AN ELEMENTARY TREATISE
ON DIFFERENTIAL EQUATIONS AND
THEIR APPLICATIONS

AN ELEMENTARY TREATISE ON
DIFFERENTIAL
EQUATIONS
AND THEIR APPLICATIONS

BY

H. T. H. PIAGGIO, M.A., D.Sc.

LATE PROFESSOR OF MATHEMATICS, UNIVERSITY OF NOTTINGHAM
FORMERLY SENIOR SCHOLAR OF ST. JOHN'S COLLEGE, CAMBRIDGE

BELL & HYMAN
LONDON

Published by
BELL & HYMAN LIMITED
Denmark House
37-39 Queen Elizabeth Street
London SE1 2QB

First published in 1920 by
G. Bell & Sons, Ltd

Reprinted four times
Revised and Enlarged Edition 1928
Reprinted fifteen times
Revised Edition 1952
Reprinted three times
Reprinted and entirely reset 1960
Reprinted 1962, 1965, 1971, 1982

British Library Cataloguing in Publication Data

Piaggio, H. T. H.
    An elementary treatise on differential equations
    and their applications.—2nd ed.
    1. Differential equations
    I. Title
    515.3′5   QA371

ISBN 0 7135 0851 5

Printed in Great Britain by
Biddles Ltd, Guildford, Surrey

PREFACE

"THE Theory of Differential Equations," said Sophus Lie, "is the most important branch of modern mathematics." The subject may be considered to occupy a central position from which different lines of development extend in many directions. If we travel along the purely analytical path, we are soon led to discuss Infinite Series, Existence Theorems and the Theory of Functions. Another leads us to the Differential Geometry of Curves and Surfaces. Between the two lies the path first discovered by Lie, leading to continuous groups of transformation and their geometrical interpretation. Diverging in another direction, we are led to the study of mechanical and electrical vibrations of all kinds and the important phenomenon of resonance. Certain partial differential equations form the starting point for the study of the conduction of heat, the transmission of electric waves, and many other branches of physics. Physical Chemistry, with its law of mass-action, is largely concerned with certain differential equations.

The object of this book is to give an account of the central parts of the subject in as simple a form as possible, suitable for those with no previous knowledge of it, and yet at the same time to point out the different directions in which it may be developed. The greater part of the text and the examples in the body of it will be found very easy. The only previous knowledge assumed is that of the elements of the differential and integral calculus and a little coordinate geometry. The miscellaneous examples at the end of the various chapters are slightly harder. They contain several theorems of minor importance, with hints that should be sufficient to enable the student to solve them. They also contain geometrical and physical applications, but great care has been taken to state the questions in such a way that no knowledge of physics is required. For instance, one question asks for a solution of a certain partial differential equation in terms of certain constants and variables. This may be regarded as a piece of pure mathematics, but it is immediately followed by a note pointing out that the work refers to a well-known experiment in heat, and giving the physical meaning of the constants and variables concerned. Finally, at the end of the book is given a set of 115 examples of much greater difficulty, most of which are taken from university examination papers. [I have to thank the Universities of London, Sheffield and Wales, and the Syndics of the Cambridge University Press for their kind permission in allowing me to use these.] The book covers the course in differential equations required for the London B.Sc. Honours or Schedule A of the Cambridge Mathematical Tripos, Part II., and also includes some of the work required for the London M.Sc. or Schedule B of the Mathematical Tripos. An appendix gives suggestions for further reading. The number of examples, both worked and unworked, is very large, and the answers to the unworked ones are given at the end of the book.

A few special points may be mentioned. The graphical method in Chapter I. (based on the MS. kindly lent me by Dr. Brodetsky of a paper he read before the Mathematical Association, and on a somewhat similar paper by Prof. Takeo Wada) has not appeared before in any text-book. The chapter dealing with numerical integration deals with the subject rather more fully than usual. It is chiefly devoted to the methods of Runge and Picard, but it also gives an account of a new method due to the present writer.

The chapter on linear differential equations with constant coefficients avoids the unsatisfactory proofs involving " infinite constants." It also points out that the use of the operator D in finding particular integrals requires more justification than is usually given. The method here adopted is at first to use the operator boldly and obtain a result, and then to verify this result by direct differentiation.

This chapter is followed immediately by one on Simple Partial Differential Equations (based on Riemann's "Partielle Differentialgleichungen"). The methods given are an obvious extension of those in the previous chapter, and they are of such great physical importance that it seems a pity to defer them until the later portions of the book, which is chiefly devoted to much more difficult subjects.

In the sections dealing with Lagrange's linear partial differential equations, two examples have been taken from M. J. M. Hill's recent paper to illustrate his methods of obtaining special integrals.

In dealing with solution in series, great prominence has been given to the method of Frobenius. One chapter is devoted to the use of the method in working actual examples. This is followed by a much harder chapter, justifying the assumptions made and dealing with the difficult questions of convergence involved. An effort has been made to state very clearly and definitely where the difficulty lies, and what are the general ideas of the somewhat complicated proofs. It is a common experience that many students when first faced by a long " epsilon-proof " are so bewildered by the details that they have very little idea of the general trend. I have to thank Mr. S. Pollard, B.A., of Trinity College, Cambridge, for his valuable help with this chapter. This is the most advanced portion of the book, and, unlike the rest of it, requires a little know- ledge of infinite series. However, references to standard text-books have been given for every such theorem used.

I have to thank Prof. W. P. Milne, the general editor of Bell's Mathematical Series, for his continual encouragement and criticism, and my colleagues Mr. J. Marshall, M.A., B.Sc., and Miss H. M. Browning, M.Sc., for their work in verifying the examples and drawing the diagrams.

I shall be very grateful for any corrections or suggestions from those who use the book.

H. T. H. PIAGGIO

UNIVERSITY COLLEGE, NOTTINGHAM
February, 1920

PREFACE

TO THE REVISED AND ENLARGED EDITION

THIS edition contains a long new chapter of a supplementary character, dealing with difficulties in the theory of singular solutions, and some little-known ideas about discriminant-loci regarded as boundaries; Riccati's equation; two additional methods for total differential equations (Mayer's general method, and the use of an integrating factor for homogeneous equations); solutions in series of linear differential equations of the second order (Fuchs' theorem, ordinary and singular points, equations of Fuchsian type, characteristic index, normal and subnormal integrals); some equations of Mathematical Physics (particularly the equation of vibrating strings and the three-dimensional Wave equation); and approximate numerical solution (Adams' method and some recent work by Remes). The other parts of the book have been revised, and a few more examples added. References have been altered when necessary.

I am deeply indebted to several friends for their valuable help and advice, particularly to Mr. H. B. Mitchell, formerly Professor at Columbia University, New York, Prof. E. H. Neville of Reading University, and my colleague, Mr. F. Underwood.

H. T. H. PIAGGIO

May, 1928

PREFACE TO THE REVISED EDITION 1952

SOME changes have been made in the treatment of Lagrange's linear partial differential equation, particularly in Arts. 124 and 125. At the end of the book, just before the index, has been added a note on limiting solutions. There are also numerous minor alterations or corrections.

H. T. H. P.

May, 1952

CONTENTS

CHAPTER IV

SIMPLE PARTIAL DIFFERENTIAL EQUATIONS

CHAPTER V

EQUATIONS OF THE FIRST ORDER, BUT NOT OF THE FIRST DEGREE

CHAPTER VI

SINGULAR SOLUTIONS

CHAPTER VII

MISCELLANEOUS METHODS FOR EQUATIONS OF THE SECOND AND HIGHER ORDERS

CHAPTER VIII

NUMERICAL APPROXIMATIONS TO THE SOLUTION OF DIFFERENTIAL EQUATIONS

CHAPTER IX

SOLUTION IN SERIES. METHOD OF FROBENIUS

CHAPTER X

EXISTENCE THEOREMS OF PICARD, CAUCHY, AND FROBENIUS

CHAPTER XI

ORDINARY DIFFERENTIAL EQUATIONS WITH THREE VARIABLES AND THE CORRESPONDING CURVES AND SURFACES

CHAPTER XII

PARTIAL DIFFERENTIAL EQUATIONS OF THE FIRST ORDER. PARTICULAR METHODS

CHAPTER XIII

PARTIAL DIFFERENTIAL EQUATIONS OF THE FIRST ORDER. GENERAL METHODS

CHAPTER XIV

PARTIAL DIFFERENTIAL EQUATIONS OF THE SECOND AND HIGHER ORDERS

CHAPTER XV

MISCELLANEOUS METHODS

APPENDIX A

Necessary and sufficient condition that the equation
$$M\,dx + N\,dy = 0$$

APPENDIX B

APPENDIX C

APPENDIX D

HISTORICAL INTRODUCTION

THE study of Differential Equations began very soon after the invention of the Differential and Integral Calculus, to which it forms a natural sequel. Newton in 1676 solved a differential equation by the use of an infinite series, only eleven years after his discovery of the fluxional form of the differential calculus in 1665. But these results were not published until 1693, the same year in which a differential equation occurred for the first time in the work of Leibniz * (whose account of the differential calculus was published in 1684).

In the next few years progress was rapid. In 1694-97 John Bernoulli † explained the method of " Separating the Variables," and he showed how to reduce a homogeneous differential equation of the first order to one in which the variables were separable. He applied these methods to problems on orthogonal trajectories. He and his brother Jacob †† (after whom " Bernoulli's Equation " is named) succeeded in reducing a large number of differential equations to forms they could solve. Integrating Factors were probably discovered by Euler (1734) and (independently of him) by Fontaine and Clairaut, though some attribute them to Leibniz. Singular Solutions, noticed by Leibniz (1694) and Brook Taylor (1715), are generally associated with the name of Clairaut (1734). The geometrical interpretation was given by Lagrange in 1774, but the theory in its present form was not given until much later by Cayley (1872) and M. J. M. Hill (1888).

The first methods of solving differential equations of the second or higher orders with constant coefficients were due to Euler. D'Alembert dealt with the case when the auxiliary equation had equal roots. Some of the symbolical methods of finding the particular integral were not given until about a hundred years later by Lobatto (1837) and Boole (1859).

The first partial differential equation to be noticed was that giving the form of a vibrating string. This equation, which is of the second order, was discussed by Euler and D'Alembert in 1747. Lagrange completed the solution of this equation, and he also dealt, in a series of memoirs from 1772 to 1785, with partial dif-

* Also spelt Leibnitz. † Also spelt Bernouilli. †† Also known as James.

ferential equations of the first order. He gave the general integral of the linear equation, and classified the different kinds of integrals possible when the equation is not linear.

These theories still remain in an unfinished state; contributions have been made by Chrystal (1892) and Hill (1917). Other methods for dealing with partial differential equations of the first order were given by Charpit (1784) and Jacobi (1836). For higher orders the most important investigations are those of Laplace 1773), Monge (1784), Ampère (1814), and Darboux (1470)

By about 1800 the subject of differential equations in its original aspect, namely the solution in a form involving only a finite number of known functions (or their integrals), was in much the same state as it is to-day. At first mathematicians had hoped to solve every differential equation in this way, but their efforts proved as fruitless as those of mathematicians of an earlier date to solve the general algebraic equation of the fifth or higher degree. The subject now became transformed, becoming closely allied to the Theory of Functions. Cauchy in 1823 proved that the infinite series obtained from a differential equation was convergent, and so really did define a function satisfying the equation. Questions of convergency (for which Cauchy was the first to give tests) are very prominent in all the investigations of this second period of the study of differential equations. Unfortunately this makes the subject very abstract and difficult for the student to grasp. In the first period the equations were not only simpler in themselves, but were studied in close connection with mechanics and physics, which indeed were often the starting point of the work.

Cauchy's investigations were continued by Briot and Bouquet (1856), and a new method, that of " Successive Approximations," was introduced by Picard (1890). Fuchs (1866) and Frobenius (1873) studied linear equations of the second and higher orders with variable coefficients. Lie's Theory of Continuous Groups (from 1884) revealed a unity underlying apparently disconnected methods. Schwarz, Klein and Goursat made their work easier to grasp by the introduction of graphical considerations, and a paper by Wada (1917) has given a graphical representation of the results of Picard and Poincaré. Runge (1895) and others dealt with numerical approximations.

Further historical notes will be found in appropriate places throughout the book. For more detailed biographies, see Rouse Ball's *Short History of Mathematics*.

CHAPTER I

INTRODUCTION AND DEFINITIONS. ELIMINATION. GRAPHICAL REPRESENTATION

1. Equations such as

$$\frac{d^2y}{dx^2} = -p^2y, \quad \dots\dots\dots\dots\dots(1)$$

$$2\frac{d^3y}{dx^3} + 3\frac{d^2y}{dx^2} + \frac{dy}{dx} - 10y = e^{-3x}\sin 5x, \quad \dots\dots\dots(2)$$

$$\left[1 + \left(\frac{dy}{dx}\right)^2\right]^{\frac{3}{2}} = 3\frac{d^2y}{dx^2}, \quad \dots\dots\dots\dots(3)$$

$$\frac{dy}{dx} = \frac{x}{y^{\frac{1}{2}}(1 + x^{\frac{1}{3}})}, \quad \dots\dots\dots\dots(4)$$

$$\frac{\partial^2y}{\partial t^2} = a^2\frac{\partial^2y}{\partial x^2}, \quad \dots\dots\dots\dots(5)$$

involving differential coefficients, are called *Differential Equations*.

2. Differential Equations arise from many problems in Algebra, Geometry, Mechanics, Physics, and Chemistry. In various places in this book we shall give examples of these, including applications to elimination, tangency, curvature, envelopes, oscillations of mechanical systems and of electric currents, bending of beams, conduction of heat, diffusion of solvents, velocity of chemical reactions, etc.

3. Definitions. Differential equations which involve only one independent variable,* like (1), (2), (3), and (4), are called *ordinary*.

Those which involve two or more independent variables and partial differential coefficients with respect to them, such as (5), are called *partial*.

---

* In equations (1), (2), (3), (4) x is the independent and y the dependent variable. In (5) x and t are the two independent variables and y the dependent.

An equation like (1), which involves a second differential co-efficient, but none of higher orders, is said to be of the second *order*. (4) is of the first order, (3) and (5) of the second, and (2) of the third.

The *degree* of an equation is the degree of the highest differential coefficient when the equation has been made rational and integral as far as the differential coefficients are concerned. Thus (1), (2), (4) and (5) are of the first degree.

(3) must be squared to rationalise it. We then see that it is of the second degree, as $\dfrac{d^2y}{dx^2}$ occurs squared.

Notice that this definition of degree does not require x or y to occur rationally or integrally.

Other definitions will be introduced when they are required.

4. Formation of differential equations by elimination. The problem of elimination will now be considered, chiefly because it gives us an idea as to what kind of solution a differential equation may have.

We shall give some examples of the elimination of arbitrary constants by the formation of *ordinary* differential equations. Later (Chap. IV.) we shall see that *partial* differential equations may be formed by the elimination of either arbitrary constants or arbitrary functions.

5. Examples

(i) Consider $x = A \cos (pt - \alpha)$, the equation of simple harmonic motion. Let us eliminate the arbitrary constants A and α.

Differentiating, $\qquad \dfrac{dx}{dt} = -pA \sin (pt - \alpha)$

and $\qquad\qquad \dfrac{d^2x}{dt^2} = -p^2A \cos (pt - \alpha) = -p^2x.$

Thus $\dfrac{d^2x}{dt^2} = -p^2x$ is the result required, an equation of the second order, whose interpretation is that the acceleration varies as the distance from the origin.

(ii) Eliminate p from the last result.

Differentiating again, $\qquad \dfrac{d^3x}{dt^3} = -p^2 \dfrac{dx}{dt}.$

Hence $\qquad \dfrac{d^3x}{dt^3} \bigg/ \dfrac{dx}{dt} = -p^2 = \dfrac{d^2x}{dt^2} \bigg/ x$, (from the last result).

Multiplying up, $x \cdot \dfrac{d^3x}{dt^3} = \dfrac{dx}{dt} \cdot \dfrac{d^2x}{dt^2}$, an equation of the third order.

(iii) Form the differential equation of all parabolas whose axis is the axis of x.

Such a parabola must have an equation of the form
$$y^2 = 4a(x - h).$$
Differentiating twice, we get
$$2y \frac{dy}{dx} = 4a,$$

$$i.e. \quad y \frac{dy}{dx} = 2a,$$

and $y \frac{d^2y}{dx^2} + \left(\frac{dy}{dx}\right)^2 = 0$, which is of the second order.

Examples for solution

Eliminate the arbitrary constants from the following equations:

(1) $y = Ae^{2x} + Be^{-2x}$.   (2) $y = A \cos 3x + B \sin 3x$.

(3) $y = Ae^{Bx}$.   (4) $y = Ax + A^3$.

(5) If $x^2 + y^2 = a^2$, prove that $\frac{dy}{dx} = -\frac{x}{y}$, and interpret the result geometrically.

(6) Prove that for any straight line through the origin $\frac{y}{x} = \frac{dy}{dx}$, and interpret this.

(7) Prove that for any straight line whatever $\frac{d^2y}{dx^2} = 0$. Interpret this.

6. To eliminate n arbitrary constants requires (in general) a differential equation of the nth order. The reader will probably have arrived at this conclusion already, from the examples of Art. 5. If we differentiate n times an equation containing n arbitrary constants, we shall obtain $(n + 1)$ equations altogether, from which the n constants can be eliminated. As the result contains an nth differential coefficient, it is of the nth order.*

* The argument in the text is that usually given, but the advanced student will notice some weak points in it. The statement that from any $(n + 1)$ equations n quantities can be eliminated, *whatever the nature of those equations*, is too sweeping. An exact statement of the necessary and sufficient conditions would be extremely complicated.

Sometimes less than $(n + 1)$ equations are required. An obvious case is $y = (A + B)x$, where the two arbitrary constants occur in such a way as to be really equivalent to one.

A less obvious case is $y^2 = 2Axy + Bx^2$. This represents two straight lines through the origin, say $y = m_1x$ and $y = m_2x$, from each of which we easily get $\frac{y}{x} = \frac{dy}{dx}$, of the first instead of the second order. The student should also obtain this result by differentiating the original equation and eliminating B. This will give
$$\left(y - x \frac{dy}{dx}\right)(y - Ax) = 0.$$

7. The most general solution of an ordinary differential equation of the nth order contains n arbitrary constants. This will probably seem obvious from the converse theorem that in general n arbitrary constants can be eliminated by a differential equation of the nth order. But a rigorous proof offers much difficulty.

If, however, we assume * that a differential equation has a solution expansible in a convergent series of ascending integral powers of x, we can easily see why the arbitrary constants are n in number.

Consider, for example, $\dfrac{d^3y}{dx^3} = \dfrac{dy}{dx}$, of order *three*.

Assume that $y = a_0 + a_1x + a_2\dfrac{x^2}{2!} + \cdots + a_n\dfrac{x^n}{n!} + \cdots$ to infinity.

Then, substituting in the differential equation, we get

$$a_3 + a_4x + a_5\frac{x^2}{2!} + \cdots + a_n\frac{x^{n-3}}{(n-3)!} + \cdots$$
$$= a_1 + a_2x + a_3\frac{x^2}{2!} + \cdots + a_n\frac{x^{n-1}}{(n-1)!} +$$

so $$a_3 = a_1, \qquad a_4 = a_2, \qquad a_5 = a_3 = a_1,$$
$$a_n = a_{n-2} = a_{n-4} = \text{etc.}$$

Hence $$y = a_0 + a_1\left(x + \frac{x^3}{3!} + \frac{x^5}{5!} + \cdots\right) + a_2\left(\frac{x^2}{2!} + \frac{x^4}{4!} + \frac{x^6}{6!} + \cdots\right)$$
$$= a_0 + a_1 \sinh x + a_2 (\cosh x - 1),$$

containing *three* arbitrary constants, a_0, a_1 and a_2.

Similar reasoning applies to the equation

$$\frac{d^ny}{dx^n} = f\left(x,\ y,\ \frac{dy}{dx},\ \frac{d^2y}{dx^2},\ \cdots,\ \frac{d^{n-1}y}{dx^{n-1}}\right).$$

In Dynamics the differential equations are usually of the second order, e.g. $\dfrac{d^2y}{dt^2} + p^2y = 0$, the equation of simple harmonic motion. To get a solution without arbitrary constants we need two conditions, such as the value of y and dy/dt when $t = 0$, giving the initial displacement and velocity.

8. Complete Primitive, Particular Integral, Singular Solution. The solution of a differential equation containing the full number of arbitrary constants is called the Complete Primitive.

Any solution derived from the Complete Primitive by giving particular values to these constants is called a Particular Integral.

* The student will see in later chapters that this assumption is not always justifiable.

Thus the Complete Primitive of $\dfrac{d^3y}{dx^3} = \dfrac{dy}{dx}$

is $\qquad y = a_0 + a_1 \sinh x + a_2 (\cosh x - 1)$,

or $\qquad y = c + a_1 \sinh x + a_2 \cosh x$, where $c = a_0 - a_2$,

or $\qquad y = c + ae^x + be^{-x}$, where $a = \frac{1}{2}(a_1 + a_2)$ and $b = \frac{1}{2}(a_2 - a_1)$

This illustrates the fact that the Complete Primitive may often be written in several different (but really equivalent) ways.

The following are Particular Integrals:

$$y = 4, \qquad\qquad \text{taking} \quad c = 4, \quad a_1 = a_2 = 0;$$
$$y = 5 \sinh x, \qquad \text{taking} \quad a_1 = 5, \quad c = a_2 = 0;$$
$$y = 6 \cosh x - 4, \quad \text{taking} \quad a_2 = 6, \quad a_1 = 0, \quad c = -4;$$
$$y = 2 + e^x - 3e^{-x}, \quad \text{taking} \quad c = 2, \quad a = 1, \quad b = -3.$$

In most equations every solution can be derived from the Complete Primitive by giving suitable values to the arbitrary constants. However, in some exceptional cases we shall find a solution, called a Singular Solution, that cannot be derived in this way. These will be discussed in Chap. VI.

Examples for solution

Solve by the method of Art. 7:

(1) $\qquad\qquad\qquad\qquad \dfrac{dy}{dx} = y.$

(2) $\qquad\qquad\qquad\qquad \dfrac{d^2y}{dx^2} = -y.$

(3) Show that the method fails for $\dfrac{dy}{dx} = \dfrac{1}{x}.$

[$\log x$ cannot be expanded in a Maclaurin series.]

(4) Verify by elimination of c that $y = cx + \dfrac{1}{c}$ is the Complete Primitive of $y = x\dfrac{dy}{dx} + 1 \Big/ \dfrac{dy}{dx}.$ Verify also that $y^2 = 4x$ is a solution of the differential equation not derivable from the Complete Primitive (*i.e.* a Singular Solution). Show that the Singular Solution is the envelope of the family of lines represented by the Complete Primitive. Illustrate by a graph.

9. Graphical representation. We shall now give some examples of a method * of sketching rapidly the general form of the family of curves representing the Complete Primitive of

$$\frac{dy}{dx} = f(x, y),$$

* Due to Dr. S. Brodetsky and Prof. Takeo Wada.

where $f(x, y)$ is a function of x and y having a perfectly definite finite value * for every pair of finite values of x and y.

The curves of the family are called the *characteristics* of the equation.

Ex. (i)
$$\frac{dy}{dx} = x(y-1).$$

Here
$$\frac{d^2y}{dx^2} = y - 1 + x\frac{dy}{dx} = (x^2 + 1)(y - 1).$$

Now a curve has its concavity upwards when the second differential coefficient is positive. Hence the characteristics will be concave up above $y = 1$, and concave down below this line. The maximum or minimum points lie on $x = 0$, since $dy/dx = 0$ there. The characteristics near $y = 1$, which is a member of the family, are flatter than those further from it.

These considerations show us that the family is of the general form shown in Fig. 1.

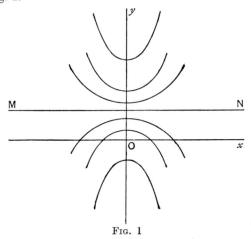

FIG. 1

Ex. (ii)
$$\frac{dy}{dx} = y + e^x.$$

Here
$$\frac{d^2y}{dx^2} = \frac{dy}{dx} + e^x = y + 2e^x.$$

We start by tracing the curve of maxima and minima $y + e^x = 0$, and the curve of inflexions $y + 2e^x = 0$. Consider the characteristic through the origin. At this point both differential coefficients are positive, so as x increases y increases also, and the curve is concave upwards. This gives us the right-hand portion of the characteristic marked 3 in Fig. 2. If we move to the left along this we get to the

* Thus excluding a function like y/x, which is indeterminate when $x = 0$ and $y = 0$.

curve of minima. At the point of intersection the tangent is parallel to Ox. After this we ascend again, so meeting the curve of inflexions. After crossing this the characteristic becomes convex upwards. It still ascends. Now the figure shows that if it cut the curve of minima again

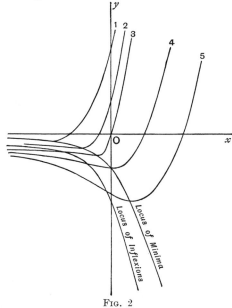

Fig. 2

the tangent could not be parallel to Ox, so it cannot cut it at all, but becomes asymptotic to it.

The other characteristics are of similar nature.

Examples for solution

Sketch the characteristics of:

(1) $$\frac{dy}{dx} = y(1 - x).$$

(2) $$\frac{dy}{dx} = x^2 y.$$

(3) $$\frac{dy}{dx} = y + x^2.$$

10. Singular points. In all examples like those in the last article, we get one characteristic, and only one, through every point of the plane. By tracing the two curves $\frac{dy}{dx} = 0$ and $\frac{d^2y}{dx^2} = 0$ we can easily sketch the system.

If, however, $f(x, y)$ becomes indeterminate for one or more points (called *singular points*), it is often very difficult to sketch the system in the neighbourhood of these points. But the following examples can be treated geometrically. In general, a complicated analytical treatment is required.*

Ex. (i). $\dfrac{dy}{dx} = \dfrac{y}{x}$. Here the origin is a singular point. The geometrical meaning of the equation is that the radius vector and the tangent have the same gradient, which can only be the case for straight

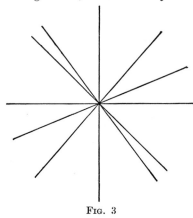

Fig. 3

lines through the origin. As the number of these is infinite, in this case an infinite number of characteristics pass through the singular point.

Ex. (ii). $\dfrac{dy}{dx} = -\dfrac{x}{y}, \quad i.e. \ \dfrac{y}{x} \cdot \dfrac{dy}{dx} = -1.$

This means that the radius vector and the tangent have gradients

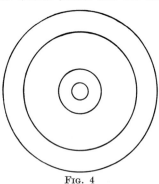

Fig. 4

whose product is -1, *i.e.* that they are perpendicular. The characteristics are therefore circles of any radius with the origin as centre.

* See a paper, "Graphical Solution," by Prof. Takeo Wada, *Memoirs of the College of Science, Kyoto Imperial University*, Vol. II. No. 3, July 1917.

In this case the singular point may be regarded as a circle of zero radius, the limiting form of the characteristics near it, but no characteristic of finite size passes through it.

Ex. (iii).
$$\frac{dy}{dx} = \frac{y - kx}{x + ky}.$$

Writing $dy/dx = \tan \psi$, $y/x = \tan \theta$, we get

$$\tan \psi = \frac{\tan \theta - k}{1 + k \tan \theta},$$

i.e. $\quad \tan \psi + k \tan \psi \tan \theta = \tan \theta - k,$

i.e. $\quad \dfrac{\tan \theta - \tan \psi}{1 + \tan \theta \tan \psi} = k,$

i.e. $\quad \tan (\theta - \psi) = k$, a constant.

The characteristics are therefore equiangular spirals, of which the singular point (the origin) is the focus.

Fig. 5

These three simple examples illustrate three typical cases. Sometimes a *finite* number of characteristics pass through a singular point, but an example of this would be too complicated to give here.*

* See Wada's paper.

MISCELLANEOUS EXAMPLES ON CHAPTER I

Eliminate the arbitrary constants from the following:

(1) $\qquad y = Ae^x + Be^{-x} + C.$

(2) $\qquad y = Ae^x + Be^{2x} + Ce^{3x}.$

[To eliminate A, B, C from the four equations obtained by successive differentiation a determinant may be used.]

(3) $\qquad y = e^x(A \cos x + B \sin x).$

(4) $\qquad y = c \cosh \dfrac{x}{c}$, (the catenary).

Find the differential equation of

(5) All parabolas whose axes are parallel to the axis of y.

(6) All circles of radius a.

(7) All circles that pass through the origin.

(8) All circles (whatever their radii or positions in the plane xOy).

[The result of Ex. 6 may be used.]

(9) Show that the results of eliminating a from

$$2y = x \frac{dy}{dx} + ax, \dotfill (1)$$

and b from

$$y = x \frac{dy}{dx} - bx^2, \dotfill (2)$$

are in each case

$$x^2 \frac{d^2y}{dx^2} - 2x \frac{dy}{dx} + 2y = 0. \dotfill (3)$$

[The complete primitive of equation (1) must satisfy equation (3), since (3) is derivable from (1). This primitive will contain a and also an arbitrary constant. Thus it is a solution of (3) containing two constants, both of which are arbitrary as far as (3) is concerned, as a does not occur in that equation. In fact, it must be the complete primitive of (3). Similarly the complete primitives of (2) and (3) are the same. Thus (1) and (2) have a common complete primitive.]

(10) Apply the method of the last example to prove that

$$y + \frac{dy}{dx} = 2ae^x$$

and

$$y - \frac{dy}{dx} = 2be^{-x}$$

have a common complete primitive.

(11) Assuming that the first two equations of Ex. 9 have a common complete primitive, find it by equating the two values of $\dfrac{dy}{dx}$ in terms of x, y, and the constants. Verify that it satisfies equation (3) of Ex. 9.

(12) Similarly obtain the common complete primitive of the two equations of Ex. 10.

(13) Prove that all curves satisfying the differential equation

$$\frac{dy}{dx} = 1 + x\left(\frac{dy}{dx}\right)^2 + x^2\frac{d^2y}{dx^2}$$

cut the axis of y at $45°$.

(14) Find the inclination to the axis of x at the point (1, 2) of the two curves which pass through that point and satisfy

$$\left(\frac{dy}{dx}\right)^2 = y^2 - 2x + x^2.$$

(15) Prove that the radius of curvature of either of the curves of Ex. 14 at the point (1, 2) is 4.

(16) Prove that in general two curves satisfying the differential equation

$$x\left(\frac{dy}{dx}\right)^2 - y\frac{dy}{dx} + 1 = 0$$

pass through any point, but that these coincide for any point on a certain parabola, which is the envelope of the curves of the system.

(17) Find the locus of a point such that the two curves through it satisfying the differential equation of Ex. (16) cut (i) orthogonally; (ii) at $45°$.

(18) Sketch (by Brodetsky and Wada's method) the characteristics of

$$\frac{dy}{dx} = x + e^y.$$

(19) Obtain solutions in series of ascending integral powers of x (as in Art. 7) of the following differential equations (in which y_1 and y_2 denote $\frac{dy}{dx}$ and $\frac{d^2y}{dx^2}$ respectively):

(i) $y_2 - xy_1 - y = 0$;
(iii) $x^2y_2 - 2xy_1 + 2y = 0$;
(v) $(x - x^2)y_2 + (1 - 5x)y_1 - 4y = 0.$

(ii) $xy_2 + xy_1 + y = 0$;
(iv) $(1 - x^2)y_2 + 2y = 0$;

[*Answers*:

(i) $y = a_0\left(1 + \frac{x^2}{2} + \frac{x^4}{2 \cdot 4} + \frac{x^6}{2 \cdot 4 \cdot 6} + \dots\right) + a_1\left(\frac{x}{1} + \frac{x^3}{1 \cdot 3} + \frac{x^5}{1 \cdot 3 \cdot 5} + \dots\right);$

(ii) $y = a_1\left(x - \frac{x^2}{1!} + \frac{x^3}{2!} - \frac{x^4}{3!} + \dots\right) = a_1xe^{-x};$ this, containing only one

arbitrary constant, is not the complete primitive, for there is another solution not of the form assumed here (see Chap. IX.);

(iii) $y = a_1x + \frac{1}{2}a_2x^2;$

(iv) $y = a_0(1 - x^2) + a_1\left(x - \frac{x^3}{1 \cdot 3} - \frac{x^5}{3 \cdot 5} - \frac{x^7}{5 \cdot 7} - \dots\right);$

(v) $y = a_0(1^2 + 2^2x + 3^2x^2 + \dots);$ see Art. 97.]

CHAPTER II

EQUATIONS OF THE FIRST ORDER AND FIRST DEGREE

11. In this chapter we shall consider equations of the form

$$M + N\frac{dy}{dx} = 0,$$

where M and N are functions of both x and y.

This equation is often written,* more symmetrically, as

$$M\,dx + N\,dy = 0.$$

Unfortunately it is not possible to solve the *general* equation of this form in terms of a finite number of known functions, but we shall discuss some special types in which this can be done.

It is usual to classify these types as

(a) Exact equations;
(b) Equations solvable by separation of the variables;
(c) Homogeneous equations;
(d) Linear equations of the first order.

The methods of this chapter are chiefly due to John Bernouilli of Bâle (1667–1748), the most inspiring teacher of his time, and to his pupil, Leonhard Euler, also of Bâle (1707–1783). Euler made great contributions to algebra, trigonometry, calculus, rigid dynamics, hydrodynamics, astronomy and other subjects.

12. Exact equations †

Ex. (i). The expression $y\,dx + x\,dy$ is an exact differential.

Thus the equation $\qquad y\,dx + x\,dy = 0,$

giving $\qquad\qquad\qquad d\,(yx) = 0,$

$\qquad\qquad\qquad i.e.\quad yx = c,$

is called an exact equation.

---

* For a rigorous justification of the use of the *differentials* dx and dy see Hardy's *Pure Mathematics*, Art. 136 [Arts. 154-155 in later editions].

† For the necessary and sufficient condition that $M\,dx + N\,dy = 0$ should be exact see Appendix A.

Ex. (ii). Consider the equation $\tan y \cdot dx + \tan x \cdot dy = 0$.

This is not exact as it stands, but if we multiply by $\cos x \cos y$ it becomes $\qquad \sin y \cos x \, dx + \sin x \cos y \, dy = 0$,

which is exact.

The solution is $\qquad \sin y \sin x = c$.

13. Integrating factors. In the last example $\cos x \cos y$ is called an *integrating factor*, because when the equation is multiplied by it we get an exact equation which can be at once integrated.

There are several rules which are usually given for determining integrating factors in particular classes of equations. These will be found in the miscellaneous examples at the end of the chapter. The proof of these rules forms an interesting exercise, but it is generally easier to solve examples without them.

14. Variables separate

Ex. (i). In the equation $\dfrac{dx}{x} = \tan y \cdot dy$, the left-hand side involves x only and the right-hand side y only, so the *variables are separate*.

Integrating, we get $\qquad \log x = -\log \cos y + c$,

$\quad$ *i.e.* $\log (x \cos y) = c$,

$\qquad x \cos y = e^c = a$, say.

Ex. (ii). $\qquad\qquad\qquad\qquad \dfrac{dy}{dx} = 2xy.$

The variables are not separate at present, but they can easily be made so. Multiply by dx and divide by y. We get

$$\frac{dy}{y} = 2x \, dx.$$

Integrating, $\qquad\qquad \log y = x^2 + c$.

As c is arbitrary, we may put it equal to $\log a$, where a is another arbitrary constant.

Thus, finally, $\qquad\qquad\qquad y = ae^{x^2}.$

Examples for solution

(1) $(12x + 5y - 9) \, dx + (5x + 2y - 4) \, dy = 0$.

(2) $\{\cos x \tan y + \cos (x+y)\} \, dx + \{\sin x \sec^2 y + \cos (x+y)\} \, dy = 0$.

(3) $(\sec x \tan x \tan y - e^x) \, dx + \sec x \sec^2 y \, dy = 0$.

(4) $(x + y)(dx - dy) = dx + dy$.

(5) $y \, dx - x \, dy + 3x^2y^2e^{x^3} \, dx = 0$.

(6) $y \, dx - x \, dy = 0$.

(7) $(\sin x + \cos x) \, dy + (\cos x - \sin x) \, dx = 0$.

(8) $\dfrac{dy}{dx} = x^3y^2.$

(9) $y \, dx - x \, dy = xy \, dx$. $\qquad\qquad$ (10) $\tan x \, dy = \cot y \, dx$.

15. Homogeneous equations. A homogeneous equation of the first order and degree is one which can be written in the form

$$\frac{dy}{dx} = f\left(\frac{y}{x}\right).$$

To test whether a function of x and y can be written in the form of the right-hand side, it is convenient to put

$$\frac{y}{x} = v \quad \text{or} \quad y = vx.$$

If the result is of the form $f(v)$, *i.e.* if the x's all cancel, the test is satisfied.

Ex. (i). $\dfrac{dy}{dx} = \dfrac{x^2 + y^2}{2x^2}$ becomes $\dfrac{dy}{dx} = \dfrac{1 + v^2}{2}$. This equation is homogeneous.

Ex. (ii). $\dfrac{dy}{dx} = \dfrac{y^3}{x^2}$ becomes $\dfrac{dy}{dx} = xv^3$. This is not homogeneous.

16. Method of solution. Since a homogeneous equation can be reduced to $\dfrac{dy}{dx} = f(v)$ by putting $y = vx$ on the right-hand side, it is natural to try the effect of this substitution on the left-hand side also. As a matter of fact, it will be found that the equation can always be solved * by this substitution (see Ex. 10 of the miscellaneous set at the end of this chapter).

Ex. (i). $$\frac{dy}{dx} = \frac{x^2 + y^2}{2x^2}.$$

Put $$y = vx,$$

i.e. $\dfrac{dy}{dx} = v + x\dfrac{dv}{dx}$, (for if y is a function of x, so is v).

The equation becomes $\quad v + x\dfrac{dv}{dx} = \dfrac{1 + v^2}{2}$,

i.e. $\quad 2x\, dv = (1 + v^2 - 2v)\, dx.$

Separating the variables, $\dfrac{2dv}{(v-1)^2} = \dfrac{dx}{x}.$

Integrating, $\quad\quad\quad \dfrac{-2}{v-1} = \log x + c.$

But $\quad v = \dfrac{y}{x}, \quad$ so $\dfrac{-2}{v-1} = \dfrac{-2}{\dfrac{y}{x} - 1} = \dfrac{-2x}{y-x} = \dfrac{2x}{x-y}.$

Multiplying by $x - y$, $\quad 2x = (x - y)(\log x + c).$

---

* By "solved" we mean reduced to an ordinary integration. Of course, this integral may not be expressible in terms of ordinary elementary functions.

Ex. (ii). $\qquad (x+y)\,dy+(x-y)\,dx=0.$

This gives $\qquad \dfrac{dy}{dx}=\dfrac{y-x}{y+x}.$

Putting $y=vx$, and proceeding as before, we get

$$v+x\frac{dv}{dx}=\frac{v-1}{v+1},$$

i.e. $\quad x\dfrac{dv}{dx}=\dfrac{v-1}{v+1}-v=-\dfrac{v^2+1}{v+1}.$

Separating the variables, $\quad -\dfrac{(v+1)\,dv}{v^2+1}=\dfrac{dx}{x},$

i.e. $\quad \dfrac{-v\,dv}{v^2+1}-\dfrac{dv}{v^2+1}=\dfrac{dx}{x}.$

Integrating, $\quad -\tfrac{1}{2}\log(v^2+1)-\tan^{-1}v=\log x+c,$

i.e. $\quad 2\log x+\log(v^2+1)+2\tan^{-1}v+2c=0,$

$\log x^2(v^2+1)+2\tan^{-1}v+a=0$, putting $2c=a.$

Substituting for v, $\log(y^2+x^2)+2\tan^{-1}\dfrac{y}{x}+a=0.$

17. Equations reducible to the homogeneous form

Ex. (i). The equation $\qquad \dfrac{dy}{dx}=\dfrac{y-x+1}{y+x+5}$

is *not* homogeneous.

This example is similar to Ex. (ii) of the last article, except that

$$\frac{y-x}{y+x}\ \text{is replaced by}\ \frac{y-x+1}{y+x+5}.$$

Now $y-x=0$ and $y+x=0$ represent two straight lines through the origin.

The intersection of $y-x+1=0$ and $y+x+5=0$ is easily found to be $(-2,-3)$.

Put $x=X-2$; $y=Y-3$. This amounts to taking new axes parallel to the old with $(-2,-3)$ as the new origin.

Then $\qquad y-x+1=Y-X$ and $y+x+5=Y+X.$

Also $\qquad\qquad dx=dX$ and $dy=dY.$

The equation becomes $\qquad \dfrac{dY}{dX}=\dfrac{Y-X}{Y+X}.$

As in the last article, the solution is

$$\log(Y^2+X^2)+2\tan^{-1}\frac{Y}{X}+a=0,$$

i.e. $\quad \log[(y+3)^2+(x+2)^2]+2\tan^{-1}\dfrac{y+3}{x+2}+a=0.$

Ex. (ii). $$\frac{dy}{dx} = \frac{y - x + 1}{y - x + 5}.$$

This equation cannot be treated as the last example, because the lines $y - x + 1 = 0$ and $y - x + 5 = 0$ are parallel.

As the right-hand side is a function of $y - x$, try putting $y - x = z$,

$$i.e. \quad \frac{dy}{dx} - 1 = \frac{dz}{dx}.$$

The equation becomes $$1 + \frac{dz}{dx} = \frac{z + 1}{z + 5},$$

$$i.e. \quad \frac{dz}{dx} = \frac{-4}{z + 5}.$$

Separating the variables, $(z + 5) \, dz = -4 \, dx.$

Integrating, $\frac{1}{2}z^2 + 5z = -4x + c,$

$$i.e. \quad z^2 + 10z + 8x = 2c.$$

Substituting for z, $\;(y - x)^2 + 10(y - x) + 8x = 2c,$

$$i.e. \quad (y - x)^2 + 10y - 2x = a, \text{ putting } 2c = a.$$

Examples for solution

(1) $(2x - y) \, dy = (2y - x) \, dx.$ [Wales.]

(2) $(x^2 - y^2) \dfrac{dy}{dx} = xy.$ [Sheffield.]

(3) $2\dfrac{dy}{dx} = \dfrac{y}{x} + \dfrac{y^2}{x^2}.$ [Math. Tripos.]

(4) $x\dfrac{dy}{dx} = y + \sqrt{(x^2 + y^2)}.$

(5) $\dfrac{dy}{dx} = \dfrac{2x + 9y - 20}{6x + 2y - 10}.$

(6) $(12x + 21y - 9) \, dx + (47x + 40y + 7) \, dy = 0.$

(7) $\dfrac{dy}{dx} = \dfrac{3x - 4y - 2}{3x - 4y - 3}.$

(8) $(x + 2y)(dx - dy) = dx + dy.$

18. Linear equations

The equation $$\frac{dy}{dx} + Py = Q,$$

where P and Q are functions of x (but not of y), is said to be *linear* of the first order.

A simple example is $\dfrac{dy}{dx} + \dfrac{1}{x} \cdot y = x^2.$

If we multiply each side of this by x, it becomes

$$x\frac{dy}{dx} + y = x^3,$$

i.e. $\quad \frac{d}{dx}(xy) = x^3.$

Hence, integrating $\quad xy = \frac{1}{4}x^4 + c.$

We have solved this example by the use of the obvious integrating factor x.

19. Let us try to find an integrating factor in the general case. If R is such a factor, then the left-hand side of

$$R\frac{dy}{dx} + RPy = RQ$$

is the differential coefficient of some product, and the first term $R\frac{dy}{dx}$ shows that the product must be Ry.

Put, therefore, $R\frac{dy}{dx} + RPy = \frac{d}{dx}(Ry) = R\frac{dy}{dx} + y\frac{dR}{dx}.$

This gives $\quad\quad\quad RPy = y\frac{dR}{dx},$

i.e. $\quad P\,dx = \frac{dR}{R},$

i.e. $\quad \int P\,dx = \log R,$

i.e. $\quad R = e^{\int P\,dx}.$

This gives the rule: *To solve* $\frac{dy}{dx} + Py = Q$, *multiply each side by* $e^{\int P\,dx}$, *which will be an integrating factor.*

20. Examples

(i) Take the example considered in Art. 18.

$$\frac{dy}{dx} + \frac{1}{x}\cdot y = x^2.$$

Here $P = \frac{1}{x}$, so $\int P\,dx = \log x$, and $e^{\log x} = x$.

Thus the rule gives the same integrating factor that we used before.

(ii) $\quad\quad\quad \frac{dy}{dx} + 2xy = 2e^{-x^2}.$

Here $P = 2x$, $\int P\,dx = x^2$, and the integrating factor is e^{x^2}.

Multiplying by this, $e^{x^2}\dfrac{dy}{dx} + 2xe^{x^2}y = 2,$

$$i.e. \quad \frac{d}{dx}\left(ye^{x^2}\right) = 2.$$

Integrating, $ye^{x^2} = 2x + c,$

$$i.e. \quad y = (2x + c)\,e^{-x^2}.$$

(iii) $\dfrac{dy}{dx} + 3y = e^{2x}.$

Here the integrating factor is e^{3x}.

Multiplying by this, $e^{3x}\dfrac{dy}{dx} + 3e^{3x}y = e^{5x},$

$$i.e. \quad \frac{d}{dx}\left(ye^{3x}\right) = e^{5x}.$$

Integrating, $ye^{3x} = \tfrac{1}{5}e^{5x} + c,$

$$i.e. \quad y = \tfrac{1}{5}e^{2x} + ce^{-3x}.$$

21. Equations reducible to the linear form

Ex. (i). $xy - \dfrac{dy}{dx} = y^3 e^{-x^2}.$

Divide by y^3, so as to free the right-hand side from y.

We get $x\cdot\dfrac{1}{y^2} - \dfrac{1}{y^3}\dfrac{dy}{dx} = e^{-x^2},$

$$i.e. \quad x\cdot\frac{1}{y^2} + \frac{1}{2}\frac{d}{dx}\left(\frac{1}{y^2}\right) = e^{-x^2}.$$

Putting $\dfrac{1}{y^2} = z,$ $2xz + \dfrac{dz}{dx} = 2e^{-x^2}.$

This is linear and, in fact, is similar to **Ex. (ii)** of the last article with z instead of y.

Hence the solution is $z = (2x + c)e^{-x^2},$

$$i.e. \quad \frac{1}{y^2} = (2x + c)\,e^{-x^2},$$

$$i.e. \quad y = \pm\,\frac{e^{\frac{1}{2}x^2}}{\sqrt{(2x + c)}}.$$

This example is a particular case of "Bernoulli's Equation"

$$\frac{dy}{dx} + Py = Qy^n,$$

where P and Q are functions of x. Jacob Bernouilli or Bernoulli of Bâle (1654-1705) studied it in 1695.

Ex. (ii). $$(2x - 10y^3)\frac{dy}{dx} + y = 0.$$

This is not linear as it stands, but if we multiply by $\frac{dx}{dy}$, we get

$$2x - 10y^3 + y\frac{dx}{dy} = 0,$$

i.e. $$\frac{dx}{dy} + \frac{2x}{y} = 10y^2.$$

This is linear, considering y as the *independent* variable.

Proceeding as before, we find the integrating factor to be y^2, and the solution

$$y^2x = 2y^5 + c,$$
i.e. $$x = 2y^3 + cy^{-2}.$$

Examples for solution

(1) $(x+a)\dfrac{dy}{dx} - 3y = (x+a)^5.$ [Wales.]

(2) $x \cos x \dfrac{dy}{dx} + y(x \sin x + \cos x) = 1.$ [Sheffield.]

(3) $x \log x \dfrac{dy}{dx} + y = 2 \log x.$ (4) $x^2y - x^3 \dfrac{dy}{dx} = y^4 \cos x.$

(5) $y + 2\dfrac{dy}{dx} = y^3(x - 1).$ (6) $(x + 2y^3)\dfrac{dy}{dx} = y.$

(7) $dx + x\, dy = e^{-y} \sec^2 y\, dy.$

22. Geometrical Problems. Orthogonal Trajectories. We shall now consider some geometrical problems leading to differential equations.

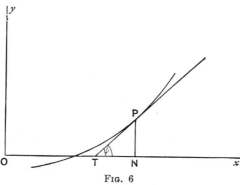

Fig. 6

Ex. (i). Find the curve whose subtangent is constant.

The subtangent $TN = PN \cot \psi = y\dfrac{dx}{dy}.$

Hence
$$y \frac{dx}{dy} = k,$$
$$dx = k \frac{dy}{y},$$
$$x + c = k \log y,$$
$$y = ae^{x/k},$$

putting the arbitrary constant c equal to $k \log a$.

Ex. (ii). Find the curve such that its length between any two points PQ is proportional to the difference of the distances of Q and P from a fixed point O.

If we keep P fixed, the arc QP will vary as OQ minus a constant.

Use polar co-ordinates, taking O as pole and OP as initial line. Then, if Q be (r, θ), we have
$$s = kr - kr_0.$$

But, as shown in treatises on the Calculus,
$$(ds)^2 = (r \, d\theta)^2 + (dr)^2.$$

Hence, in our problem,
$$k^2 (dr)^2 = (r \, d\theta)^2 + (dr)^2,$$
$$i.e. \quad d\theta = \pm \sqrt{(k^2 - 1)} \frac{dr}{r}$$
$$= \frac{1}{a} \frac{dr}{r}, \text{ say,}$$

giving $r = ce^{a\theta}$, the equiangular spiral.

Ex. (iii). Find the Orthogonal Trajectories of the family of semi-cubical parabolas $ay^2 = x^3$, where a is a variable parameter.

Two families of curves are said to be orthogonal trajectories when every member of one family cuts every member of the other at right angles.

We first obtain the differential equation of the given family by eliminating a.

Differentiating
$$ay^2 = x^3.$$

we get
$$2ay \frac{dy}{dx} = 3x^2,$$

whence, by division,
$$\frac{2}{y} \frac{dy}{dx} = \frac{3}{x}. \quad \dots\dots\dots\dots\dots\dots\dots\dots(1)$$

Now $\frac{dy}{dx} = \tan \psi$, where ψ is the inclination of the tangent to the axis of x. The value of ψ for the trajectory, say ψ', is given by
$$\psi = \psi' \pm \tfrac{1}{2}\pi,$$
$$i.e. \quad \tan \psi = -\cot \psi',$$

i.e. $\frac{dy}{dx}$ for the given family is to be replaced by $-\frac{dx}{dy}$ for the trajectory.

Making this change in (1), we get

$$-\frac{2}{y}\frac{dx}{dy}=\frac{3}{x},$$

$$2x\,dx+3y\,dy=0,$$

$$2x^2+3y^2=c,$$

a family of similar and similarly situated ellipses.

Ex. (iv). Find the family of curves that cut the family of spirals $r=a\theta$ at a constant angle α.

As before, we start by eliminating a.

This gives $$\frac{r\,d\theta}{dr}=\theta.$$

Now $\dfrac{r\,d\theta}{dr}=\tan\phi$, where ϕ is the angle between the tangent and the radius vector. If ϕ' is the corresponding angle for the second family,

$$\phi'=\phi\pm\alpha,$$

$$\tan\phi'=\frac{\tan\phi\pm\tan\alpha}{1\mp\tan\phi\tan\alpha}=\frac{\theta+k}{1-k\theta},$$

putting in the value found for $\tan\phi$ and writing k instead of $\pm\tan\alpha$.

Thus, for the second family,

$$\frac{r\,d\theta}{dr}=\frac{\theta+k}{1-k\theta}.$$

The solution of this will be left as an exercise for the student. The result will be found to be

$$r=c(\theta+k)^{k^2+1}e^{-k\theta}.$$

Examples for solution

(1) Find the curve whose subnormal is constant.

(2) The tangent at any point P of a curve meets the axis of x in T. Find the curve for which $OP=PT$, O being the origin.

(3) Find the curve for which the angle between the tangent and radius vector at any point is twice the vectorial angle.

(4) Find the curve for which the projection of the ordinate on the normal is constant.

Find the orthogonal trajectories of the following families of curves:

(5) $x^2-y^2=a^2$.   (6) $x^{4/3}+y^{4/3}=a^{4/3}$.

(7) $px^2+qy^2=a^2$, (p and q constant).

(8) $r\theta=a$.   (9) $r=\dfrac{a\theta}{1+\theta}$.

(10) Find the family of curves that cut a family of concentric circles at a constant angle α.

MISCELLANEOUS EXAMPLES ON CHAPTER II

(1) $(3y^2 - x)\dfrac{dy}{dx} = y$. $\qquad\qquad$ (2) $x\dfrac{dy}{dx} = y + 2\sqrt{(y^2 - x^2)}$.

(3) $\tan x \cos y \, dy + \sin y \, dx + e^{\sin x} dx = 0$.

(4) $x^3 \dfrac{dy}{dx} + 3y^2 = xy^2$. [Sheffield.]

(5) $x^3 \dfrac{dy}{dx} = y^3 + y^2\sqrt{(y^2 - x^2)}$.

(6) Show that $\qquad\dfrac{dy}{dx} = -\dfrac{ax + hy + g}{hx + by + f}$

represents a family of conics.

⟨7) Show that $\qquad y\,dx - 2x\,dy = 0$
represents a system of parabolas with a common axis and tangent at the vertex.

(8) Show that $\qquad (4x + 3y + 1)\,dx + (3x + 2y + 1)\,dy = 0$
represents a family of hyperbolas having as asymptotes the lines
$$x + y = 0 \quad \text{and} \quad 2x + y + 1 = 0.$$

(9) If $\qquad\dfrac{dy}{dx} + 2y \tan x = \sin x$

and $y = 0$ when $x = \frac{1}{3}\pi$, show that the maximum value of y is $\frac{1}{8}$.
[Math. Tripos.]

(10) Show that the solution of the general homogeneous equation of the first order and degree $\dfrac{dy}{dx} = f\left(\dfrac{y}{x}\right)$ is
$$\log x = \int \frac{dv}{f(v) - v} + c,$$
where $v = y/x$.

(11) Prove that $x^h y^k$ is an integrating factor of
$$py\,dx + qx\,dy + x^m y^n (ry\,dx + sx\,dy) = 0$$
if $\qquad\dfrac{h+1}{p} = \dfrac{k+1}{q}$ and $\dfrac{h+m+1}{r} = \dfrac{k+n+1}{s}$.

Use this method to solve
$$3y\,dx - 2x\,dy + x^2 y^{-1}(10y\,dx - 6x\,dy) = 0.$$

(12) By differentiating the equation
$$\int \frac{f(xy) + F(xy)}{f(xy) - F(xy)} \frac{d(xy)}{xy} + \log \frac{x}{y} = c$$
verify that
$$\frac{1}{xy\{f(xy) - F(xy)\}}$$

is an integrating factor of

$$f(xy)\, y\, dx + F(xy)\, x\, dy = 0.$$

Hence solve $(x^2 y^2 + xy + 1)\, y\, dx - (x^2 y^2 - xy + 1)\, x\, dy = 0.$

(13) Prove that if the equation $M\, dx + N\, dy = 0$ is exact,

$$\frac{\partial N}{\partial x} = \frac{\partial M}{\partial y}.$$

[For a proof of the converse see Appendix A.]

(14) Verify that the condition for an exact equation is satisfied by

$$(P\, dx + Q\, dy)\, e^{\int f(x)\, dx} = 0$$

if

$$\frac{\partial P}{\partial y} = \frac{\partial Q}{\partial x} + Q f(x).$$

Hence show that an integrating factor can always be found for

$$P\, dx + Q\, dy = 0$$

if

$$\frac{1}{Q}\left[\frac{\partial P}{\partial y} - \frac{\partial Q}{\partial x}\right]$$

is a function of x only.

Solve by this method

$$(x^3 + xy^4)\, dx + 2y^3\, dy = 0.$$

(15) Find the curve (i) whose polar subtangent is constant;
(ii) whose polar subnormal is constant.

(16) Find the curve which passes through the origin and is such that the area included between the curve, the ordinate, and the axis of x is k times the cube of that ordinate.

(17) The normal PG to a curve meets the axis of x in G. If the distance of G from the origin is twice the abscissa of P, prove that the curve is a rectangular hyperbola.

(18) Find the curve which is such that the portion of the axis of x cut off between the origin and the tangent at any point is proportional to the ordinate of that point.

(19) Find the orthogonal trajectories of the following families of curves:

(i) $(x-1)^2 + y^2 + 2ax = 0,$

(ii) $r = a\theta,$

(iii) $r = a + \cos n\theta,$

and interpret the first result geometrically.

(20) Obtain the differential equation of the system of confocal conics

$$\frac{x^2}{a^2 + \lambda} + \frac{y^2}{b^2 + \lambda} = 1,$$

and hence show that the system is its own orthogonal trajectory.

(21) Find the family of curves cutting the family of parabolas $y^2 = 4ax$ at $45°$.

(22) If $u + iv = f(x + iy)$, where u, v, x and y are all real, prove that the families $u = $ constant, $v = $ constant are orthogonal trajectories.

Also prove that $\quad \dfrac{\partial^2 u}{\partial x^2} + \dfrac{\partial^2 u}{\partial y^2} = 0 = \dfrac{\partial^2 v}{\partial x^2} + \dfrac{\partial^2 v}{\partial y^2}.$

[This theorem is of great use in obtaining lines of force and lines of constant potential in Electrostatics or stream lines in Hydrodynamics. u and v are called Conjugate Functions.]

(23) The rate of·decay of radium is proportional to the amount remaining. Prove that the amount at any time t is given by

$$A = A_0 e^{-kt}.$$

(24) If $\dfrac{dv}{dt} = g\left(1 - \dfrac{v^2}{k^2}\right)$ and $v = 0$ if $t = 0$, prove that

$$v = k \tanh \frac{gt}{k}.$$

[This gives the velocity of a falling body in air, taking the resistance of the air as proportional to v^2. As t increases, v approaches the limiting value k. A similar equation gives the ionisation of a gas after being subjected to an ionising influence for time t.]

(25) Two liquids are boiling in a vessel. It is found that the ratio of the quantities of each passing off as vapour at any instant is proportional to the ratio of the quantities still in the liquid state. Prove that these quantities (say x and y) are connected by a relation of the form

$$y = cx^k.$$

[From Partington's *Higher Mathematics for Students of Chemistry*, p. 220.]

CHAPTER III

LINEAR EQUATIONS WITH CONSTANT COEFFICIENTS

23. The equations to be discussed in this chapter are of the form

$$p_0 \frac{d^n y}{dx^n} + p_1 \frac{d^{n-1}y}{dx^{n-1}} + \dots + p_{n-1}\frac{dy}{dx} + p_n y = f(x), \quad \dots\dots\dots(1)$$

where $f(x)$ is a function of x, but the p's are all constant.

These equations are most important in the study of vibrations of all kinds, mechanical, acoustical, and electrical. This will be illustrated by the miscellaneous examples at the end of the chapter. The methods to be given below are chiefly due to Euler and D'Alembert.*

We shall also discuss systems of simultaneous equations of this form, and equations reducible to this form by a simple transformation.

24. The simplest case; equations of the first order. If we take $n = 1$ and $f(x) = 0$, equation (1) becomes

$$p_0 \frac{dy}{dx} + p_1 y = 0, \quad \dots\dots\dots\dots\dots(2)$$

$$\textit{i.e.} \quad p_0 \frac{dy}{y} + p_1\, dx = 0,$$

or $\qquad\qquad p_0 \log y + p_1 x = \text{constant},$

so $\qquad\qquad \log y = -p_1 x/p_0 + \text{constant}$

$$= -p_1 x/p_0 + \log A, \text{ say,}$$

giving $\qquad\qquad y = A e^{-p_1 x/p_0}.$

25. Equations of the second order. If we take $n = 2$ and $f(x) = 0$, equation (1) becomes

$$p_0 \frac{d^2 y}{dx^2} + p_1 \frac{dy}{dx} + p_2 y = 0. \quad \dots\dots\dots\dots(3)$$

---

* Jean-le-Rond D'Alembert of Paris (1717-1783) is best known by "D'Alembert's Principle" in Dynamics. The application of this principle to the motion of fluids led him to partial differential equations.

The solution of equation (2) suggests that $y = Ae^{mx}$, where m is some constant, may satisfy (3).

With this value of y, equation (3) reduces to

$$Ae^{mx}(p_0 m^2 + p_1 m + p_2) = 0.$$

Thus, if m is a root of

$$p_0 m^2 + p_1 m + p_2 = 0, \quad\dots\dots\dots\dots\dots\dots(4)$$

$y = Ae^{mx}$ is a solution of equation (3), whatever the value of A.

Let the roots of equation (4) be α and β. Then, if α and β are unequal, we have two solutions of equation (3), namely

$$y = Ae^{\alpha x} \text{ and } y = Be^{\beta x}.$$

Now, if we substitute $y = Ae^{\alpha x} + Be^{\beta x}$ in equation (3), we shall get

$$Ae^{\alpha x}(p_0 \alpha^2 + p_1 \alpha + p_2) + Be^{\beta x}(p_0 \beta^2 + p_1 \beta + p_2) = 0,$$

which is obviously true as α and β are the roots of equation (4).

Thus the sum of two solutions gives a third solution (this might have been seen at once from the fact that equation (3) was *linear*). As this third solution contains two arbitrary constants, equal in number to the order of the equation, we shall regard it as the general solution.

Equation (4) is known as the " auxiliary equation."

Example

To solve $2\dfrac{d^2 y}{dx^2} + 5\dfrac{dy}{dx} + 2y = 0$ put $y = Ae^{mx}$ as a trial solution. This leads to $\qquad Ae^{mx}(2m^2 + 5m + 2) = 0,$
which is satisfied by $m = -2$ or $-\frac{1}{2}$.

The general solution is therefore

$$y = Ae^{-2x} + Be^{-\frac{1}{2}x}.$$

26. Modification when the auxiliary equation has imaginary or complex roots. When the auxiliary equation (4) has roots of the form $p + iq$, $p - iq$, where $i^2 = -1$, it is best to modify the solution

$$y = Ae^{(p+iq)x} + Be^{(p-iq)x}, \quad\dots\dots\dots\dots\dots\dots(5)$$

so as to present it without imaginary quantities.

To do this we use the theorems (given in any book on Analytical Trigonometry)

$$e^{iqx} = \cos qx + i \sin qx,$$
$$e^{-iqx} = \cos qx - i \sin qx.$$

Equation (5) becomes

$$y = e^{px}\{A(\cos qx + i\sin qx) + B(\cos qx - i\sin qx)\}$$
$$= e^{px}\{E\cos qx + F\sin qx\},$$

writing E for $A + B$ and F for $i(A - B)$. E and F are arbitrary

constants, just as A and B are. It looks at first sight as if F must be imaginary, but this is not necessarily so. Thus, if
$$A = 1 + 2i \text{ and } B = 1 - 2i, \quad E = 2 \text{ and } F = -4.$$

Example
$$\frac{d^2y}{dx^2} - 6\frac{dy}{dx} + 13y = 0$$
leads to the auxiliary equation
$$m^2 - 6m + 13 = 0,$$
whose roots are $m = 3 \pm 2i$.

The solution may be written as
$$y = Ae^{(3+2i)x} + Be^{(3-2i)x},$$
or in the preferable form
$$y = e^{3x}(E \cos 2x + F \sin 2x),$$
or again as
$$y = Ce^{3x} \cos(2x - \alpha),$$
where
$$C \cos \alpha = E \text{ and } C \sin \alpha = F,$$
so that
$$C = \sqrt{(E^2 + F^2)} \text{ and } \tan \alpha = F/E.$$

27. Peculiarity of the case of equal roots. When the auxiliary equation has equal roots $\alpha = \beta$, the solution
$$y = Ae^{\alpha x} + Be^{\beta x}$$
reduces to
$$y = (A + B)e^{\alpha x}.$$
Now $A + B$, the sum of two arbitrary constants, is really only a single arbitrary constant. Thus the solution cannot be regarded as the most general one.

We shall prove later (Art. 34) that the general solution is
$$y = (A + Bx)e^{\alpha x}.$$

28. Extension to orders higher than the second. The methods of Arts. 25 and 26 apply to equation (1) whatever the value of n, as long as $f(x) = 0$.

Ex. (i).
$$\frac{d^3y}{dx^3} - 6\frac{d^2y}{dx^2} + 11\frac{dy}{dx} - 6y = 0.$$

The auxiliary equation is
$$m^3 - 6m^2 + 11m - 6 = 0,$$
giving $m = 1, 2,$ or 3.

Thus
$$y = Ae^x + Be^{2x} + Ce^{3x}.$$

Ex. (ii).
$$\frac{d^3y}{dx^3} - 8y = 0.$$

The auxiliary equation is $m^3 - 8 = 0$,
$$i.e. \quad (m - 2)(m^2 + 2m + 4) = 0,$$
giving
$$m = 2 \text{ or } -1 \pm i\sqrt{3}.$$

Thus
$$y = Ae^{2x} + e^{-x}(E \cos x\sqrt{3} + F \sin x\sqrt{3}),$$
or
$$y = Ae^{2x} + Ce^{-x} \cos(x\sqrt{3} - \alpha).$$

Examples for solution

Solve

(1) $\dfrac{d^2y}{dx^2} + 4\dfrac{dy}{dx} + 3y = 0.$ (2) $\dfrac{d^2y}{dx^2} + 4y = 0.$

(3) $\dfrac{d^2y}{dx^2} + 7\dfrac{dy}{dx} + 12y = 0.$ (4) $\dfrac{d^2y}{dx^2} - 4\dfrac{dy}{dx} + 5y = 0.$

(5) $\dfrac{d^2s}{dt^2} + 4\dfrac{ds}{dt} + 13s = 0.$ (6) $\dfrac{d^2s}{dt^2} + 4\dfrac{ds}{dt} = 0.$

(7) $\dfrac{d^3y}{dx^3} + 2\dfrac{d^2y}{dx^2} - \dfrac{dy}{dx} - 2y = 0.$

(8) What does the solution to the last example become if the initial conditions are

$$y = 1, \frac{dy}{dx} = 0 \text{ when } x = 0,$$

and if y is to remain finite when $x = +\infty$?

Solve

(9) $\dfrac{d^4y}{dx^4} + 13\dfrac{d^2y}{dx^2} + 36y = 0.$

(10) $\dfrac{d^4y}{dx^4} - 13\dfrac{d^2y}{dx^2} + 36y = 0.$

(11) $\dfrac{d^3y}{dx^3} + 8y = 0.$ (12) $\dfrac{d^6y}{dx^6} - 64y = 0.$

(13) $l\dfrac{d^2\theta}{dt^2} + g\theta = 0$, given that $\theta = \alpha$ and $\dfrac{d\theta}{dt} = 0$ when $t = 0$.

[The approximate equation for small oscillations of a simple pendulum of length l, starting from rest in a position inclined at α to the vertical.]

(14) Find the condition that trigonometrical terms should appear in the solution of

$$m\frac{d^2s}{dt^2} + k\frac{ds}{dt} + cs = 0.$$

[The equation of motion of a particle of mass m, attracted to a fixed point in its line of motion by a force of c times its distance from that point, and damped by a frictional resistance of k times its velocity. The condition required expresses that the motion should be oscillatory, *e.g.* a tuning fork vibrating in air where the elastic force tending to restore it to the equilibrium position is proportional to the displacement and the resistance of the air is proportional to the velocity.]

(15) Prove that if k is so small that k^2/mc is negligible, the solution of the equation of Ex. (14) is approximately $e^{-kt/2m}$ times what it would be if k were zero.

[This shows that slight damping leaves the frequency practically unaltered, but causes the amplitude of successive vibrations to diminish in a geometric progression.]

(16) Solve $L\dfrac{d^2Q}{dt^2} + R\dfrac{dQ}{dt} + \dfrac{Q}{C} = 0$, given that $Q = Q_0$ and $\dfrac{dQ}{dt} = 0$ when $t = 0$, and that $CR^2 < 4L$.

[Q is the charge at time t on one of the coatings of a Leyden jar of capacity C, whose coatings are connected when $t = 0$ by a wire of resistance R and coefficient of self-induction L.]

29. The Complementary Function and the Particular Integral. So far we have dealt only with examples where the $f(x)$ of equation (1) has been equal to zero. We shall now show the relation between the solution of the equation when $f(x)$ is not zero and the solution of the simpler equation derived from it by replacing $f(x)$ by zero. To start with a simple example, consider the equation

$$2\frac{d^2y}{dx^2} + 5\frac{dy}{dx} + 2y = 5 + 2x.$$

It is obvious that $y = x$ is one solution. Such a solution, containing no arbitrary constants, is called a *Particular Integral*.

Now if we write $y = x + v$, the differential equation becomes

$$2\frac{d^2v}{dx^2} + 5\left(1 + \frac{dv}{dx}\right) + 2(x + v) = 5 + 2x,$$

$$i.e. \quad 2\frac{d^2v}{dx^2} + 5\frac{dv}{dx} + 2v = 0,$$

giving $\qquad\qquad v = Ae^{-2x} + Be^{-\frac{1}{2}x},$

so that $\qquad\qquad y = x + Ae^{-2x} + Be^{-\frac{1}{2}x}.$

The terms containing the arbitrary constants are called the *Complementary Function*.

This can easily be generalised.

If $y = u$ is a particular integral of

$$p_0\frac{d^ny}{dx^n} + p_1\frac{d^{n-1}y}{dx^{n-1}} + \ldots + p_{n-1}\frac{dy}{dx} + p_ny = f(x), \quad \ldots\ldots\ldots(6)$$

so that $\quad p_0\dfrac{d^nu}{dx^n} + p_1\dfrac{d^{n-1}u}{dx^{n-1}} + \ldots + p_{n-1}\dfrac{du}{dx} + p_nu = f(x), \quad \ldots\ldots\ldots(7)$

put $y = u + v$ in equation (6) and subtract equation (7). This gives

$$p_0\frac{d^nv}{dx^n} + p_1\frac{d^{n-1}v}{dx^{n-1}} + \ldots + p_{n-1}\frac{dv}{dx} + p_nv = 0\ldots\ldots\ldots\ldots(8)$$

If the solution of (8) be $v = F(x)$, containing n arbitrary constants, the general solution of (6) is

$$y = u + F(x),$$

and $F(x)$ is called the Complementary Function.

Thus the general solution of a linear differential equation with constant coefficients is the sum of a Particular Integral and the Complementary Function, the latter being the solution of the equation obtained by substituting zero for the function of x occurring.

Examples for solution

Verify that the given functions are particular integrals of the following equations, and find the general solutions:

(1) e^x; $\quad \dfrac{d^2y}{dx^2} - 2\dfrac{dy}{dx} + 2y = e^x$.

(2) 3; $\quad \dfrac{d^2y}{dx^2} - 13\dfrac{dy}{dx} + 12y = 36$.

(3) $2 \sin 3x$; $\quad \dfrac{d^2y}{dx^2} + 4y = -10 \sin 3x$.

For what values of the constants are the given functions particular integrals of the following equations?

(4) ae^{bx}; $\quad \dfrac{d^2y}{dx^2} + 13\dfrac{dy}{dx} + 42y = 112e^x$.

(5) ae^{bt}; $\quad \dfrac{d^2s}{dt^2} + 9s = 60e^{-t}$.

(6) $a \sin px$; $\quad \dfrac{d^2y}{dx^2} + y = 12 \sin 2x$.

(7) $a \sin px + b \cos px$; $\quad \dfrac{d^2y}{dx^2} + 4\dfrac{dy}{dx} + 3y = 8 \cos x - 6 \sin x$.

(8) a; $\quad \dfrac{d^2y}{dx^2} + 5\dfrac{dy}{dx} + 6y = 12$.

Obtain, by trial, particular integrals of the following:

(9) $\dfrac{d^2y}{dx^2} + 2\dfrac{dy}{dx} + 5y = 80e^{3x}$.

(10) $\dfrac{d^2y}{dx^2} + 2\dfrac{dy}{dx} + 37y = 300e^{7x}$.

(11) $\dfrac{d^2y}{dx^2} + 9y = 40 \sin 5x$.

(12) $\dfrac{d^2y}{dx^2} - 8\dfrac{dy}{dx} + 9y = 40 \sin 5x$.

(13) $\dfrac{d^2y}{dx^2} + 8\dfrac{dy}{dx} + 25y = 50$.

30. The operator D and the fundamental laws of algebra. When a particular integral is not obvious by inspection, it is convenient to employ certain methods involving the operator D, which stands for $\dfrac{d}{dx}$. This operator is also useful in establishing the form of the complementary function when the auxiliary equation has equal roots.

D^2 will be used for $\dfrac{d^2}{dx^2}$, D^3 for $\dfrac{d^3}{dx^3}$, and so on.

The expression $2\dfrac{d^2y}{dx^2} + 5\dfrac{dy}{dx} + 2y$ may then be written

$$2D^2y + 5Dy + 2y,$$
or
$$(2D^2 + 5D + 2)y.$$

We shall even write this in the factorised form

$$(2D + 1)(D + 2)y,$$

factorising the expression in D as if it were an **ordinary** algebraic quantity. Is this justifiable?

The operations performed in ordinary algebra **are based upon** three laws:

    I. The Distributive Law
$$m(a + b) = ma + mb;$$
    II. The Commutative Law
$$ab = ba;$$
    III. The Index Law $\quad a^m \cdot a^n = a^{m+n}.$

Now D satisfies the first and third of these laws, for

$$D(u + v) = Du + Dv,$$
and
$$D^m \cdot D^n u = D^{m+n} \cdot u$$

                                      (m and n positive integers).

As for the second law, $D(cu) = c(Du)$ is true if c is a constant, but not if c is a variable.

    Also $\qquad D^m(D^n u) = D^n(D^m u)$

                                       (m and n *positive* integers).

Thus D satisfies the fundamental laws of algebra except in that it is not commutative with variables. **In** what follows we shall write $\qquad F(D) \equiv p_0 D^n + p_1 D^{n-1} + \dots + p_{n-1}D + p_n,$

where the p's are constants and n is a positive integer. We are justified in factorising this or performing any other operations depending on the fundamental laws of algebra. For an example in which the index law for operators ceases to hold when *negative* powers of D occur, see Ex. (iii) of Art. 37.

    31. $F(D)e^{ax} = e^{ax}F(a)$. Since

$$De^{ax} = ae^{ax},$$
$$D^2 e^{ax} = a^2 e^{ax},$$
and so on,
$$\begin{aligned}
F(D)e^{ax} &= (p_0 D^n + p_1 D^{n-1} + \dots + p_{n-1}D + p_n)e^{ax} \\
&= (p_0 a^n + p_1 a^{n-1} + \dots + p_{n-1}a + p_n)e^{ax} \\
&= e^{ax}F(a).
\end{aligned}$$

32. $F(D)\{e^{ax}V\} = e^{ax}F(D+a)V$, where V is any function of x. By Leibniz's theorem for the nth differential coefficient of a product,

$$D^n\{e^{ax}V\} = (D^n e^{ax})V + n(D^{n-1}e^{ax})(DV)$$
$$+ \tfrac{1}{2}n(n-1)(D^{n-2}e^{ax})(D^2V) + \ldots + e^{ax}(D^n V)$$
$$= a^n e^{ax}V + na^{n-1}e^{ax}DV + \tfrac{1}{2}n(n-1)a^{n-2}e^{ax}D^2V + \ldots + e^{ax}D^n V$$
$$= e^{ax}\{a^n + na^{n-1}D + \tfrac{1}{2}n(n-1)a^{n-2}D^2 + \ldots + D^n\}V$$
$$= e^{ax}(D+a)^n V.$$

Similarly $D^{n-1}\{e^{ax}V\} = e^{ax}(D+a)^{n-1}V$, and so on.

Therefore

$$F(D)\{e^{ax}V\} = (p_0 D^n + p_1 D^{n-1} + \ldots + p_{n-1}D + p_n)\{e^{ax}V\}$$
$$= e^{ax}\{p_0(D+a)^n + p_1(D+a)^{n-1} + \ldots + p_{n-1}(D+a) + p_n\}V$$
$$= e^{ax}F(D+a)V.$$

33. $F(D^2)\cos ax = F(-a^2)\cos ax$. Since
$$D^2\cos ax = -a^2\cos ax,$$
$$D^4\cos ax = (-a^2)^2\cos ax,$$

and so on,

$$F(D^2)\cos ax = (p_0 D^{2n} + p_1 D^{2n-2} + \ldots + p_{n-1}D^2 + p_n)\cos ax$$
$$= \{p_0(-a^2)^n + p_1(-a^2)^{n-1} + \ldots + p_{n-1}(-a^2) + p_n\}\cos ax$$
$$= F(-a^2)\cos ax.$$

Similarly $\qquad F(D^2)\sin ax = F(-a^2)\sin ax.$

34. Complementary Function when the auxiliary equation has equal roots. When the auxiliary equation has equal roots α and α, it may be written $\qquad m^2 - 2m\alpha + \alpha^2 = 0.$

The original differential equation will then be

$$\frac{d^2y}{dx^2} - 2\alpha\frac{dy}{dx} + \alpha^2 y = 0,$$
$$i.e. \quad (D^2 - 2\alpha D + \alpha^2)y = 0,$$
$$i.e. \quad (D-\alpha)^2 y = 0. \qquad\ldots\ldots\ldots\ldots\ldots\ldots(9)$$

We have already found that $y = Ae^{\alpha x}$ is one solution. To find a more general one put $y = e^{\alpha x}V$, where V is a function of x.

By Art. 32,

$$(D-\alpha)^2\{e^{\alpha x}V\} = e^{\alpha x}(D-\alpha+\alpha)^2 V = e^{\alpha x}D^2 V.$$

Thus equation (9) becomes

$$D^2 V = 0,$$
$$i.e. \quad V = A + Bx,$$
so that $\qquad\qquad y = e^{\alpha x}(A + Bx).$

Similarly the equation $(D-\alpha)^p y = 0$
reduces to $$D^p V = 0,$$
giving $$V = (A_1 + A_2 x + A_3 x^2 + \ldots + A_p x^{p-1}),$$
and $$y = e^{\alpha x}(A_1 + A_2 x + A_3 x^2 + \ldots + A_p x^{p-1}).$$

When there are several repeated roots, as in
$$(D-\alpha)^p (D-\beta)^q (D-\gamma)^r y = 0, \quad \ldots\ldots\ldots\ldots\ldots\ldots(10)$$
we note that as the operators are commutative we may rewrite the equation in the form
$$(D-\beta)^q (D-\gamma)^r \{(D-\alpha)^p y\} = 0,$$
which is therefore satisfied by any solution of the simpler equation
$$(D-\alpha)^p y = 0. \quad \ldots\ldots\ldots\ldots\ldots\ldots\ldots(11)$$
Similarly equation (10) is satisfied by any solution of
$$(D-\beta)^q y = 0, \quad \ldots\ldots\ldots\ldots\ldots\ldots\ldots(12)$$
or of $$(D-\gamma)^r y = 0. \quad \ldots\ldots\ldots\ldots\ldots\ldots\ldots(13)$$

The general solution of (10) is the sum of the general solutions of (11), (12), and (13), containing together $(p+q+r)$ arbitrary constants.

Ex. (i). Solve $\qquad (D^4 - 8D^2 + 16)y = 0,$
$$i.e. \quad (D^2 - 4)^2 y = 0.$$
The auxiliary equation is $(m^2 - 4)^2 = 0,$
$$m = 2 \text{ (twice)} \quad \text{or} \quad -2 \text{ (twice)}.$$
Thus by the rule the solution is
$$y = (A + Bx)e^{2x} + (E + Fx)e^{-2x}.$$
Ex. (ii). Solve $\qquad (D^2 + 1)^2 y = 0.$
The auxiliary equation is $(m^2 + 1)^2 = 0,$
$$m = i \text{ (twice)} \quad \text{or} \quad -i \text{ (twice)}.$$
Thus $$y = (A + Bx)e^{ix} + (E + Fx)e^{-ix},$$
or better $$y = (P + Qx)\cos x + (R + Sx)\sin x.$$

Examples for solution

(1) $(D^4 + 2D^3 + D^2)y = 0.$ $\qquad$ (2) $(D^6 + 3D^4 + 3D^2 + 1)y = 0.$
(3) $(D^4 - 2D^3 + 2D^2 - 2D + 1)y = 0.$ $\qquad$ (4) $(4D^5 - 3D^3 - D^2)y = 0.$
(5) Show that
$$F(D^2)(P \cosh ax + Q \sinh ax) = F(a^2)(P \cosh ax + Q \sinh ax).$$
(6) Show that $(D-a)^{4n}(e^{ax} \sin px) = p^{4n} e^{ax} \sin px.$

35. Symbolical methods of finding the Particular Integral when $f(x) = e^{ax}$. The following methods are a development of the idea of treating the operator D as if it were an ordinary algebraic quantity. We shall proceed tentatively, at first performing any operations that seem plausible, and then, when a result has been obtained in this manner, verifying it by direct differentiation. We shall use the notation $\dfrac{1}{F(D)} f(x)$ to denote a particular integral of the equation

$$F(D) y = f(x).$$

(i) If $f(x) = e^{ax}$, the result of Art. 31,

$$F(D) e^{ax} = e^{ax} F(a)$$

suggests that, as long as $F(a) \neq 0$, $\dfrac{1}{F(a)} e^{ax}$ may be a value of $\dfrac{1}{F(D)} e^{ax}$.

This suggestion is easily verified, for

$$F(D) \left\{ \frac{1}{F(a)} e^{ax} \right\} = \frac{e^{ax} F(a)}{F(a)}, \quad \text{by Art. 31,}$$
$$= e^{ax}.$$

(ii) If $F(a) = 0$, $(D - a)$ must be a factor of $F(D)$.
Suppose that $F(D) = (D - a)^p \phi(D)$, where $\phi(a) \neq 0$.
Then the result of Art. 32,

$$F(D) \{ e^{ax} V \} = e^{ax} F(D + a) V,$$

suggests that the following may be true, if V is 1,

$$\frac{1}{F(D)} e^{ax} = \frac{1}{(D-a)^p \phi(D)} e^{ax} = \frac{1}{(D-a)^p} \left\{ \frac{e^{ax} \cdot 1}{\phi(a)} \right\} = \frac{e^{ax}}{\phi(a)} \frac{1}{D^p} \cdot 1$$
$$= \frac{e^{ax}}{\phi(a)} \frac{x^p}{p!}.$$

adopting the very natural suggestion that $\dfrac{1}{D}$ is the operator inverse to D, that is the operator that integrates with respect to x, while $\dfrac{1}{D^p}$ integrates p times. Again the result obtained in this tentative manner is easily verified, for

$$F(D) \left\{ \frac{e^{ax}}{\phi(a)} \frac{x^p}{p!} \right\} = (D-a)^p \phi(D) \left\{ \frac{e^{ax}}{\phi(a)} \frac{x^p}{p!} \right\}$$
$$= \phi(D) \left[(D-a)^p \left\{ \frac{e^{ax}}{\phi(a)} \frac{x^p}{p!} \right\} \right]$$
$$= \phi(D) \left[\frac{e^{ax}}{\phi(a)} D^p \frac{x^p}{p!} \right], \quad \text{by Art. 32,}$$
$$= \phi(D) \left[\frac{e^{ax}}{\phi(a)} \cdot 1 \right]$$
$$= e^{ax}, \quad \text{by Art. 31.}$$

In working numerical examples it will not be necessary to repeat the verification of our tentative methods.

Ex. (i). $$(D+3)^2 y = 50e^{2x}.$$
The particular integral is

$$\frac{1}{(D+3)^2} \cdot 50e^{2x} = \frac{50e^{2x}}{(2+3)^2} = 2e^{2x}.$$

Adding the complementary function, we get

$$y = 2e^{2x} + (A + Bx)e^{-3x}.$$

Ex. (ii). $$(D-2)^2 y = 50e^{2x}.$$

If we substitute 2 for D in $\dfrac{1}{(D-2)^2}$ $50e^{2x}$, we get infinity

But using the other method,

$$\frac{1}{(D-2)^2} \cdot 50e^{2x} = 50e^{2x} \frac{1}{D^2} \cdot 1 = 50e^{2x} \cdot \tfrac{1}{2}x^2 = 25x^2 e^{2x}.$$

Adding the complementary function, we get

$$y = 25x^2 e^{2x} + (A + Bx)e^{2x}.$$

Examples for solution

Solve

(1) $(D^2 + 6D + 25)\, y = 104e^{3x}.$ (2) $(D^2 + 2pD + p^2 + q^2)\, y = e^{ax}.$

(3) $(D^2 - 9)\, y = 54e^{3x}.$ (4) $(D^3 - D)\, y = e^x + e^{-x}.$

(5) $(D^2 - p^2)\, y = a \sinh px.$ (6) $(D^3 + 4D^2 + 4D)\, y = 8e^{-2x}.$

36. Particular Integral when f(x) = cos ax.

From Art. 33,

$$\phi(D^2) \cos ax = \phi(-a^2) \cos ax.$$

This suggests that we may obtain the particular integral by writing $-a^2$ for D^2 wherever it occurs.

Ex. (i). $$(D^2 + 3D + 2)\, y = \cos 2x.$$

$$\frac{1}{D^2 + 3D + 2} \cdot \cos 2x = \frac{1}{-4 + 3D + 2} \cdot \cos 2x = \frac{1}{3D - 2} \cdot \cos 2x.$$

To get D^2 in the denominator, try the effect of writing

$$\frac{1}{3D - 2} = \frac{3D + 2}{9D^2 - 4},$$

suggested by the usual method of dealing with surds.
This gives

$$\frac{3D + 2}{-36 - 4} \cos 2x = -\tfrac{1}{40}(3D \cos 2x + 2 \cos 2x)$$

$$= -\tfrac{1}{40}(-6 \sin 2x + 2 \cos 2x)$$

$$= \tfrac{1}{20}(3 \sin 2x - \cos 2x).$$

Ex. (ii). $(D^3 + 6D^2 + 11D + 6)\, y = 2 \sin 3x.$

$$\frac{1}{D^3 + 6D^2 + 11D + 6}\, 2 \sin 3x = 2\frac{1}{-9D - 54 + 11D + 6} \sin 3x$$

$$= \frac{1}{D - 24} \sin 3x$$

$$= \frac{D + 24}{D^2 - 576} \sin 3x$$

$$= -\tfrac{1}{585}(3 \cos 3x + 24 \sin 3x)$$

$$= -\tfrac{1}{195}(\cos 3x + 8 \sin 3x).$$

We may now show, by direct differentiation, that the results obtained are correct.

If this method is applied to

$$[\phi\, (D^2) + D\psi\, (D^2)]y = P \cos ax + Q \sin ax,$$

where P, Q and a are constants, we obtain

$$\frac{\phi\, (-a^2)\, .\, (P \cos ax + Q \sin ax) + a\psi\, (-a^2)\, .\, (P \sin ax - Q \cos ax)\, .}{\{\phi\, (-a^2)\}^2 + a^2\{\psi(-a^2)\}^2.}$$

It is quite easy to show that this is really a particular integral, provided that the denominator does not vanish. This exceptional case is treated later (Art. 38).

Examples for solution

Solve

(1) $(D + 1)\, y = 10 \sin 2x.$ (2) $(D^2 - 5D + 6)\, y = 100 \sin 4x.$

(3) $(D^2 + 8D + 25)\, y = 48 \cos x - 16 \sin x.$

(4) $(D^2 + 2D + 401)\, y = \sin 20x + 40 \cos 20x.$

(5) Prove that the particular integral of

$$\frac{d^2s}{dt^2} + 2k\, \frac{ds}{dt} + p^2s = a \cos qt$$

may be written in the form $b \cos (qt - \epsilon),$

where $b = a/\{(p^2 - q^2)^2 + 4k^2q^2\}^{\frac{1}{2}}$ and $\tan \epsilon = 2kq/(p^2 - q^2).$

Hence prove that if q is a variable and k, p and a constants, b is greatest when $q = \sqrt{(p^2 - 2k^2)} = p$ approx. if k is very small, and then $\epsilon = \pi/2$ approx. and $b = a/2kp$ approx.

[This differential equation refers to a vibrating system damped by a force proportional to the velocity and acted upon by an external periodic force. The particular integral gives the forced vibrations and the complementary function the free vibrations, which are soon damped out (see Ex. 15 following Art. 28). The forced vibrations have the greatest amplitude if the period $2\pi/q$ of the external force is very nearly equal to that of the free vibrations (which is

$2\pi/\sqrt{(p^2 - k^2)} = 2\pi/p$ approx.), and then ϵ the difference in phase between the external force and the response is approx. $\pi/2$. This is the important phenomenon of *Resonance*, which has important applications to Acoustics, Engineering and Wireless Telegraphy.]

37. Particular integral when f(x) = x^m, where m is a positive integer

In this case the tentative method is to expand $\dfrac{1}{F(D)}$ in a series of ascending powers of D.

Ex. (i).

$$\frac{1}{D^2 + 4} x^2 = \tfrac{1}{4}(1 + \tfrac{1}{4}D^2)^{-1}x^2$$
$$= \tfrac{1}{4}(1 - \tfrac{1}{4}D^2 + \tfrac{1}{16}D^4 \ldots) x^2$$
$$= \tfrac{1}{4}(x^2 - \tfrac{1}{2}).$$

Hence, adding the complementary function, the solution suggested for

$$(D^2 + 4) y = x^2$$

is $\qquad y = \tfrac{1}{4}(x^2 - \tfrac{1}{2}) + A \cos 2x + B \sin 2x.$

Ex. (ii).

$$\frac{1}{D^2 - 4D + 3} x^3 = \tfrac{1}{2}\left(\frac{1}{1 - D} - \frac{1}{3 - D}\right) x^3, \quad \text{by partial fractions,}$$

$$= \tfrac{1}{2}\left\{(1 + D + D^2 + D^3 + D^4 + \ldots) - \tfrac{1}{3}\left(1 + \frac{D}{3} + \frac{D^2}{9} + \frac{D^3}{27} + \frac{D^4}{81} + \ldots\right)\right\}x^3$$

$$= \{\tfrac{1}{3} + \tfrac{4}{9}D + \tfrac{13}{27}D^2 + \tfrac{40}{81}D^3 + \tfrac{121}{243}D^4 + \ldots\} x^3$$

$$= \tfrac{1}{3}x^3 + \tfrac{4}{3}x^2 + \tfrac{26}{9}x + \tfrac{80}{27}.$$

Adding the complementary function, the solution suggested for

$$(D^2 - 4D + 3) y = x^3$$

is $\qquad y = \tfrac{1}{3}x^3 + \tfrac{4}{3}x^2 + \tfrac{26}{9}x + \tfrac{80}{27} + Ae^x + Be^{3x}.$

Ex. (iii).

$$\frac{1}{D^2(D^2 + 4)} 96x^2 = 96 \cdot \frac{1}{D^2}\left\{\frac{1}{D^2 + 4} x^2\right\}$$

$$= 96 \cdot \frac{1}{D^2} \cdot \tfrac{1}{4}\left(x^2 - \frac{1}{2}\right), \quad \text{from Ex. (i),}$$

$$= 96 \cdot \frac{1}{4}\left(\frac{x^4}{12} - \frac{x^2}{4}\right)$$

$$= 2x^4 - 6x^2.$$

Hence the solution of $D^2(D^2 + 4) y = 96x^2$ should be

$$y = 2x^4 - 6x^2 + A \cos 2x + B \sin 2x + E + Fx.$$

Alternative method.

$$\frac{1}{D^2(D^2 + 4)} \cdot 96x^2 = \frac{96}{D^2} \cdot \tfrac{1}{4}(1 - \tfrac{1}{4}D^2 + \tfrac{1}{16}D^4 - \ldots) x^2$$

$$= (24D^{-2} - 6 + \tfrac{3}{2}D^2 - \ldots) x^2$$

$$= 2x^4 - 6x^2 + 3.$$

This gives an extra term 3, which is, however, included in the complementary function.

* The method adopted in Exs. (i) and (ii), where $F(D)$ does not contain D as a factor, may be justified as follows. Suppose the expansions have been obtained by ordinary long division. This is always possible, although the use of partial fractions may be more convenient in practice. If the division is continued until the quotient contains D^m, the remainder will have D^{m+1} as a factor. Call it $\phi(D) . D^{m+1}$. Then

$$\frac{1}{F(D)} = c_0 + c_1 D + c_2 D^2 + \dots + c_m D^m + \frac{\phi(D) . D^{m+1}}{F(D)} . \quad \dots\dots(1)$$

This is an algebraical identity, leading to

$$1 = F(D)\{c_0 + c_1 D + c_2 D^2 + \dots + c_m D^m\} + \phi(D) . D^{m+1}. \quad \dots\dots\dots(2)$$

Now equation (2), which is true when D is an algebraical quantity, is of the simple form depending only on the elementary laws of algebra which have been shown to apply to the operator D, and it does not involve the difficulties which arise when division by functions of D is concerned. Therefore equation (2) is also true when each side of the equation is regarded as an operator. Operating on x^m we get, since $D^{m+1} x^m = 0$,

$$x^m = F(D)\{(c_0 + c_1 D + c_2 D^2 + \dots + c_m D^m) x^m\}, \quad \dots\dots\dots(3)$$

which proves that the expansion obtained in (1), *disregarding the remainder*, supplies a particular integral of $F(D) y = x^m$.

It is interesting to note that this method holds good even if the expansion would be divergent for algebraical values of D.

To verify the first method in cases like Ex. (iii), we have to prove that

$$\frac{1}{D^r} . \{(c_0 + c_1 D + c_2 D^2 + \dots + c_m D^m) x^m\},$$

i.e. $\quad (c_0 D^{-r} + c_1 D^{-r+1} + c_2 D^{-r+2} + \dots + c_m D^{-r+m}) x^m,$

is a particular integral of $\{F(D) . D^r\} y = x^m,$

i.e. that $\{F(D) . D^r\} \{(c_0 D^{-r} + c_1 D^{-r+1} + c_2 D^{-r+2}$
$$+ \dots + c_m D^{-r+m}) x^m\} = x^m. \quad \dots\dots\dots(4)$$

Now $\qquad\qquad \{F(D) . D^r\} u = F(D) . \{D^r u\},$

also $\qquad\qquad D^r\{(c_s D^{-r+s}) x^m\} = (c_s D^s) x^m;$

hence the expression on the left-hand side of (4) becomes

$$F(D)\{c_0 + c_1 D + c_2 D^2 + \dots + c_m D^m) x^m\} = x^m, \text{ by (3)},$$

which is what was to be proved.

In the alternative method we get r extra terms in the particular integral, say

$$(c_{m+1} D^{-r+m+1} + \dots + c_{m+r} D^m) x^m.$$

These give terms involving the $(r-1)^{\text{th}}$ and lower powers of x. But these all occur in the complementary function. Hence the first method is preferable.

Note that if $D^{-1}u$ denotes the *simplest* form of the integral of u, without any arbitrary constant,

$$D^{-1}(D \cdot 1) = D^{-1} \cdot 0 = 0,$$

while
$$D(D^{-1} \cdot 1) = D \cdot x = 1,$$

so that
$$D(D^{-1} \cdot 1) \neq D^{-1} \cdot (D \cdot 1).$$

Similarly $D^m(D^{-m} \cdot x^n) \neq D^{-m}(D^m \cdot x^n)$, if m is greater than n.

So when *negative* powers of D concerned, the laws of algebra are *not* always obeyed. This explains why the two different methods adopted in Ex. (iii) give different results.

Examples for solution

Solve (1) $(D+1)y = x^3$.   (2) $(D^2+2D)y = 24x$.

   (3) $(D^2-6D+9)y = 54x+18$. (4) $(D^4-6D^3+9D^2)y = 54x+18$.

   (5) $(D^2-D-2)y = 44-76x-48x^2$.

   (6) $(D^3-D^2-2D)y = 44-76x-48x^2$.

38. Particular integrals in other simple cases. We shall now give some typical examples of the evaluation of particular integrals in simple cases which have not been dealt with in the preceding articles. The work is tentative, as before. For the sake of brevity the verification is omitted, as it is very similar to the verifications already given.

Ex. (i). $(D^2+4)y = \sin 2x$.

We cannot evaluate $\dfrac{1}{D^2+4}\sin 2x$ by writing -2^2 for D^2, as in Art. 36, for this gives zero in the denominator.

But $i \sin 2x$ is the imaginary part of e^{2ix}, and

$$\frac{1}{D^2+4}e^{2ix} = e^{2ix}\frac{1}{(D+2i)^2+4} \cdot 1, \qquad \text{as in Art. 35,}$$

$$= e^{2ix}\frac{1}{D(D+4i)} \cdot 1$$

$$= e^{2ix}\frac{1}{4iD} \cdot \left(1+\frac{D}{4i}\right)^{-1} \cdot 1$$

$$= e^{2ix}\frac{1}{4iD} \cdot \left\{\left(1-\frac{D}{4i}+\frac{D^2}{4^2i^2}-\cdots\right) \cdot 1\right\} \qquad \ldots\ldots\ldots(1)$$

$$= e^{2ix}\frac{1}{4iD} \cdot 1 = e^{2ix}\frac{x}{4i}$$

$$= -\tfrac{1}{4}ix(\cos 2x + i\sin 2x)$$

$$\left[otherwise \quad \frac{1}{D^2+4}e^{2ix} = \frac{1}{D-2i}\left(\frac{1}{D+2i}e^{2ix}\right) = \frac{1}{D-2i}\left(\frac{1}{4i}e^{2ix}\right)\right.$$

$$\left. = e^{2ix}\frac{1}{D} \cdot \frac{1}{4i} = e^{2ix}\frac{x}{4i}\right];$$

hence, picking out the imaginary part,

$$\frac{1}{D^2+4}\sin 2x = -\tfrac{1}{4}x\cos 2x.$$

Adding the complementary function, we get

$$y = A\cos 2x + B\sin 2x - \tfrac{1}{4}x\cos 2x.$$

Ex. (ii). $(D^2-5D+6)\,y = e^{2x}x^3.$

$$\frac{1}{(D^2-5D+6)}\cdot e^{2x}x^3 = \left(\frac{1}{2-D}-\frac{1}{3-D}\right)\cdot e^{2x}x^3$$

$$= e^{2x}\left(-\frac{1}{D}-\frac{1}{1-D}\right)x^3$$

$$= e^{2x}\left(-\frac{1}{D}-1-D-D^2-D^3-D^4-\ldots\right)x^3$$

$$= e^{2x}(-\tfrac{1}{4}x^4-x^3-3x^2-6x-6).$$

Adding the complementary function, we get

$$y = Ae^{3x} - e^{2x}(\tfrac{1}{4}x^4+x^3+3x^2+6x-B),$$

including the term $-6e^{2x}$ in $Be^{2x}.$

Ex. (iii). $(D^2-6D+13)\,y = 8e^{3x}\sin 2x.$

$$\frac{1}{(D^2-6D+13)}\cdot 8e^{3x}\sin 2x = 8e^{3x}\frac{1}{\{(D+3)^2-6(D+3)+13\}}\cdot\sin 2x$$

$$= 8e^{3x}\frac{1}{D^2+4}\sin 2x$$

$$= 8e^{3x}(-\tfrac{1}{4}x\cos 2x)\qquad\text{(see Ex. (i))}$$

$$= -2xe^{3x}\cos 2x.$$

Adding the complementary function, we get

$$y = e^{3x}(A\cos 2x + B\sin 2x - 2x\cos 2x).$$

These methods are sufficient to evaluate nearly all the particular integrals that the student is likely to meet. All other cases may be dealt with on the lines indicated in (33) and (34) of the miscellaneous examples at the end of this chapter.

Examples for solution

Solve

(1) $(D^2+1)\,y = 4\cos x.$ (2) $(D-1)\,y = (x+3)\,e^{2x}.$

(3) $(D^3-3D-2)\,y = 540x^3e^{-x}.$ (4) $(D^2+2D+2)\,y = 2e^{-x}\sin x.$

5) $(D^2+1)^2\,y = 24x\cos x.$ (6) $(D^5-D)\,y = 12e^x+8\sin x - 2x.$

(7) $(D^2-6D+25)\,y = 2e^{3x}\cos 4x + 8e^{3x}(1-2x)\sin 4x.$

39. The Homogeneous Linear Equation. This is the name given to the form $(p_0x^nD^n + p_1x^{n-1}D^{n-1} + \ldots + p_n)y = f(x).$

It reduces to the type considered before if we put $x = e^t.$

Ex. $\qquad\qquad (x^3D^3 + 3x^2D^2 + xD) \, y = 24x^2.$

Put $\qquad\qquad\qquad\qquad x = e^t,$

$$\frac{dx}{dt} = e^t = x,$$

so that $\quad D = \dfrac{d}{dx} = \dfrac{dt}{dx}\dfrac{d}{dt} = \dfrac{1}{x}\dfrac{d}{dt};$

$$D^2 = D\left(\frac{1}{x}\frac{d}{dt}\right) = -\frac{1}{x^2}\frac{d}{dt} + \frac{1}{x}D\frac{d}{dt} = \frac{1}{x^2}\left(-\frac{d}{dt} + \frac{d^2}{dt^2}\right);$$

$$D^3 = D\frac{1}{x^2}\left(-\frac{d}{dt} + \frac{d^2}{dt^2}\right) = -\frac{2}{x^3}\left(-\frac{d}{dt} + \frac{d^2}{dt^2}\right) + \frac{1}{x^2}D\left(-\frac{d}{dt} + \frac{d^2}{dt^2}\right)$$

$$= -\frac{2}{x^3}\left(-\frac{d}{dt} + \frac{d^2}{dt^2}\right) + \frac{1}{x^3}\left(-\frac{d^2}{dt^2} + \frac{d^3}{dt^3}\right)$$

$$= \frac{1}{x^3}\left(2\frac{d}{dt} - 3\frac{d^2}{dt^2} + \frac{d^3}{dt^3}\right);$$

thus the given differential equation reduces to $\dfrac{d^3y}{dt^3} = 24e^{2t},$

giving $\qquad\qquad y = A + Bt + Ct^2 + 3e^{2t}$

$$= A + B \log x + C \, (\log x)^2 + 3x^2.$$

Another method is indicated in (28)-(30) of the miscellaneous examples at the end of this chapter.

The equation

$$p_0(a+bx)^n D^n y + p_1(a+bx)^{n-1}D^{n-1}y + \ldots + p_n y = f(x)$$

can be reduced to the homogeneous linear form by putting $z = a + bx$, giving

$$Dy = \frac{dy}{dx} = \frac{dy}{dz}\frac{dz}{dx} = b\frac{dy}{dz}.$$

Examples for solution

(1) $x^2\dfrac{d^2y}{dx^2} - 2x\dfrac{dy}{dx} + 2y = 4x^3.$ $\qquad$ (2) $x^2\dfrac{d^2y}{dx^2} + 9x\dfrac{dy}{dx} + 25y = 50.$

(3) $x^3\dfrac{d^3y}{dx^3} + 3x^2\dfrac{d^2y}{dx^2} + x\dfrac{dy}{dx} + 8y = 65 \cos (\log x).$

(4) $x^4\dfrac{d^4y}{dx^4} + 2x^3\dfrac{d^3y}{dx^3} + x^2\dfrac{d^2y}{dx^2} - x\dfrac{dy}{dx} + y = \log x.$

(5) $(1+2x)^2\dfrac{d^2y}{dx^2} - 6(1+2x)\dfrac{dy}{dx} + 16y = 8(1+2x)^2.$

(6) $(1+x)^2\dfrac{d^2y}{dx^2} + (1+x)\dfrac{dy}{dx} + y = 4 \cos \log (1+x).$

40. Simultaneous linear equations with constant coefficients. The method will be illustrated by an example. We have two dependent variables, y and z, and one independent variable x. D stands for $\dfrac{d}{dx}$, as before.

Consider
$$(5D + 4)y - (2D + 1)z = e^{-x}, \quad \dots \dots \dots \dots (1)$$
$$(D + 8)y - 3z = 5e^{-x}. \quad \dots \dots \dots \dots (2)$$

Eliminate z, as in simultaneous linear equations of elementary algebra. To do this we multiply equation (1) by 3 and operate on equation (2) by $(2D + 1)$.

Subtracting the results, we get
$$\{3(5D + 4) - (2D + 1)(D + 8)\}y = 3e^{-x} - (2D + 1)5e^{-x},$$
$$i.e. \quad (-2D^2 - 2D + 4)y = 8e^{-x},$$
or
$$(D^2 + D - 2)y = -4e^{-x}.$$

Solving this in the usual way, we get
$$y = 2e^{-x} + Ae^x + Be^{-2x}.$$

The easiest way to get z in this particular example is to use equation (2), which does not involve any differential coefficients of z. Substituting for y in (2), we get
$$14e^{-x} + 9Ae^x + 6Be^{-2x} - 3z = 5e^{-x},$$
so that
$$z = 3e^{-x} + 3Ae^x + 2Be^{-2x}.$$

However, when the equations do not permit of such a simple method of finding z, we may eliminate y. (But see p. 48.)

In our case this gives
$$\{-(D + 8)(2D + 1) + 3(5D + 4)\}z = (D + 8)e^{-x} - (5D + 4)5e^{-x},$$
$$i.e. \quad (-2D^2 - 2D + 4)z = 12e^{-x},$$
giving
$$z = 3e^{-x} + Ee^x + Fe^{-2x}.$$

To find the relation between the four constants A, B, E, and F, substitute in either of the original equations, say (2). This gives
$$(D + 8)(2e^{-x} + Ae^x + Be^{-2x}) - 3(3e^{-x} + Ee^x + Fe^{-2x}) = 5e^{-x},$$
$$i.e. \quad (9A - 3E)e^x + (6B - 3F)e^{-2x} = 0,$$
whence
$$E = 3A \quad \text{and} \quad F = 2B,$$
so
$$z = 3e^{-x} + Ee^x + Fe^{-2x} = 3e^{-x} + 3Ae^x + 2Be^{-2x}, \quad \text{as before.}$$

Examples for solution

(1) $Dy - z = 0,$
$(D - 1)y - (D + 1)z = 0.$

(2) $(D - 17)y + (2D - 8)z = 0,$
$(13D - 53)y - 2z = 0.$

(3) $(2D^2 - D + 9)y - (D^2 + D + 3)z = 0,$
$(2D^2 + D + 7)y - (D^2 - D + 5)z = 0.$

(4) $(D+1)y = z + e^x$,   (5) $(D^2+5)y - 4z = -36 \cos 7x$,

     $(D+1)z = y + e^x$. $y + D^2z = 99 \cos 7x$.

(6) $(2D+1)y + (D+32)z = 91e^{-x} + 147 \sin 2x + 135 \cos 2x$,

       $y - (D-8)z = 29e^{-x} + 47 \sin 2x + 23 \cos 2x$.

MISCELLANEOUS EXAMPLES ON CHAPTER III

Solve

(1) $(D-1)^3 y = 16e^{3x}$.   (2) $(4D^2 + 12D + 9)y = 144xe^{-\frac{3}{2}x}$.

(3) $(D^4 + 6D^3 + 11D^2 + 6D)y = 20e^{-2x} \sin x$.

(4) $(D^3 - D^2 + 4D - 4)y = 68e^x \sin 2x$.

(5) $(D^4 - 6D^2 - 8D - 3)y = 256(x+1)e^{3x}$.

(6) $(D^4 - 8D^2 - 9)y = 50 \sinh 2x$.   (7) $(D^4 - 2D^2 + 1)y = 40 \cosh x$.

(8) $(D-2)^2 y = 8(x^2 + e^{2x} + \sin 2x)$.   (9) $(D-2)^2 y = 8x^2 e^{2x} \sin 2x$.

(10) $(D^2+1)y = 3 \cos^2 x + 2 \sin^3 x$.

(11) $(D^4 + 10D^2 + 9)y = 96 \sin 2x \cos x$.

(12) $(D-a)^a y = a^x$, where a is a positive integer.

(13) $\dfrac{d^2y}{dx^2} + \dfrac{1}{x}\dfrac{dy}{dx} = \dfrac{12 \log x}{x^2}$.   (14) $\dfrac{d^2y}{dx^2} + \dfrac{2}{x}\dfrac{dy}{dx} = 10$.

(15) $\dfrac{d^3y}{dx^3} = \dfrac{6y}{x^3}$.   (16) $(x+1)^2 \dfrac{d^2y}{dx^2} + (x+1)\dfrac{dy}{dx} = (2x+3)(2x+4)$.

(17) $\dfrac{d^2x}{dt^2} - 4\dfrac{dx}{dt} + 4x = y$,

    $\dfrac{d^2y}{dt^2} + 4\dfrac{dy}{dt} + 4y = 25x + 16e^t$.

(18) $\dfrac{dx}{dt} = 2y$; $\dfrac{dy}{dt} = 2z$; $\dfrac{dz}{dt} = 2x$.   (19) $t\dfrac{dx}{dt} + y = 0$; $t\dfrac{dy}{dt} + x = 0$.

(20) $t^2 \dfrac{d^2x}{dt^2} + t\dfrac{dx}{dt} + 2y = 0$,

    $t^2 \dfrac{d^2y}{dt^2} + t\dfrac{dy}{dt} - 2x = 0$.

(21) Show that the solution of $(D^{2n+1} - 1)y = 0$ consists of Ae^x and n pairs of terms of the form

$$e^{cx}(B_r \cos sx + C_r \sin sx),$$

where $c = \cos \dfrac{2\pi r}{2n+1}$ and $s = \sin \dfrac{2\pi r}{2n+1}$,

r taking the values 1, 2, 3 ... n successively.

(22) If $(D-a)u = 0$,

                  $(D-a)v = u$,

and $(D-a)y = v$,

find successively u, v, and y, and hence solve $(D-a)^3 y = 0$.

(23) Show that the solution of
$$(D-a)(D-a-h)(D-a-2h)\,y=0$$
can be written $Ae^{ax}+Be^{ax}\dfrac{(e^{hx}-1)}{h}+Ce^{ax}\dfrac{(e^{2hx}-2e^{hx}+1)}{h^2}$.

Hence deduce the solution of $(D-a)^3\,y=0$.

[This method is due to D'Alembert. The advanced student will notice that it is not quite satisfactory without further discussion. It is obvious that the second differential equation is the limit of the first, but it is *not* obvious that the *solution* of the second is the limit of the *solution* of the first.]

(24) If $(D-a)^3 e^{mx}$ is denoted by z, prove that z, $\dfrac{\partial z}{\partial m}$, and $\dfrac{\partial^2 z}{\partial m^2}$ all vanish when $m=a$.

Hence prove that e^{ax}, xe^{ax}, and $x^2 e^{ax}$ are all solutions of $(D-a)^3 y=0$.

[Note that the operators $(D-a)^3$ and $\dfrac{\partial}{\partial m}$ are commutative.]

(25) Show that
$$\frac{\cos ax-\cos (a+h)\,x}{(a+h)^2-a^2}$$
is a solution of $(D^2+a^2)\,y=\cos (a+h)\,x$.

Hence deduce the Particular Integral of $(D^2+a^2)\,y=\cos ax$.

[This is open to the same objection as Example 23.]

(26) Prove that if V is a function of x and $F(D)$ has its usual meaning

(i) $D^n[xV] = xD^n V+nD^{n-1}V$;

(ii) $F(D)\,[xV] = xF(D)\,V+F'(D)\,V$;

(iii) $\dfrac{1}{F(D)}\Big[xV\Big] = \Big\{x-\dfrac{1}{F(D)}\cdot F'(D)\Big\}\dfrac{1}{F(D)}\,V$;

(iv) $\dfrac{1}{F(D)}\Big[x^n V\Big] = \Big\{x-\dfrac{1}{F(D)}\cdot F'(D)\Big\}^n\dfrac{1}{F(D)}\,V$.

The operators must be used in the proper order. The work is sometimes laborious.

(27) Obtain the Particular Integrals of (i) $(D-1)\,y=xe^{2x}$,

(ii) $(D+1)\,y=x^2\cos x$,

by using the results (iii) and (iv) of the last example.

(28) Prove, by induction or otherwise, that if θ stands for $x\dfrac{d}{dx}$,
$$x^n\frac{d^n y}{dx^n}=\theta(\theta-1)(\theta-2)\ldots(\theta-n+1)\,y.$$

(29) Prove that

(i) $F(\theta)x^m = x^m F(m)$;

(ii) $\dfrac{1}{F(\theta)}x^m = \dfrac{x^m}{F(m)}$, provided $F(m)\neq0$;

(iii) $\dfrac{1}{F(\theta)}[x^m V]=x^m\cdot\dfrac{1}{F(\theta+m)}\,V$,

where V is a function of x.

(30) By using the results of the last question, prove that the solution of

$$x^2 \frac{d^2y}{dx^2} - 4x\frac{dy}{dx} + 6y = x^5 \text{ is } \tfrac{1}{6}x^5 + Ax^a + Bx^b,$$

where a and b are the roots of $m(m-1) - 4m + 6 = 0$,

i.e. 2 and 3.

(31) Given that $(D-1)y = e^{2x}$,

prove that $(D-1)(D-2)y = 0$.

By writing down the general solution of the second differential equation (involving two unknown constants) and substituting in the first, obtain the value of one of these constants, hence obtaining the solution of the first equation.

(32) Solve $\dfrac{d^2y}{dx^2} + p^2y = \sin ax$ by the method of the last question.

(33) If u_1 denotes $e^{ax} \displaystyle\int ue^{-ax}\, dx$,

u_2 denotes $e^{bx} \displaystyle\int u_1 e^{-bx}\, dx$,

etc.,

prove the solution of $F(D)y = u$, where $F(D)$ is the product of n factors

$$(D-a)(D-b) \ldots$$

may be written $y = u_n$.

This is true even if the factors of $F(D)$ are not all different.

Hence solve $(D-a)(D-b)y = e^{ax} \log x$.

(34) By putting $\dfrac{1}{F(D)}$ into partial fractions, prove the solution of

$F(D)y = u$ may be expressed in the form

$$\Sigma \frac{1}{F'(a)} e^{ax} \int ue^{-ax}\, dx,$$

provided the factors of $F(D)$ are all different.

[If the factors of $F(D)$ are not all different, we get repeated integrations.]

Theoretically the methods of this example and the last enable us to solve any linear equation with constant coefficients. Unfortunately, unless u is one of the simple functions (products of exponentials, sines and cosines, and polynomials) discussed in the text, we are generally left with an indefinite integration which cannot be performed.

If $u = f(x)$, we can rewrite $e^{ax} \displaystyle\int ue^{-ax}\, dx$

in the form $\displaystyle\int_k^x f(t) e^{a(x-t)}\, dt$,

where the lower limit k is an arbitrary constant.

(35) (i) Verify that

$$y = \frac{1}{p} \int_k^x f(t) \sin p (x - t) \, dt$$

is a Particular Integral of

$$\frac{d^2y}{dx^2} + p^2 y = f(x).$$

[Remember that if a and b are functions of x,

$$\frac{d}{dx} \int_a^b F(x, t) \, dt = F(x, b) \frac{db}{dx} - F(x, a) \frac{da}{dx} \quad \int_a^b \frac{dF(x, t)}{dx} \, dt.]$$

(ii) Obtain this Particular Integral by using the result of the last example.

(iii) Hence solve $\quad (D^2 + 1) y = \text{cosec } x$.

(iv) Show that this method will also give the solution of

$$(D^2 + 1) \, y = f(x)$$

(in a form free from signs of integration, if $f(x)$ is any one of the functions tan x, cot x, sec x).

(36) Show that the Particular Integral of $\dfrac{d^2y}{dt^2} + p^2 y = k \cos pt$ represents an oscillation with an indefinitely increasing amplitude.

[This is the phenomenon of RESONANCE, which we have mentioned before (see Ex. 5 following Art. 36). Of course the physical equations of this type are only approximate, so it must not be assumed that the oscillation really becomes infinite. Still it may become too large for safety. It is for this reason that soldiers break step on crossing a bridge, in case their steps might be in tune with the natural oscillation of the structure.]

(37) Show that the Particular Integral of

$$\frac{d^2y}{dt^2} + 2h \frac{dy}{dt} + (h^2 + p^2) \, y = ke^{-ht} \cos pt$$

represents an oscillation with a variable amplitude $\dfrac{k}{2p} \, te^{-ht}$.

Find the maximum value of this amplitude, and show that it is very large if h is very small. What is the value of the amplitude after an infinite time?

[This represents the forced vibration of a system which is *in resonance* with the forcing agency, when both are *damped* by friction. The result shows that if this friction is small the forced vibrations soon become large, though not infinite as in the last example. This is an advantage in some cases. If the receiving instruments of wireless telegraphy were not in resonance with the Hertzian waves, the effects would be too faint to be detected.]

(38) Solve $\dfrac{d^4y}{dx^4} - n^4y = 0.$

[This equation gives the lateral displacement y of any portion of a thin vertical shaft in rapid rotation, x being the vertical height of the portion considered.]

(39) If, in the last example,

$$\frac{dy}{dx} = y = 0 \quad \text{when} \quad x = 0 \quad \text{and} \quad x = l,$$

prove that $\qquad y = E\,(\cos nx - \cosh nx) + F\,(\sin nx - \sinh nx)$

and $\qquad\qquad\qquad\qquad \cos nl \cosh nl = 1.$

[This means that the shaft is supported at two points, one a height l above the other, and is compelled to be vertical at these points. The last equation gives n when l is known.]

(40) Prove that the Complementary Function of

$$\frac{d^3y}{dt^3} + 3\frac{d^2y}{dt^2} + 4\frac{dy}{dt} + 2y = 40$$

becomes negligible when t increases sufficiently, while that of

$$\frac{d^3y}{dt^3} - \frac{d^2y}{dt^2} + 2y = 40$$

oscillates with indefinitely increasing amplitude.

[An equation of this type holds approximately for the angular velocity of the governor of a steam turbine. The first equation corresponds to a *stable* motion of revolution, the second to *unstable* motion or "*hunting*." See the Appendix to Perry's *Steam Engine*.]

(41) Prove that the general solution of the simultaneous equations:

$$m\frac{d^2x}{dt^2} = Ve - He\frac{dy}{dt},$$

$$m\frac{d^2y}{dt^2} = He\frac{dx}{dt},$$

where m, V, H, and e are constants, is

$$x = A + B\cos(\omega t - \alpha),$$

$$y = \frac{V}{H}t + C + B\sin(\omega t - \alpha),$$

where $\omega = \dfrac{He}{m}$ and A, B, C, α are arbitrary constants.

Given that $\dfrac{dx}{dt} = \dfrac{dy}{dt} = x = y = 0$ when $t = 0$, show that these reduce to

$$x = \frac{V}{\omega H}\,(1 - \cos \omega t),$$

$$y = \frac{V}{\omega H}\,(\omega t - \sin \omega t), \text{ the equations of a cycloid.}$$

[These equations give the path of a corpuscle of mass m and charge e repelled from a negatively-charged sheet of zinc illuminated with ultra-violet light, under a magnetic field H parallel to the surface. V is the electric intensity due to the charged surface. By finding experimentally the greatest value of x, Sir J. J. Thomson determined $\dfrac{2V}{\omega H}$, from which the important ratio $\dfrac{m}{e}$ is calculated when V and H are known. See *Phil. Mag.* Vol. 48, p. 547, 1899.]

(42) Given the simultaneous equations,

$$L_1 \frac{d^2 I_1}{dt^2} + M \frac{d^2 I_2}{dt^2} + \frac{I_1}{c_1} = Ep \cos pt,$$

$$L_2 \frac{d^2 I_2}{dt^2} + M \frac{d^2 I_1}{dt^2} + \frac{I_2}{c_2} = 0$$

where L_1, L_2, M, c_1, c_2, E and p are constants, prove that I_1 is of the form

$$a_1 \cos pt + A_1 \cos (mt - \alpha) + B_1 \cos (nt - \beta),$$

and I_2 of the form

$$a_2 \cos pt + A_2 \cos (mt - \alpha) + B_2 \cos (nt - \beta),$$

where

$$a_1 = \frac{E}{k} pc_1(1 - p^2 c_2 L_2),$$

$$a_2 = \frac{EM}{k} p^3 c_1 c_2,$$

k denoting the expression

$$(L_1 L_2 - M^2)c_1 c_2 p^4 - (L_1 c_1 + L_2 c_2)p^2 + 1\ ;$$

m and n are certain *definite* constants; A_1, B_1, α and β are arbitrary constants; and A_2 is expressible in terms of A_1 and B_2 in terms of B_1.

Prove further that m and n are *real* if L_1, L_2, M, c_1, and c_2 are real and positive, and $L_1 L_2 > M^2$.

[These equations give the primary and secondary currents I_1 and I_2 in a transformer when the circuits contain condensers of capacities c_1 and c_2. L_1 and L_2 are the coefficients of self-induction and M that of mutual induction. The resistances (which are usually very small) have been neglected. $E \sin pt$ is the impressed E.M.F. of the primary.]

Alternative methods for simultaneous equations. In Ex. 3, p. 42, having found y, we can find z without integration by operating on the given equations by D and $(D+2)$ respectively and subtracting. Given $f(D)$, $F(D)$, any two polynomials in D with no common factor containing D, we can find other polynomials $\phi(D)$, $\psi(D)$, such that

$$\phi(D)f(D) - \psi(D)\cdot F(D) = 1. \quad \text{(Cf. Smith's *Algebra*, Art. 100.)}$$

In simple cases we can obtain $\phi(D)$, $\psi(D)$ by inspection.

Alternatively, we may replace the given equations of Ex. 3 by their sum and difference. Proceeding similarly in Ex. 4, we may take $y + z$ and $y - z$ as new variables.

CHAPTER IV

SIMPLE PARTIAL DIFFERENTIAL EQUATIONS

41. In this chapter we shall consider some of the ways in which partial differential equations arise, the construction of simple particular solutions, and the formation of more complex solutions from infinite series of the particular solutions. We shall also explain the application of Fourier's Series, by which we can make these complex solutions satisfy given conditions.

The equations considered include those that occur in problems on the conduction of heat, the vibrations of strings, electrostatics and gravitation, telephones, electro-magnetic waves, and the diffusion of solvents.

The methods of this chapter are chiefly due to Euler, D'Alembert, and Lagrange.*

42. Elimination of arbitrary functions. In Chapter I. we showed how to form *ordinary* differential equations by the elimination of arbitrary constants. Partial differential equations can often be formed by the elimination of arbitrary functions.

Ex. (i). Eliminate the arbitrary functions f and F from

$$y = f(x - at) + F(x + at). \quad \dots\dots\dots\dots\dots\dots\dots\dots(1)$$

We get
$$\frac{\partial y}{\partial x} = f'(x - at) + F'(x + at)$$

and
$$\frac{\partial^2 y}{\partial x^2} = f''(x - at) + F''(x + at). \quad \dots\dots\dots\dots\dots\dots(2)$$

Similarly
$$\frac{\partial y}{\partial t} = -af'(x - at) + aF'(x + at)$$

and
$$\frac{\partial^2 y}{\partial t^2} = a^2 f''(x - at) + a^2 F''(x + at). \quad \dots\dots\dots\dots\dots(3)$$

---

* Joseph Louis Lagrange of Turin (1736-1813), the greatest mathematician of the eighteenth century, contributed largely to every branch of Mathematics. He created the Calculus of Variations and much of the subject of Partial Differential Equations, and he greatly developed Theoretical Mechanics and Infinitesimal Calculus.

From (2) and (3),
$$\frac{\partial^2 y}{\partial x^2} = \frac{1}{a^2}\frac{\partial^2 y}{\partial t^2}, \quad\dots\dots\dots\dots\dots\dots\dots(4)$$

a partial differential equation of the second order.*

Ex. (ii). Eliminate the arbitrary function f from
$$z = f\left(\frac{y}{x}\right).$$

We get
$$\frac{\partial z}{\partial x} = -\frac{y}{x^2}f'\left(\frac{y}{x}\right)$$

and
$$\frac{\partial z}{\partial y} = \frac{1}{x}f'\left(\frac{y}{x}\right),$$

so
$$x\frac{\partial z}{\partial x} + y\frac{\partial z}{\partial y} = 0.$$

Examples for solution

Eliminate the arbitrary functions from the following equations:

(1) $z = f(x + ay)$.   (2) $z = f(x + iy) + F(x - iy)$, where $i^2 = -1$.

(3) $z = f(x\cos\alpha + y\sin\alpha - at) + F(x\cos\alpha + y\sin\alpha + at)$.

(4) $z = f(x^2 - y^2)$.   (5) $z = e^{ax+by}f(ax - by)$.

(6) $z = x^n f\left(\frac{y}{x}\right)$.

43. Elimination of arbitrary constants. We have seen in Chapter I. how to eliminate arbitrary constants by ordinary differential equations. This can also be effected by partials.

Ex. (i). Eliminate A and p from $z = Ae^{pt}\sin px$.

We get
$$\frac{\partial^2 z}{\partial x^2} = -p^2 Ae^{pt}\sin px,$$

and
$$\frac{\partial^2 z}{\partial t^2} = p^2 Ae^{pt}\sin px;$$

therefore
$$\frac{\partial^2 z}{\partial x^2} + \frac{\partial^2 z}{\partial t^2} = 0.$$

Ex. (ii). Eliminate a, b, and c from
$$z = a(x + y) + b(x - y) + abt + c.$$

We get
$$\frac{\partial z}{\partial x} = a + b,$$

$$\frac{\partial z}{\partial y} = a - b,$$

$$\frac{\partial z}{\partial t} = ab.$$

---

* This equation holds for the transverse vibrations of a stretched string. The most general solution of it is equation (1), which represents two waves travelling with speed *a*, one to the right and the other to the left. See pp. 61, 218, 256.

But $$(a+b)^2 - (a-b)^2 = 4ab.$$

Therefore $$\left(\frac{\partial z}{\partial x}\right)^2 - \left(\frac{\partial z}{\partial y}\right)^2 = 4\frac{\partial z}{\partial t}.$$

Examples for solution

Eliminate the arbitrary constants from the following equations:

(1) $z = Ae^{-p^2t} \cos px.$ (2) $z = Ae^{-pt} \cos qx \sin ry,$ where $p^2 = q^2 + r^2.$

(3) $z = ax + (1-a)y + b.$ (4) $z = ax + by + a^2 + b^2.$

(5) $z = (x-a)^2 + (y-b)^2.$ (6) $az + b = a^2x + y.$

44. Special difficulties of partial differential equations. As we have already stated in Chapter I., every *ordinary* differential equation of the nth order may be regarded as derived from a solution containing n arbitrary *constants*.* It might be supposed that every *partial* differential equation of the nth order was similarly derivable from a solution containing n arbitrary *functions*. However, this is not true. In general it is impossible to express the eliminant of n arbitrary functions as a partial differential equation of order n. An equation of a higher order is required, and the result is not unique.†

In this chapter we shall content ourselves with finding particular solutions. By means of these we can solve such problems as most commonly arise from physical considerations.‡ We may console ourselves for our inability to find the most general solutions by the reflection that in those cases when they have been found it is often extremely difficult to apply them to any particular problem.§

* It will be shown later (Chap. VI.) that in certain exceptional cases an ordinary differential equation admits of Singular Solutions in addition to the solution with arbitrary constants. These Singular Solutions are not derivable from the ordinary solution by giving the constants particular values, but are of quite a different form.

† See Edwards' *Differential Calculus*, Arts. 512 and 513, or Williamson's *Differential Calculus*, Art. 317.

‡ The physicist will take it as obvious that every such problem *has* a solution, and moreover that this solution is *unique*. From the point of view of pure mathematics, it is a matter of great difficulty to prove the first of these facts: this proof has only been given quite recently by the aid of the Theory of Integral Equations (see Heywood and Fréchet's *L'Equation de Fredholm et ses applications à la Physique Mathématique*). The second fact is easily proved by the aid of Green's Theorem (see Carslaw's *Conduction of Heat*, 2nd ed. p. 14).

§ For example, Whittaker has proved that the most general solution of Laplace's equation

$$\frac{\partial^2 V}{\partial x^2} + \frac{\partial^2 V}{\partial y^2} + \frac{\partial^2 V}{\partial z^2} = 0$$

is $$V = \int_0^{2\pi} f(x \cos t + y \sin t + iz, t)\, dt,$$

but if we wish to find a solution satisfying certain given conditions on a given surface, we generally use a solution in the form of an infinite series.

45. Simple particular solutions

Ex. (i). Consider the equation $\dfrac{\partial^2 z}{\partial x^2} = \dfrac{1}{a^2} \dfrac{\partial z}{\partial t}$ (which gives the conduction of heat in one dimension). This equation is *linear*. Now, in the treatment of ordinary linear equations we found exponentials very useful. This suggests $z = e^{mx+nt}$ as a trial solution. Substituting in the differential equation, we get

$$m^2 e^{mx+nt} = \frac{1}{a^2} n e^{mx+nt},$$

which is true if $\qquad n = m^2 a^2.$

Thus $e^{mx+m^2a^2t}$ is a solution.

Changing the sign of m, $e^{-mx+m^2a^2t}$ is also a solution.

Ex. (ii). Find a solution of the same equation that vanishes when $t = +\infty$.

In the previous solutions t occurs in $e^{m^2a^2t}$. This increases with t, since m^2a^2 is positive if m and a are real. To make it decrease, put $m = ip$, so that $m^2a^2 = -p^2a^2$.

This gives $e^{ipx - p^2a^2t}$ as a solution.

Similarly $e^{-ipx - p^2a^2t}$ is a solution.

Hence, as the differential equation is linear, $e^{-p^2a^2t}(Ae^{ipx} + Be^{-ipx})$ is a solution, which we replace, as usual, by

$$e^{-p^2a^2t}(E \cos px + F \sin px).$$

Ex. (iii). Find a solution of $\dfrac{\partial^2 z}{\partial x^2} + \dfrac{\partial^2 z}{\partial y^2} = 0$ which shall vanish when $y = +\infty$, and also when $x = 0$.

Putting $z = e^{mx+ny}$, we get $(m^2 + n^2)e^{mx+ny} = 0$, so $m^2 + n^2 = 0$.

The condition when $y = +\infty$ demands that n should be real and negative, say $n = -p$.

Then $\qquad\qquad\qquad m = \pm ip.$

Hence $\qquad\quad e^{-py}(Ae^{ipx} + Be^{-ipx})$ is a solution,

$\qquad$ *i.e.* $\quad e^{-py}(E \cos px + F \sin px)$ is a solution.

But $\qquad\qquad\quad z = 0$ if $x = 0$, so $E = 0$.

The solution required is therefore $Fe^{-py} \sin px$.

Examples for solution

(1) $\dfrac{\partial^2 y}{\partial x^2} = \dfrac{\partial^2 y}{\partial t^2}$, given that $y = 0$ when $x = +\infty$ and also when $t = +\infty$.

(2) $\dfrac{\partial^2 z}{\partial x^2} = \dfrac{1}{a^2} \dfrac{\partial^2 z}{\partial y^2}$, given that z is never infinite (for any real values of x or y), and that $z = 0$ when $x = 0$ or $y = 0$.

(3) $\dfrac{\partial z}{\partial x} + a \dfrac{\partial z}{\partial y} = 0$, given that z is never infinite, and that $\dfrac{\partial z}{\partial x} = 0$ when $x = y = 0$.

(4) $\dfrac{\partial^2 V}{\partial x^2} + \dfrac{\partial^2 V}{\partial y^2} + \dfrac{\partial^2 V}{\partial z^2} = 0$, given that $V = 0$ when $x = +\infty$, when $y = -\infty$, and also when $z = 0$.

(5) $\dfrac{\partial^2 V}{\partial x^2} = \dfrac{\partial^2 V}{\partial y \, \partial z}$, given that V is never infinite, and that $V = C$ and $\dfrac{\partial V}{\partial x} = \dfrac{\partial V}{\partial y} = \dfrac{\partial V}{\partial z} = 0$ when $x = y = z = 0$.

(6) $\dfrac{\partial^2 V}{\partial x^2} + \dfrac{\partial^2 V}{\partial y^2} \quad \dfrac{\partial V}{\partial t}$, given that $V = 0$ when $t = +\infty$, when $x = 0$ or l, and when $y = 0$ or l.

46. More complicated initial and boundary conditions.* In Ex. (iii) of Art. 45, we found $Fe^{-py} \sin px$ as a solution of

$$\frac{\partial^2 z}{\partial x^2} + \frac{\partial^2 z}{\partial y^2} = 0,$$

satisfying the conditions that $z = 0$ if $y = +\infty$ or if $x = 0$.

Suppose that we impose two extra conditions, † say $z = 0$ if $x = l$ and $z = lx - x^2$ if $y = 0$ for all values of x between 0 and l.

The first condition gives $\sin pl = 0$,

    i.e. $pl = n\pi$, where n is any integer.

For simplicity we will at first take $l = \pi$, giving $p = n$, any integer.

The second condition gives $F \sin px = \pi x - x^2$ for *all* values of x between 0 and π. This is impossible.

However, instead of the solution consisting of a single term, we may take

$$F_1 e^{-y} \sin x + F_2 e^{-2y} \sin 2x + F_3 e^{-3y} \sin 3x + \dots ,$$

since the equation is linear (if this is not clear, cf. Chap. III. Art. 25) giving p the values 1, 2, 3, ... and adding the results.

By putting $y = 0$ and equating to $\pi x - x^2$ we get

$$F_1 \sin x + F_2 \sin 2x + F_3 \sin 3x + \dots$$
$$= \pi x - x^2 \text{ for } \textit{all} \text{ values of } x \text{ between 0 and } \pi.$$

The student will possibly think this equation as impossible to satisfy as the other, but it is a remarkable fact that we *can* choose values of the F's that make this true.

This is a particular case of a more general theorem, which we now enunciate.

---

* As t usually denotes time and x and y rectangular coordinates, a condition such as $z = 0$ when $t = 0$ is called an *initial* condition, while one such as $z = 0$ if $x = 0$, or if $x = l$, or if $y = x$, is called a *boundary* condition.

† This is the problem of finding the steady distribution of temperature in a semi-infinite rectangular strip of metal of breadth l, when the infinite sides are kept at $0°$ and the base at $(lx - x^2)°$.

47. Fourier's Half-Range Series. Every function of x which satisfies certain conditions can be expanded in a convergent series of the form

$$f(x) = a_1 \sin x + a_2 \sin 2x + a_3 \sin 3x + \dots \text{ to inf.}$$

for all values of x between 0 and π (but not necessarily for the extreme values $x = 0$ and $x = \pi$).

This is called Fourier's * half-range sine series.

The conditions alluded to are satisfied in practically every physical problem.†

Similarly, under the same conditions $f(x)$ may be expanded in a half-range cosine series

$$b_0 + b_1 \cos x + b_2 \cos 2x + b_3 \cos 3x + \dots \text{ to inf.}$$

These are called half-range series as against the series valid between 0 and 2π, which contains both sine and cosine terms.

The proofs of these theorems are very long and difficult.‡ However, if it be *assumed that these expansions are possible*, it is easy to find the values of the coefficients.

Multiply the sine series by $\sin nx$, and integrate term by term,§ giving

$$\int_0^\pi f(x) \sin nx \, dx = a_1 \int_0^\pi \sin x \sin nx \, dx + a_2 \int_0^\pi \sin 2x \sin nx \, dx + \dots .$$

The term with a_n as a factor is

$$a_n \int_0^\pi \sin^2 nx \, dx$$

$$= \frac{a_n}{2} \int_0^\pi (1 - \cos 2nx) \, dx = \frac{a_n}{2} \left[x - \frac{1}{2n} \sin 2nx \right]_0^\pi$$

$$= \tfrac{1}{2} a_n \pi.$$

---

* Jean Baptiste Joseph Fourier of Auxerre (1768-1830) is best known as the author of *La Théorie analytique de la chaleur*. His series arose in the solution of problems on the conduction of heat.

† It is *sufficient* for $f(x)$ to be single-valued, finite, and continuous, and have only a limited number of maxima and minima between $x = 0$ and $x = \pi$. However, these conditions are not necessary. The necessary and sufficient set of conditions has not yet been discovered.

‡ For a full discussion of Fourier's Series, see Carslaw's *Fourier's Series and Integrals* and Hobson's *Theory of Functions*.

§ The assumption that this is legitimate is another point that requires justification.

The term involving any other coefficient, say a_r, is

$$a_r \int_0^\pi \sin rx \sin nx \, dx$$

$$= \frac{a_r}{2} \int_0^\pi \{\cos (n-r)x - \cos (n+r)x\} \, dx$$

$$= \frac{a_r}{2} \left[\frac{\sin (n-r)x}{n-r} - \frac{\sin (n+r)x}{n+r} \right]_0^\pi = 0.$$

So all the terms on the right vanish except one.

Thus

$$\int_0^\pi f(x) \sin nx \, dx = \tfrac{1}{2} a_n \pi,$$

or

$$a_n = \frac{2}{\pi} \int_0^\pi f(x) \sin nx \, dx.$$

Similarly, it is easy to prove that if

$$f(x) = b_0 + b_1 \cos x + b_2 \cos 2x + \ldots$$

for values of x between 0 and π, then

$$b_0 = \frac{1}{\pi} \int_0^\pi f(x) \, dx$$

and

$$b_n = \frac{2}{\pi} \int_0^\pi f(x) \cos nx \, dx$$

for values of n other than 0.

48. Examples of Fourier's Series

(i) Expand $\pi x - x^2$ in a half-range sine series, valid between $x = 0$ and $x = \pi$.

It is better not to quote the formula established in the last article.

Let $\qquad \pi x - x^2 = a_1 \sin x + a_2 \sin 2x + a_3 \sin 3x + \ldots$.

Multiply by $\sin nx$ and integrate from 0 to π, giving

$$\int_0^\pi (\pi x - x^2) \sin nx \, dx = a_n \int_0^\pi \sin^2 nx \, dx = \frac{\pi}{2} a_n, \text{ as before.}$$

Now, integrating by parts,

$$\int_0^\pi (\pi x - x^2) \sin nx \, dx = \left[-\frac{1}{n} (\pi x - x^2) \cos nx \right]_0^\pi + \frac{1}{n} \int_0^\pi (\pi - 2x) \cos nx \, dx$$

$$= 0 + \left[\frac{1}{n^2} (\pi - 2x) \sin nx \right]_0^\pi + \frac{2}{n^2} \int_0^\pi \sin nx \, dx$$

$$= 0 - \frac{2}{n^3} \left[\cos nx \right]_0^\pi = \frac{4}{n^3} \text{ if } n \text{ is odd or 0 if } n \text{ is even.}$$

Thus $a_n = \dfrac{8}{\pi n^3}$ if n is odd or 0 if n is even, giving finally

$$\pi x - x^2 = \frac{8}{\pi} (\sin x + \tfrac{1}{27} \sin 3x + \tfrac{1}{125} \sin 5x + \ldots).$$

(ii) Expand $f(x)$ in a half-range series valid from $x = 0$ to $x = \pi$, where

$$f(x) = mx \text{ between } x = 0 \text{ and } x = \frac{\pi}{2}$$

and $\qquad\qquad f(x) = m(\pi - x) \text{ between } x = \frac{\pi}{2} \text{ and } x = \pi.$

In this case $f(x)$ is given by different analytical expressions in different parts of the range.* The only novelty lies in the evaluation of the integrals.

In this case

$$\int_0^\pi f(x) \sin nx \, dx = \int_0^{\pi/2} f(x) \sin nx \, dx + \int_{\pi/2}^\pi f(x) \sin nx \, dx$$

$$= \int_0^{\pi/2} mx \sin nx \, dx + \int_{\pi/2}^\pi m(\pi - x) \sin nx \, dx.$$

We leave the rest of the work to the student. The result is

$$\frac{4m}{\pi} \left(\sin x - \tfrac{1}{9} \sin 3x + \tfrac{1}{25} \sin 5x - \tfrac{1}{49} \sin 7x + \ldots \right).$$

The student should draw the graph of the given function, and compare it with the graph of the first term and of the sum of the first two terms of this expansion.†

Examples for solution

Expand the following functions in half-range sine series, valid between $x = 0$ and $x = \pi$:

(1) 1.   (2) x.   (3) x^3.   (4) $\cos x$.   (5) e^x.

(6) $f(x) = 0$ from $x = 0$ to $x = \frac{\pi}{4}$, and from $x = \frac{3\pi}{4}$ to π,

$\qquad f(x) = (4x - \pi)(3\pi - 4x)$ from $x = \frac{\pi}{4}$ to $x = \frac{3\pi}{4}$.

(7) Which of these expansions hold good (a) for $x = 0$?
$\qquad\qquad\qquad\qquad\qquad\qquad\qquad\qquad$ (b) for $x = \pi$?

49. Application of Fourier's series to satisfy boundary conditions. We can now complete the solution of the problem of Art. 46.

We found in Art. 46 that

$$F_1 e^{-y} \sin x + F_2 e^{-2y} \sin 2x + F_3 e^{-3y} \sin 3x + \ldots$$

satisfied all the conditions, if

$$F_1 \sin x + F_2 \sin 2x + F_3 \sin 3x + \ldots = \pi x - x^2$$

for all values of x between 0 and π.

---

* Fourier's theorem applies even if $f(x)$ is given by a graph with no analytical expression at all, if the conditions given in the footnote to Art. 47 are satisfied.

For a function given graphically, these integrals are determined by arithmetical approximation or by an instrument known as a Harmonic Analyser.

† Several of the graphs will be found in Carslaw's *Fourier's Series and Integrals*, 2nd ed., Chap. VII. More elaborate ones are given in the *Phil. Mag.*, Vol. 45 (1898).

In Ex. (i) of Art. 48 we found that, between 0 and π,

$$\frac{8}{\pi}(\sin x + \tfrac{1}{27}\sin 3x + \tfrac{1}{125}\sin 5x + \ldots) = \pi x - x^2.$$

Thus the solution required is

$$\frac{8}{\pi}(e^{-y}\sin x + \tfrac{1}{27}e^{-3y}\sin 3x + \tfrac{1}{125}e^{-5y}\sin 5x + \ldots).$$

50. In the case when the boundary condition involved l instead of π, we found $Fe^{-py}\sin px$ as a solution of the differential equation, and the conditions showed that p, instead of being a positive integer n, must be of the form $n\pi/l$.

Thus $\qquad F_1 e^{-\pi y/l}\sin \pi x/l + F_2 e^{-2\pi y/l}\sin 2\pi x/l + \ldots$

satisfies all the conditions if

$$F_1\sin \pi x/l + F_2\sin 2\pi x/l + \ldots = lx - x^2$$

for all values of x between 0 and l.

Put $\pi x/l = z$. Then $lx - x^2 = \dfrac{l^2}{\pi^2}(\pi z - z^2)$. The F's are thus $\dfrac{l^2}{\pi^2}$ times as much as before. The solution is therefore

$$\frac{8l^2}{\pi^3}(e^{-\pi y/l}\sin \pi x/l + \tfrac{1}{27}e^{-3\pi y/l}\sin 3\pi x/l + \tfrac{1}{125}e^{-5\pi y/l}\sin 5\pi x/l + \ldots).$$

MISCELLANEOUS EXAMPLES ON CHAPTER IV

(1) Verify that $V = \dfrac{1}{\sqrt{t}}e^{-x^2/4Kt}$ is a solution of

$$\frac{\partial^2 V}{\partial x^2} = \frac{1}{K}\frac{\partial V}{\partial t}.$$

(2) Eliminate A and p from $V = Ae^{-px}\sin(2p^2Kt - px)$.

(3) Transform $\qquad \dfrac{\partial V}{\partial t} = K\dfrac{\partial^2 V}{\partial x^2} - hV$

to $\qquad \dfrac{\partial W}{\partial t} = K\dfrac{\partial^2 W}{\partial x^2}$

by putting $V = e^{-ht}W$.

[The first equation gives the temperature of a conducting rod whose surface is allowed to radiate heat into air at temperature zero. The given transformation reduces the problem to one without radiation.]

(4) Transform

$$\frac{\partial V}{\partial t} = \frac{K}{r^2}\frac{\partial}{\partial r}\left(r^2\frac{\partial V}{\partial r}\right) \quad \text{to} \quad \frac{\partial W}{\partial t} = K\frac{\partial^2 W}{\partial r^2}$$

by putting $W = rV$.

[The first equation gives the temperature of a sphere, when heat flows radially.]

(5) Eliminate the arbitrary functions from

$$V = \frac{1}{r}[f(r-at) + F(r+at)].$$

(6) (i) Show that if e^{mx+int} is a solution of

$$\frac{\partial V}{\partial t} = K\frac{\partial^2 V}{\partial x^2} - hV,$$

where n and h are real, then m must be complex.

(ii) Hence, putting $m = -g - if$, show that $V_0 e^{-gx}\sin(nt-fx)$ is a solution that reduces to $V_0\sin nt$ for $x=0$, provided $K(g^2-f^2)=h$ and $n=2Kfg$.

(iii) If $V=0$ when $x=+\infty$, show that if K and n are positive so are g and f.

[In Ångström's method of measuring K (the " diffusivity "), one end of a very long bar is subjected to a periodic change of temperature $V_0\sin nt$. This causes heat waves to travel along the bar. By measuring their velocity and rate of decay n/f and g are found. K is then calculated from $K = n/2fg$.]

(7) Find a solution of $\dfrac{\partial V}{\partial t} = K\dfrac{\partial^2 V}{\partial x^2}$ reducing to $V_0\sin nt$ for $x=0$ and to zero for $x=+\infty$.

[This is the problem of the last question when no radiation takes place. The bar may be replaced by a semi-infinite solid bounded by a plane face, if the flow is always perpendicular to that face. Kelvin found K for the earth by this method.]

(8) Prove that the simultaneous equations

$$-\frac{\partial V}{\partial x} = RI + L\frac{\partial I}{\partial t},$$

$$-\frac{\partial I}{\partial x} = KV + C\frac{\partial V}{\partial t},$$

are satisfied by

$$V = V_0 e^{-(g+if)x+int},$$

$$I = I_0 e^{-(g+if)x+int},$$

if

$$g^2 - f^2 = RK - n^2 LC,$$

$$2fg = n(RC + LK),$$

and

$$I_0^2(R + iLn) = V_0^2(K + iCn).$$

[These are Heaviside's equations for a telephone cable with resistance R, capacity C, inductance L, and leakance K, all measured per unit length. I is the current and V the electromotive force.]

(9) Show that in the last question g is independent of n if $RC = KL$.

[The attenuation of the wave depends upon g, which in general depends upon n. Thus, if a sound is composed of harmonic waves of different frequencies, these waves are transmitted with different degrees of attenuation. The sound received at the other end is therefore distorted. Heaviside's device of increasing L and K to make $RC = KL$ prevents this distortion.]

(10) In question (8), if $L = K = 0$, show that both V and I are propagated with velocity $\sqrt{(2n/RC)}$.
[The velocity is given by n/f.]

(11) Show that the simultaneous equations

$$\frac{k}{c}\frac{\partial P}{\partial t} = \frac{\partial \gamma}{\partial y} - \frac{\partial \beta}{\partial z}; \qquad -\frac{\mu}{c}\frac{\partial \alpha}{\partial t} = \frac{\partial R}{\partial y} - \frac{\partial Q}{\partial z};$$

$$\frac{k}{c}\frac{\partial Q}{\partial t} = \frac{\partial \alpha}{\partial z} - \frac{\partial \gamma}{\partial x}; \qquad -\frac{\mu}{c}\frac{\partial \beta}{\partial t} = \frac{\partial P}{\partial z} - \frac{\partial R}{\partial x};$$

$$\frac{k}{c}\frac{\partial R}{\partial t} = \frac{\partial \beta}{\partial x} - \frac{\partial \alpha}{\partial y}; \qquad -\frac{\mu}{c}\frac{\partial \gamma}{\partial t} = \frac{\partial Q}{\partial x} - \frac{\partial P}{\partial y};$$

are satisfied by

$$P = 0; \qquad\qquad \alpha = 0;$$
$$Q = 0; \qquad\qquad \beta = \beta_0 \sin p(x - vt);$$
$$R = R_0 \sin p(x - vt); \qquad \gamma = 0;$$

provided that $v = c/\sqrt{k\mu}$ and $\beta_0 = -\sqrt{(k/\mu)}\, R_0$.

[These are Maxwell's electromagnetic equations for a dielectric of specific inductive capacity k and permeability μ. P, Q, R are the components of the electric intensity and α, β, γ those of the magnetic intensity. c is the ratio of the electromagnetic to the electrostatic units (which is equal to the velocity of light in free ether). The solution shows that plane electromagnetic waves travel with the velocity $c/\sqrt{k\mu}$, and that the electric and magnetic intensities are perpendicular to the direction of propagation and to each other.]

(12) Find a solution of $\dfrac{\partial V}{\partial t} = K \dfrac{\partial^2 V}{\partial x^2}$ such that

$V \neq \infty$ if $t = +\infty$;

$V = 0$ if $x = 0$ or π, for all values of t;

$V = \pi x - x^2$ if $t = 0$, for values of x between 0 and π.

[N.B. Before attempting this question read again Arts. 46 and 49. V is the temperature of a non-radiating rod of length π whose ends are kept at $0°$, the temperature of the rod being initially $(\pi x - x^2)°$ at a distance x from an end.]

(13) What does the solution of the last question become if the length of the rod is l instead of π?
[N.B. Proceed as in Art. 50.]

(14) Solve question (12) if the condition $V = 0$ for $x = 0$ or π is replaced by $\dfrac{\partial V}{\partial x} = 0$ for $x = 0$ or π.

[Instead of the ends being at a constant temperature, they are here treated so that no heat can pass through them.]

(15) Solve equation (12) if the expression $\pi x - x^2$ is replaced by 100.

(16) Find a solution of $\dfrac{\partial V}{\partial t} = K \dfrac{\partial^2 V}{\partial x^2}$ such that

$V \neq \infty$ if $t = +\infty$;

$V = 100$ if $x = 0$ or π for all values of t;

$V = 0$ if $t = 0$ for all values of x between 0 and π.

[Here the initially ice-cold rod has its ends in boiling water.]

(17) Solve question (15) if the length is l instead of π. If l increases indefinitely, show that the infinite series becomes the integral

$$\frac{200}{\pi} \int_0^\infty \frac{1}{\alpha} e^{-K\alpha^2 t} \sin \alpha x \, d\alpha.$$

[*N.B.* This is called a Fourier's Integral. To obtain this result put $(2r + 1)\pi/l = \alpha$ and $2\pi/l = d\alpha$.

Kelvin used an integral in his celebrated estimate of the age of the earth from the observed rate of increase of temperature underground. (See example (107) of the miscellaneous set at the end of the book.) Strutt's later discovery that heat is continually generated within the earth by radio-active processes shows that Kelvin's estimate was too small.]

(18) Find a solution of $\dfrac{\partial V}{\partial t} = K \dfrac{\partial^2 V}{\partial x^2}$ such that

V is finite when $t = +\infty$;

$\left.\begin{array}{l} \dfrac{\partial V}{\partial x} = 0 \text{ when } x = 0, \\[2mm] V = 0 \text{ when } x = l, \end{array}\right\}$ for all values of t;

$V = V_0$ when $t = 0$, for all values of x between 0 and l.

[If a small test-tube containing a solution of salt is completely submerged in a very large vessel full of water, the salt diffuses up out of the test-tube into the water of the large vessel. If V_0 is the initial concentration of the salt and l the length of test-tube it fills, V gives the concentration at any time at a height x above the bottom of the test-tube. The condition $\dfrac{\partial V}{\partial x} = 0$ when $x = 0$ means that no diffusion takes place at the closed end. $V = 0$ when $x = l$ means that at the top of the test-tube we have nearly pure water.]

(19) Find a solution of $\dfrac{\partial^2 y}{\partial t^2} = v^2 \dfrac{\partial^2 y}{\partial x^2}$ such that

y involves x trigonometrically;

$y = 0$ when $x = 0$ or π, for all values of t;

$\dfrac{\partial y}{\partial t} = 0$ when $t = 0$, for all values of x;

$\left.\begin{array}{l} y = mx \text{ between } x = 0 \text{ and } \dfrac{\pi}{2}, \\[3mm] y = m(\pi - x) \text{ between } x = \dfrac{\pi}{2} \text{ and } \pi, \end{array}\right\}$ when $t = 0$.

[*N.B.* See the second worked example of Art. 48.

y is the transverse displacement of a string stretched between two points a distance π apart. The string is plucked aside a distance $m\pi/2$ at its middle point and then released.]

* (20) Writing the solution of $\dfrac{d^2y}{dx^2} = D^2y$, where D is a constant, in the form

$$y = e^{xD}A + e^{-xD}B,$$

deduce the solution of $\dfrac{\partial^2y}{\partial x^2} = \dfrac{\partial^2y}{\partial t^2}$ in the form

$$y = f(t+x) + F(t-x)$$

by substituting $\dfrac{\partial}{\partial t}$ for D, $f(t)$ and $F(t)$ for A and B respectively, and using Taylor's theorem in its symbolical form

$$f(t+x) = e^{xD}f(t).$$

[The results obtained by these symbolical methods should be regarded merely as *probably* correct. Unless they can be verified by other means, a very careful examination of the argument is necessary to see if it can be taken backwards from the result to the differential equation.

Heaviside has used symbolical methods to solve some otherwise insoluble problems. See his *Electromagnetic Theory*.]

* (21) From the solution of $\dfrac{dy}{dx} = D^2y$, where D is a constant, deduce that of $\dfrac{\partial y}{\partial x} = \dfrac{\partial^2y}{\partial t^2}$ in the form

$$y = f(t) + x\frac{\partial^2f}{\partial t^2} + \frac{x^2}{2!}\frac{\partial^4f}{\partial t^4} + \cdots .$$

[This is not a solution unless the series is convergent.]

General solution of $\dfrac{\partial^2y}{\partial x^2} = \dfrac{1}{a^2}\dfrac{\partial^2y}{\partial t^2}$.

As a trial solution put $y = f(x+mt)$, where m is constant.

This gives $\qquad f''(x+mt) = \dfrac{m^2}{a^2} \cdot f''(x+mt),$

which is satisfied if $\qquad m = \pm a.$

Thus $y = f(x-at)$ and $y = F(x+at)$ are two solutions, and as the differential equation is linear, a third solution is

$$y = f(x-at) + F(x+at),$$

containing a number of arbitrary functions equal to the order (two) of the differential equation, so no more general solution can be expected. (Cf. pp. 218 and 256.)

[Arts. 178-181 form a supplement to this chapter. They deal chiefly with the equation of vibrating strings and with the three-dimensional wave equation. At the end of Art. 181 is a list of some important works on the differential equations of Mathematical Physics.]

* To be omitted on a first reading.

CHAPTER V

EQUATIONS OF THE FIRST ORDER BUT NOT OF THE FIRST DEGREE

51. In this chapter we shall deal with some special types of equations of the first order and of degree higher than the first for which the solution can sometimes be obtained without the use of infinite series. For brevity dy/dx will be denoted by p.

These special types are:

        (a) Those solvable for p.

        (b) Those solvable for y.

        (c) Those solvable for x.

52. Equations solvable for p. If we can solve for p, the equation of the nth degree is reduced to n equations of the first degree, to which we apply the methods of Chap. II.

Ex. (i). The equation $p^2 + px + py + xy = 0$ gives

$$p = -x \quad \text{or} \quad p = -y;$$

from which $\qquad 2y = -x^2 + c_1 \quad \text{or} \quad x = -\log y + c_2;$

or, expressed as one equation,

$$(2y + x^2 - c_1)(x + \log y - c_2) = 0. \qquad \dots\dots\dots\dots\dots(1)$$

At this point we meet with a difficulty; the complete primitive apparently contains *two* arbitrary constants, whereas we expect only *one*, as the equation is of the first order.

But consider the solution

$$(2y + x^2 - c)(x + \log y - c) = 0. \qquad \dots\dots\dots\dots\dots(2)$$

If we are considering only one value of each of the constants c, c_1, and c_2, these equations each represent a pair of curves, and of course not the same pair (unless $c = c_1 = c_2$). But if we consider the infinite set of pairs of curves obtained by giving the constants all possible values from $-\infty$ to $+\infty$, we shall get the same infinite set when taken altogether, though possibly in a different order. Thus (2) can be taken as the complete primitive.

Ex. (ii). $$p^2 + p - 2 = 0.$$

Here $$p = 1 \quad \text{or} \quad p = -2,$$

giving $$y = x + c_1 \quad \text{or} \quad y = -2x + c_2.$$

As before, we take the complete primitive as

$$(y - x - c)(y + 2x - c) = 0,$$

not $$(y - x - c_1)(y + 2x - c_2) = 0.$$

Each of these equations represents all lines parallel either to $y = x$ or to $y = -2x$.

Examples for solution

(1) $p^2 + p - 6 = 0.$ (2) $p^2 + 2xp = 3x^2.$ (3) $p^2 = x^5.$

(4) $x + yp^2 = p(1 + xy).$ (5) $p^3 - p(x^2 + xy + y^2) + xy(x + y) = 0.$

(6) $p^2 - 2p \cosh x + 1 = 0.$

53. Equations solvable for y. If the equation is solvable for y, we differentiate the solved form with respect to x.

Ex. (i). $$p^2 - py + x = 0.$$

Solving for y, $$y = p + \frac{x}{p}.$$

Differentiating, $$p = \frac{dp}{dx} + \frac{1}{p} - \frac{x}{p^2}\frac{dp}{dx},$$

$$i.e. \quad \left(p - \frac{1}{p}\right)\frac{dx}{dp} + \frac{x}{p^2} = 1.$$

This is a linear equation of the first order, considering p as the independent variable. Proceeding as in Art. 19, the student will obtain

$$x = p(c + \cosh^{-1} p)(p^2 - 1)^{-\frac{1}{2}}.$$

Hence, as $y = p + \dfrac{x}{p}$, $\quad y = p + (c + \cosh^{-1} p)(p^2 - 1)^{-\frac{1}{2}}.$

These two equations for x and y in terms of p give the parametric equations of the solution of the differential equation. For any given value of c, to each value of p correspond one definite value of x and one of y, defining a point. As p varies, the point moves, tracing out a curve. In this example we can eliminate p and get the equation connecting x and y, but for tracing the curve the parametric forms are as good, if not better.

Ex. (ii). $$3p^5 - py + 1 = 0.$$

Solving for y, $$y = 3p^4 + p^{-1}.$$

Differentiating, $$p = 12p^3\frac{dp}{dx} - p^{-2}\frac{dp}{dx},$$

$$i.e. \quad dx = (12p^2 - p^{-3})\,dp.$$

Integrating, $$x = 4p^3 + \tfrac{1}{2}p^{-2} + c,$$

and from above, $$y = 3p^4 + p^{-1}.$$

The student should trace the graph of this for some particular value of c, say $c = 0$.

54. Equations solvable for x. If the equation is solvable for x, we differentiate the solved form with respect to y, and rewrite $\dfrac{dx}{dy}$ in the form $\dfrac{1}{p}$.

Ex. $p^2 - py + x = 0$. This was solved in the last article by solving for y.

Solving for x, $\qquad\qquad x = py - p^2$.

Differentiating with respect to y,

$$\frac{1}{p} = p + y\frac{dp}{dy} - 2p\frac{dp}{dy},$$

$$i.e. \quad \left(p - \frac{1}{p}\right)\frac{dy}{dp} + y = 2p,$$

which is a linear equation of the first order, considering p as the independent and y as the dependent variable. This may be solved as in Art. 19. The student will obtain the result found in the last article.

Examples for solution

(1) $x = 4p + 4p^3$.

(2) $p^2 - 2xp + 1 = 0$.

(3) $y = p^2x + p$.

(4) $y = x + p^3$.

(5) $p^3 + p = e^y$.

(6) $2y + p^2 + 2p = 2x(p+1)$.

(7) $p^3 - p(y+3) + x = 0$.

(8) $y = p\sin p + \cos p$.

(9) $y = p\tan p + \log\cos p$.

(10) $e^{p-y} = p^2 - 1$.

(11) $p = \tan\left(x - \dfrac{p}{1+p^2}\right)$.

(12) Prove that all curves of the family given by the solution of Ex. 1 cut the axis of y at right angles. Find the value of c for that curve of the family that goes through the point $(0, 1)$.

Trace this curve on squared paper.

(13) Trace the curve given by the solution of Ex. 9 with $c = 0$. Draw the tangents at the points given by $p = 0$, $p = \cdot 1$, $p = \cdot 2$ and $p = \cdot 3$, and verify, by measurement, that the gradients of these tangents are respectively 0, $\cdot 1$, $\cdot 2$ and $\cdot 3$.

CHAPTER VI

SINGULAR SOLUTIONS *

55. We know from coordinate geometry that the straight line $y = mx + \dfrac{a}{m}$ touches the parabola $y^2 = 4ax$, whatever the value of m.

Consider the point of contact P of any particular tangent. At P the tangent and parabola have the same direction, so they have a common value of $\dfrac{dy}{dx}$, as well as of x and y.

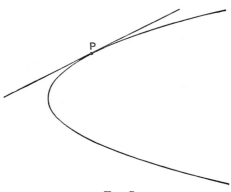

Fɪɢ. 7

But for the tangent $m = \dfrac{dy}{dx} = p$ say, so the tangent satisfies the differential equation $y = px + \dfrac{a}{p}$.

Hence the equation holds also for the parabola at P, where x, y, and p are the same as for the tangent. As P may be any point on the parabola, the equation of the parabola $y^2 = 4ax$ must be a solution of the differential equation, as the student will easily verify.

* The arguments of this chapter will be based upon geometrical intuition. The results therefore cannot be considered to be *proved*, but merely suggested as *probably true in certain cases*. The analytical theory presents grave difficulties (see M. J. M. Hill, *Proc. Lond. Math. Soc.*, 1918).

In general, if we have any singly infinite system of curves which all touch a fixed curve, which we will call their *envelope*,* and if this family represents the complete primitive of a certain differential equation of the first order, then the envelope represents a solution of the differential equation. For at every point of the envelope x, y, and p have the same value for the envelope and the curve of the family that touches it there.

Such a solution is called a Singular Solution. It does not contain any arbitrary constant, and is not deducible from the Complete Primitive by giving a particular value to the arbitrary constant in it, save in exceptional cases (Art. 160).

Examples for solution

Prove that the straight line $y = x$ is the envelope of the family of parabolas $y = x + \frac{1}{4}(x - c)^2$. Prove that the point of contact is (c, c), and that $p = 1$ for the parabola and envelope at this point. Obtain the differential equation of the family of parabolas in the form $y = x + (p - 1)^2$, and verify that the equation of the envelope satisfies this.

Trace the envelope and a few parabolas of the family, taking c as 0, 1, 2, etc.

56. We shall now consider how to obtain singular solutions. It has been shown that the envelope of the curves represented by the complete primitive gives a singular solution, so we shall commence by examining the method of finding envelopes.

The general method † is to eliminate the parameter c between $(x, y, c) = 0$, the equation of the family of curves, and

$$\frac{\partial f}{\partial c} = 0.$$

E.g.   if $f(x, y, c) = 0$ is $y - cx - \dfrac{1}{c} = 0$, ...(1)

$$\frac{\partial f}{\partial c} = 0 \quad \text{is} \quad -x + \frac{1}{c^2} = 0, \quad \text{.....................(2)}$$

giving $c = \pm 1/\sqrt{x}.$

---

* In Lamb's *Infinitesimal Calculus*, 2nd ed., Art. 155, the envelope of a family is defined as the locus of ultimate intersection of consecutive curves of the family. As thus defined it may include node- or cusp-loci in addition to or instead of what we have called envelopes. (We shall give a geometrical reason for this in Art. 56; see Lamb for an analytical proof.)

† See Lamb's *Infinitesimal Calculus*, 2nd ed., Art. 156. If $f(x, y, c,)$ is of the form $Lc^2 + Mc + N$, the result comes to $M^2 = 4LN$. Thus, for

$$y - cx - \frac{1}{c} = 0,$$

$$\textit{i.e.} \quad c^2 x - cy + 1 = 0,$$

the result is $y^2 = 4x.$

[Arts. 155–156, 2nd ed., become Arts. 138–139 in the 3rd ed.]

Substituting in (1), $\qquad y = \pm 2\sqrt{x},$

or $\qquad\qquad\qquad y^2 = 4x.$

This method is equivalent to finding the locus of intersection of

$$f(x, y, c) = 0,$$

and $\qquad\qquad\qquad f(x, y, c + h) = 0,$

two curves of the family with parameters that differ by a small quantity h, and proceeding to the limit when h approaches zero. The result is called the *c-discriminant* of $f(x, y, c) = 0.$

57. Now consider the diagrams 8, 9, 10, 11.

Fig. 8 shows the case where the curves of the family have no special singularity. The locus of the ultimate intersections

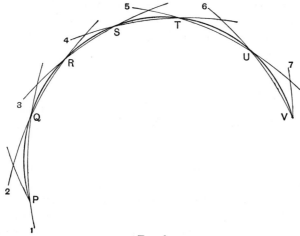

Fig. 8

$PQRSTUV$ is a curve which has two points in common with each of the curves of the family (*e.g.* Q and R lie on the locus and also on the curve marked 2). In the limit the locus $PQRSTUV$ therefore touches each curve of the family, and is what we have defined as the envelope.

In Fig. 9 each curve of the family has a node. Two consecutive curves intersect in three points (*e.g.* curves 2 and 3 in the points P, Q, and R).

The locus of such points consists of three distinct parts EE', AA', and BB'.

When we proceed to the limit, taking the consecutive curves ever closer and closer, AA' and BB' will move up to coincidence with the node-locus NN', while EE' will become an envelope. So in this case we expect the c-discriminant to contain the square of the equation of the node-locus, as well as the equation of the envelope.

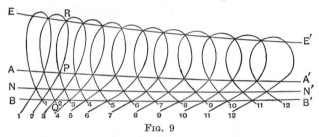

FIG. 9

As Fig. 10 shows, the direction of the node-locus NN' at any point P on it is in general not the same as that of either branch of the curve with the node at P. The node-locus has x and y in common with the curve at P, but not p, so *the node-locus is not a solution of the differential equation of the curves of the family.*

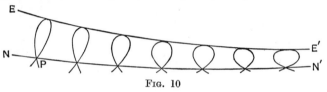

FIG. 10

If the node shrinks into a cusp, the loci EE' and NN' of Fig. 10 move up to coincidence, forming the cusp-locus CC' of Fig. 11. Now NN' was shown to be the coincidence of two loci AA' and BB' of Fig. 9, so CC' is really the coincidence of three loci, and its equation must be expected to occur cubed in the c-discriminant.

Fig. 11 shows that the cusp-locus, like the node-locus, is *not* (in general) a solution of the differential equation.

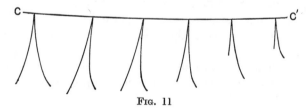

FIG. 11

To sum up, we may expect the c-discriminant to contain:

       (i) *the envelope,*
      (ii) *the node-locus squared,*
     (iii) *the cusp-locus cubed.*

The envelope is a singular solution, but the node- and cusp-loci are not (in general *) solutions at all.

58. The following examples will illustrate the preceding results:

Ex. (i). $y = p^2.$

The complete primitive is easily found to be $4y = (x - c)^2,$

$$i.e. \quad c^2 - 2cx + x^2 - 4y = 0.$$

As this is a quadratic in c, we can write down the discriminant at once as $(2x)^2 = 4(x^2 - 4y),$

i.e. $y = 0$, representing the envelope of the family of equal parabolas given by the complete primitive, and occurring to the first degree only, as an envelope should.

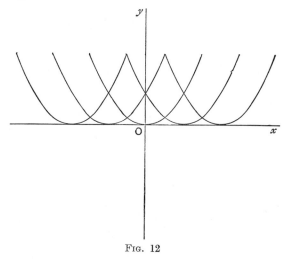

Fig. 12

Ex. (ii). $3y = 2px - 2\dfrac{p^2}{x}.$

Proceeding as in the last chapter, we get

$$3p = 2p + 2\frac{p^2}{x^2} + \left(2x - 4\frac{p}{x}\right)\frac{dp}{dx},$$

$$i.e. \quad px^2 - 2p^2 = (2x^3 - 4px)\frac{dp}{dx},$$

$$i.e. \quad x^2 - 2p = 0 \quad \text{or} \quad p = 2x\frac{dp}{dx}. \quad \dots\dots\dots\dots\dots\dots\dots(\text{A})$$

$$ie.. \quad \frac{dx}{x} = 2\frac{dp}{p},$$

---

* We say *in general*, because it is conceivable that in some special example a node- or cusp-locus may coincide with an envelope or with a curve of the family.

$$i.e. \quad \log x = 2 \log p - \log c,$$
$$i.e. \quad cx = p^2,$$

whence
$$3y = \pm 2c^{\frac{1}{2}}x^{\frac{3}{2}} - 2c,$$

i.e. $(3y + 2c)^2 = 4cx^3$, a family of semi-cubical parabolas with their cusps on the axis of y.

The c-discriminant is
$$(3y - x^3)^2 = 9y^2,$$
$$i.e. \quad x^3(6y - x^3) = 0.$$

The cusp-locus appears cubed, and the other factor represents the envelope.

It is easily verified that $6y = x^3$ is a solution of the differential equation, while $x = 0$ (giving $p = \infty$) is not.

If we take the first alternative of the equations (A),
$$i.e. \quad x^2 - 2p = 0,$$

we get by substitution for p in the differential equation
$$3y = \tfrac{1}{2}x^3,$$

i.e. the envelope.

This illustrates another method of finding singular solutions.

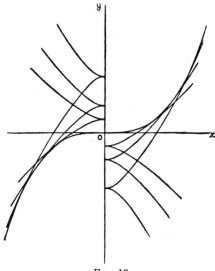

Fig. 13

Examples for solution

Find the complete primitives and singular solutions (if any) of the following differential equations. Trace the graphs for Examples 1-4:

(1) $4p^2 - 9x = 0$.

(2) $4p^2(x - 2) = 1$.

(3) $xp^2 - 2yp + 4x = 0$.

(4) $p^2 + y^2 - 1 = 0$.

(5) $p^2 + 2xp - y = 0$.

(6) $xp^2 - 2yp + 1 = 0$.

(7) $4xp^2 + 4yp - 1 = 0$.

59. The p-discriminant. We shall now consider how to obtain the singular solutions of a differential equation directly from the equation itself, without having to find the complete primitive.

Consider the equation $x^2p^2 - yp + 1 = 0$.

If we give x and y any definite numerical values, we get a quadratic for p. For example, if

$$x = \sqrt{2}, \quad y = 3, \quad 2p^2 - 3p + 1 = 0,$$

$$p = \tfrac{1}{2} \quad \text{or} \quad 1.$$

Thus there are two curves of the family satisfying this equation through every point. These two curves will have the same tangent at all points where the equation has equal roots in p, *i.e.* where the discriminant $y^2 - 4x^2 = 0$.

Similar conclusions hold for the quadratic $Lp^2 + Mp + N = 0$, where L, M, N are any functions of x and y. There are two curves through every point in the plane, but these curves have the same direction at all points on the locus $M^2 - 4LN = 0$.

More generally, the differential equation

$$f(x, y, p) \equiv L_0p^n + L_1p^{n-1} + L_2p^{n-2} + \ldots + L_n = 0,$$

where the L's are functions of x and y, gives n values of p for a given pair of values of x and y, corresponding to n curves through any point. Two of these n curves have the same tangent at all points on the locus given by eliminating p from

$$\begin{cases} f(x, y, p) = 0, \\ \dfrac{\partial f}{\partial p} = 0, \end{cases}$$

for this is the condition given in books on theory of equations for the existence of a repeated root.

We are thus led to the p-discriminant, and we must now investigate the properties of the loci represented by it.

60. The Envelope. The p-discriminant of the equation

$$y = px + \frac{1}{p}$$

or

$$p^2x - py + 1 = 0$$

is

$$y^2 = 4x.$$

We have already found that the complete primitive consists of the tangents to the parabola, which is the singular solution. Two of these tangents pass through every point P in the plane, and these tangents coincide for points on the envelope.

This is an example of the p-discriminant representing an envelope. Fig. 15 shows a more general case of this.

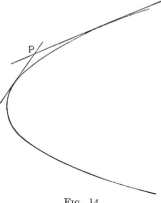

Fig. 14

Consider the curve SQP as moving up to coincidence with the curve PRT, always remaining in contact with the envelope QRU. The point P will move up towards R, and the tangents to the two curves through P will finally coincide with each other and with the tangent to the envelope at R. Thus R is a point for which the p's of the two curves of the system through the point coincide, and consequently the p-discriminant vanishes.

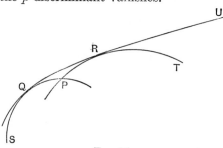

Fig. 15

Thus the p-discriminant may be an envelope of the curves of the system, and if so, as shown in Art. 55, is a singular solution.

61. The tac-locus. The envelope is thus the locus of points where two *consecutive* curves of the family have the same value of p. But it is quite possible for two non-consecutive curves to touch.

Consider a family of circles, all of equal radius, whose centres lie on a straight line.

Fig. 16 shows that the line of centres is the locus of the point of contact of pairs of circles. This is called a *tac-locus*. Fig. 17

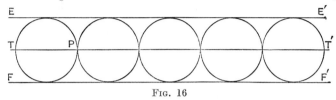

FIG. 16

shows circles which do not quite touch, but cut in pairs of neighbouring points, lying on two neighbouring loci AA', BB'. When we proceed to the limiting case of contact these two loci coincide in the tac-locus TT'. Thus the p-discriminant may be expected to contain the equation of the tac-locus squared.

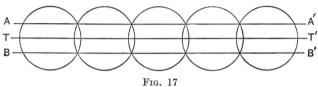

FIG. 17

It is obvious that at the point P in Fig. 16 the direction of the tac-locus is *not* the direction of the two circles. Thus the relation between x, y, and p satisfied by the circles will not be satisfied by the tac-locus, which has the same x and y but a different p at P. In general, *the tac-locus does not furnish a solution of the differential equation*.

62. The circles of the last article are represented by

$$(x+c)^2 + y^2 = r^2,$$

if the line of centres is Ox.

This gives $\qquad x+c = \pm\sqrt{r^2 - y^2},$

or $\qquad 1 = \mp\, yp/\sqrt{r^2 - y^2},$

i.e. $\qquad y^2 p^2 + y^2 - r^2 = 0.$

The p-discriminant of this is $y^2(y^2 - r^2) = 0.$

The line $y = 0$ (occurring squared, as we expected) is the tac-locus, $y = \pm r$ are the envelopes EE' and FF' of Fig. 16; $y = \pm r$, giving $p = 0$, are singular solutions of the differential equation, but $y = 0$ does not satisfy it.

63. The cusp-locus. The contact that gives rise to the equal roots in p may be between two branches of the same curve instead of between two different curves, *i.e.* the *p*-discriminant vanishes at a cusp.

As shown in Fig. 18, the direction of the cusp-locus at any point *P* on it is in general not the same as that of the tangent to the cusp, so *the cusp-locus is not a solution of the differential equation.*

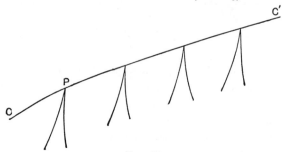

Fɪɢ. 18

It is natural to enquire if the equation of the cusp-locus will appear cubed in the *p*-discriminant, as in the *c*-discriminant. To decide this, consider the locus of points for which the two *p*'s are nearly but not quite equal, when the curves have very flat nodes. This will be the locus *NN'* of Fig. 19. In the limit, when the nodes

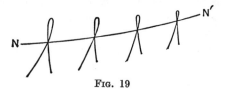

Fɪɢ. 19

contract into cusps, we get the cusp-locus, and as in this case there is no question of two or more loci coinciding, we expect the *p*-discriminant to contain the equation of the cusp-locus to the first power only.

64. Summary of results. The *p*-discriminant therefore may be expected to contain

> (i) the envelope,
> (ii) the tac-locus squared,
> (iii) the cusp-locus,

and the *c*-discriminant to contain

> (i) the envelope,
> (ii) the node-locus squared,
> (iii) the cusp-locus cubed.

Of these only the envelope is a solution of the differential equation.

65. Examples

Ex. (i). $$p^2(2-3y)^2 = 4(1-y).$$

Writing this in the form

$$\frac{dx}{dy} = \pm \frac{2-3y}{2\sqrt{(1-y)}},$$

we easily find the complete primitive in the form

$$(x-c)^2 = y^2(1-y).$$

The c-discriminant and p-discriminant are respectively

$$y^2(1-y) = 0 \quad \text{and} \quad (2-3y)^2(1-y) = 0.$$

$1-y=0$, which occurs in both to the first degree, gives an envelope; $y=0$, which occurs squared in the c-discriminant and not at all in the p-discriminant, gives a node-locus; $2-3y=0$, which occurs squared in the p-discriminant and not at all in the c-discriminant, gives a tac-locus.

It is easily verified that of these three loci only the equation of the envelope satisfies the differential equation.

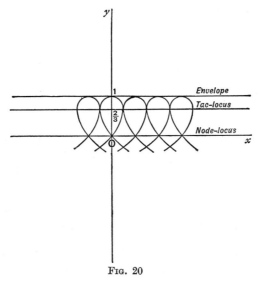

FIG. 20

Ex. (ii). Consider the family of circles

$$x^2 + y^2 + 2cx + 2c^2 - 1 = 0.$$

By eliminating c (by the methods of Chap. I.), we obtain the differential equation

$$2y^2p^2 + 2xyp + x^2 + y^2 - 1 = 0.$$

The c- and p-discriminants are respectively

$$x^2 - 2(x^2 + y^2 - 1) = 0 \quad \text{and} \quad x^2 y^2 - 2y^2(x^2 + y^2 - 1) = 0,$$
$$i.e. \quad x^2 + 2y^2 - 2 = 0 \quad \text{and} \quad y^2(x^2 + 2y^2 - 2) = 0.$$

$x^2 + 2y^2 - 2 = 0$ gives an envelope as it occurs to the first degree in both discriminants, while $y = 0$ gives a tac-locus, as it occurs squared in the p-discriminant and not at all in the c-discriminant. The circle given by the original equation touches the envelope at the points

$$\{-2c, \ \pm\sqrt{(1 - 2c^2)}\},$$

which are imaginary when c is numerically greater than $\frac{1}{2}\sqrt{2}$.

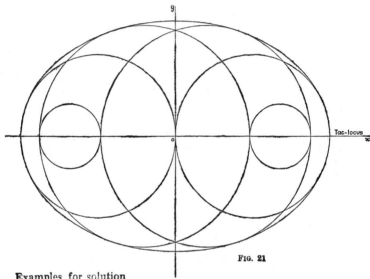

FIG. 21

Examples for solution

In the following examples find the complete primitive if the differential equation is given or the differential equation if the complete primitive is given. Find the singular solutions (if any). Trace the graphs.

(1) $4x(x - 1)(x - 2)p^2 - (3x^2 - 6x + 2)^2 = 0.$ (2) $4xp^2 - (3x - 1)^2 = 0.$

(3) $yp^2 - 2xp + y = 0.$ (4) $3xp^2 - 6yp + x + 2y = 0.$

(5) $p^2 + 2px^3 - 4x^2 y = 0.$ (6) $p^3 - 4xyp + 8y^2 = 0.$

(7) $x^2 + y^2 - 2cx + c^2 \cos^2 \alpha = 0.$ (8) $c^2 + 2cy - x^2 + 1 = 0.$

(9) $c^2 + (x + y)c + 1 - xy = 0.$ (10) $x^2 + y^2 + 2cxy + c^2 - 1 = 0.$

66. Clairaut's Form.* We commenced this chapter by considering the equation

$$y = px + \frac{a}{p}.$$

---

* Alexis Claude Clairaut, of Paris (1713-1765), although best known in connection with differential equations, wrote chiefly on astronomy.

This is a particular case of Clairaut's Form

$$y = px + f(p). \qquad \ldots\ldots\ldots\ldots\ldots\ldots(1)$$

To solve, differentiate with respect to x.

$$p = p + \{x + f'(p)\}\frac{dp}{dx};$$

therefore $\qquad \dfrac{dp}{dx} = 0, \quad p = c, \qquad \ldots\ldots\ldots\ldots\ldots\ldots\ldots(2)$

or $\qquad\qquad 0 = x + f'(p). \qquad \ldots\ldots\ldots\ldots\ldots\ldots\ldots(3)$

Using (1) and (2) we get the complete primitive, the family of straight lines, $\qquad y = cx + f(c). \qquad \ldots\ldots\ldots\ldots\ldots\ldots(4)$

If we eliminate p from (1) and (3) we shall simply get the p-discriminant.

To find the c-discriminant we eliminate c from (4) and the result of differentiating (4) partially with respect to c, *i.e.*

$$0 = x + f'(c). \qquad \ldots\ldots\ldots\ldots\ldots\ldots(5)$$

Equations (4) and (5) differ from (1) and (3) only in having c instead of p. The eliminants are therefore the same. Thus both discriminants must represent the envelope.*

Of course it is obvious that a family of straight lines cannot have node-, cusp-, or tac-loci.

Equation (4) gives the important result that *the complete primitive of a differential equation of Clairaut's Form may be written down immediately by simply writing c in place of p.*

67. Example

Find the curve such that OT varies as $\tan \psi$, where T is the point in which the tangent at any point cuts the axis of x, ψ is its inclination to this axis, and O is the origin.

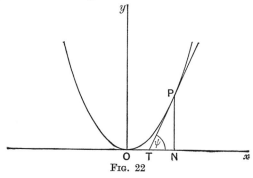

Fig. 22

---

* But in some cases the discriminants represent not only the envelope, but also its inflexional tangents (Art. 161).

From the figure, $\qquad OT = ON - TN$

$$= x - y \cot \psi$$

$$= x - \frac{y}{p},$$

since $\qquad \tan \psi = p$;

therefore $\qquad x - \dfrac{y}{p} = kp,$

$i.e.\quad y = px - kp^2.$

This is of Clairaut's Form, so the complete primitive is

$$y = cx - kc^2,$$

and the singular solution is the discriminant of this,

$i.e.\quad x^2 = 4ky.$

The curve required is the parabola represented by this singular solution. The complete primitive represents the family of straight lines tangent to this parabola.

Examples for solution

Find the complete primitive and singular solutions of the following differential equations. Trace the graphs for Examples (1), (2), (4), (7), (8) and (9).

(1) $y = px + p^2$.

(2) $y = px + p^3$.

(3) $y = px + \cos p$.

(4) $y = px + \sqrt{(a^2p^2 + b^2)}$.

(5) $p = \log (px - y)$.

(6) $\sin px \cos y = \cos px \sin y + p$.

(7) Find the differential equation of the curve such that the tangent makes with the co-ordinate axes a triangle of constant area k^2, and hence find the equation of the curve in integral form.

(8) Find the curve such that the tangent cuts off intercepts from the axes whose sum is constant.

(9) Find the curve such that the part of the tangent intercepted between the axes is of constant length.

MISCELLANEOUS EXAMPLES ON CHAPTER VI

Illustrate the solutions by a graph whenever possible.

(1) Examine for singular solutions $p^2 + 2xp = 3x^2$.

(2) Reduce $\qquad xyp^2 - (x^2 + y^2 - 1)p + xy = 0$

to Clairaut's Form by the substitution $X = x^2;\ Y = y^2$.

Hence show that the equation represents a family of conics touching the four sides of a square.

(3) Show that $xyp^2 + (x^2 - y^2 - h^2)p - xy = 0$
represents a family of confocal conics, with the foci at $(\pm h, 0)$, touching the four imaginary lines joining the foci to the circular points at infinity.

(4) Show by geometrical reasoning or otherwise that the substitution
$$x = aX + bY, \quad y = a'X + b'Y,$$
converts any differential equation of Clairaut's form to another equation of Clairaut's form.

(5) Show that the complete primitive of $8p^3x = y(12p^2 - 9)$ is $(x + c)^3 = 3y^2c$, the p-discriminant $y^2(9x^2 - 4y^2) = 0$, and the c-discriminant $y^4(9x^2 - 4y^2) = 0$. Interpret these discriminants.

(6) Reduce the differential equation
$$x^2p^2 + yp(2x + y) + y^2 = 0, \quad \text{where } p = \frac{dy}{dx}$$
to Clairaut's form by the substitution $\xi = y, \eta = xy$.

Hence, or otherwise, solve the equation.

Prove that $y + 4x = 0$ is a singular solution; and that $y = 0$ is both part of the envelope and part of an ordinary solution. [London.]

(7) Solve $y^2\left(y - x\frac{dy}{dx}\right) = x^4\left(\frac{dy}{dx}\right)^2$, which can be transformed to Clairaut's form by suitable substitutions. [London.]

(8) Integrate the differential equations:

(i) $3(p + x)^2 = (p - x)^3$.

(ii) $y^2(1 + 4p^2) - 2pxy - 1 = 0$.

In (ii) find the singular solution and explain the significance of any factors that occur. [London.]

(9) Show that the curves of the family
$$y^2 - 2cx^2y + c^2(x^4 - x^3) = 0$$
all have a cusp at the origin, touching the axis of x.

By eliminating c obtain the differential equation of the family in the form
$$4p^2x^2(x - 1) - 4pxy(4x - 3) + (16x - 9)y^2 = 0.$$

Show that both discriminants take the form $x^3y^2 = 0$, but that $x = 0$ is not a solution, while $y = 0$ is a particular integral.

[This example shows that our theory does not apply without modification to families of curves with a cusp at a fixed point.]

(10) Show that the complete primitive of
$$r^4 + r^2\left(\frac{dr}{d\theta}\right)^2 = a^4$$
represents the family of equal lemniscates of Bernoulli
$$r^2 = a^2\cos 2(\theta - \alpha),$$
inscribed in the circle $r = a$, which is the singular solution, with the point $r = 0$ as a node-locus.

(11) Obtain and interpret the complete primitive and singular solution of

$$\left(\frac{dr}{d\theta}\right)^2 + r^2 - 2ra = 0.$$

(12) Show that $r = c\theta - c^2$ is the complete primitive and $4r = \theta^2$ the singular solution of

$$r = \theta\frac{dr}{d\theta} - \left(\frac{dr}{d\theta}\right)^2.$$

Verify that the singular solution touches the complete primitive at the point $(c^2, 2c)$, the common tangent there making an angle $\tan^{-1} c$ with the radius vector.

[For a supplementary discussion of singular solutions, including difficulties concerning their definition and the definition of an envelope, the occurrence of particular solutions in the discriminants, the idea of boundaries, and the methods of calculating discriminants, see Arts. 160-161. These will throw additional light on Exs. 7 and 9 above.]

CHAPTER VII

MISCELLANEOUS METHODS FOR EQUATIONS OF THE SECOND AND HIGHER ORDERS

68. In this chapter we shall be concerned chiefly with the reduction of equations of the second order to those of the first order. We shall show that the order can always be so reduced if the equation

	(i)	does not contain y explicitly;
or	(ii)	does not contain x explicitly;
or	(iii)	is homogeneous.

A special form of equation, of some importance in Dynamics, may be reduced by using an integrating factor.

The remainder of the chapter will be devoted to the linear equation, excluding the simple case, already fully discussed in Chapter III., where the coefficients are merely constants. It will be found that the linear equation of the second order can be reduced to one of the first order if

	(i)	the operator can be factorised,
or	(ii)	any one integral belonging to the complementary function is known.

If the complete complementary function is known, the equation may be solved by the method of *Variation of Parameters*. This elegant method (due to Lagrange) is applicable to linear equations of any order.

Further information on linear equations, such as the condition for exact equations, the normal form, the invariantive condition of equivalence, and the Schwarzian derivative, will be found in the form of problems among the miscellaneous examples at the end of the chapter, with hints sufficient to enable the student to work them out for himself.

We shall use suffixes to denote differentiations with respect to x, e.g. y_2 for $\dfrac{d^2y}{dx^2}$, but when the independent variable is any other than x the differential coefficients will be written in full.

69. y absent. If y does not occur explicitly in an equation of the second order, write p for y_1 and $\dfrac{dp}{dx}$ for y_2.

We obtain an equation containing only $\dfrac{dp}{dx}$, p, and x, and so of the first order.

Consider, for example, $xy_2 + y_1 = 4x$.

This transforms into $x\dfrac{dp}{dx} + p = 4x$,

which can be integrated at once
$$xp = 2x^2 + a,$$
$$\text{i.e.} \quad p = 2x + \frac{a}{x}.$$

By integrating, $y = x^2 + a \log x + b$,
where a and b are arbitrary constants.

This method may be used to reduce an equation of the nth order not containing y explicitly to one of the $(n-1)^{\text{th}}$.

70. x absent. If x is the absent letter, we may still write p for y_1, but for y_2 we now write $p\dfrac{dp}{dy}$, since $p\dfrac{dp}{dy} = \dfrac{dy}{dx}\dfrac{dp}{dy} = \dfrac{dp}{dx} = y_2$. The procedure reduces an equation of the second order without x to one of the first order in the variables p and y.

For example, $yy_2 = y_1^2$

transforms into $yp\dfrac{dp}{dy} = p^2$,

from which the student will easily obtain
$$p = by \quad \text{and} \quad y = ae^{bx}.$$

Examples for solution

(1) $y_2 \cos^2 x = 1$.   (2) $yy_2 + y_1^2 = y_1$.   (3) $yy_2 + 1 = y_1^2$.

(4) Reduce to the previous example, and hence solve
$$y_1 y_3 + y_1^2 = 2y_2^2.$$

(5) $xy_3 + y_2 = 12x$.   (6) $y_n - 2y_{n-1} = e^x$.

(7) Integrate and interpret geometrically
$$\frac{(1 + y_1^2)^{\frac{3}{2}}}{y_2} = k.$$

(8) The radius of curvature of a certain curve is equal to the length of the normal between the curve and the axis of x. Prove that the curve is a catenary or a circle, according as it is convex or concave to the axis of x.

(9) Find and solve the differential equation of the curve the length of whose arc, measured from a fixed point A to a variable point P, is proportional to the tangent of the angle between the tangent at P and the axis of x.

* **71. Homogeneous equations.** If x and y are regarded as of dimension 1,

$$y_1 \text{ is of dimension } \quad 0,$$
$$y_2 \text{ is of dimension } \quad -1,$$
$$y_3 \text{ is of dimension } \quad -2,$$

and so on.

We define a homogeneous equation as one in which all the terms are of the same dimensions. We have already in Chap. II. dealt with homogeneous equations of the first order and degree, and in Chap. III. with the homogeneous linear equation

$$x^n y_n + A x^{n-1} y_{n-1} + B x^{n-2} y_{n-2} + \ldots + H x y_1 + K y = 0$$

(where A, B, ... H, K are merely constants), for which we used the substitution $x = e^t$ or $t = \log x$.

Let us make the same substitution in the homogeneous equation

$$x y y_2 + x y_1{}^2 = 3 y y_1. \quad \ldots\ldots\ldots\ldots\ldots\ldots\ldots\ldots(1)$$

Now
$$y_1 = \frac{dt}{dx}\frac{dy}{dt} = \frac{1}{x}\frac{dy}{dt},$$

$$y_2 = \frac{dy_1}{dx} = -\frac{1}{x^2}\frac{dy}{dt} + \frac{1}{x}\frac{d}{dx}\frac{dy}{dt}$$

$$= -\frac{1}{x^2}\frac{dy}{dt} + \frac{1}{x}\frac{dt}{dx}\frac{d^2y}{dt^2}$$

$$= -\frac{1}{x^2}\frac{dy}{dt} + \frac{1}{x^2}\frac{d^2y}{dt^2}.$$

Substituting in (1) and multiplying by x, we get

$$y\left(\frac{d^2y}{dt^2} - \frac{dy}{dt}\right) + \left(\frac{dy}{dt}\right)^2 = 3y\frac{dy}{dt},$$

i.e. $\quad y\dfrac{d^2y}{dt^2} + \left(\dfrac{dy}{dt}\right)^2 = 4y\dfrac{dy}{dt}.$

This is an equation, with t absent, similar to those in the last article with x absent.

By putting $\dfrac{dy}{dt} = q$, the student will easily obtain

$$yq = 2(y^2 + b),$$

giving
$$t + c = \tfrac{1}{4} \log (y^2 + b).$$

Hence $y^2 + b = e^{4(t+c)}$

$$= ax^4, \text{ replacing } e^{4c} \text{ by another arbitrary constant } a.$$

72. The example of Art. 71 came out easily because it had no superfluous x's left after associating x^2 with y_2 and x with y_1. In fact, it could have been written

$$y(x^2 y_2) + (xy_1)^2 = 3y(xy_1).$$

But
$$(x^2 + y^2)(y - xy_1) + x^2 y^2 y_2 = 0 \quad\ldots\ldots\ldots\ldots\ldots\ldots(2)$$

cannot be so written. To reduce this to a form similar to that of the last example, put $y = vx$, a substitution used for homogeneous equations in Chap. II.

(2) becomes
$$(x^2 + x^2 v^2)(vx - v_1 x^2 - vx) + x^4 v^2 (xv_2 + 2v_1) = 0,$$
$$i.e. \quad -(1 + v^2) v_1 + v^2 (xv_2 + 2v_1) = 0,$$

which may be written $v^2 x^2 v_2 = (1 - v^2) xv_1. \quad\ldots\ldots\ldots\ldots\ldots\ldots(3)$

We now proceed as before and put $x = e^t$, giving

$$xv_1 = \frac{dv}{dt},$$

and
$$x^2 v_2 = \frac{d^2 v}{dt^2} - \frac{dv}{dt}.$$

(3) becomes $v^2 \left(\dfrac{d^2 v}{dt^2} - \dfrac{dv}{dt} \right) = (1 - v^2) \dfrac{dv}{dt},$

$$i.e. \quad v^2 \frac{d^2 v}{dt^2} = \frac{dv}{dt}, \quad\ldots\ldots\ldots\ldots\ldots\ldots\ldots\ldots\ldots(4)$$

an equation with t absent.

As before, put $\dfrac{dv}{dt} = q, \quad \dfrac{d^2 v}{dt^2} = q \dfrac{dq}{dv}.$

(4) becomes $v^2 q \dfrac{dq}{dv} = q,$

$$i.e. \quad \frac{dq}{dv} = \frac{1}{v^2} \text{ (unless } q = 0, \text{ giving } y = cx),$$

$$i.e. \quad \frac{dv}{dt} = q = \frac{1}{a} - \frac{1}{v},$$

$$i.e. \quad dt = \frac{av\,dv}{v - a} = \left(a + \frac{a^2}{v - a} \right) dv,$$

$$i.e. \quad t = av + a^2 \log (v - a) + b,$$

and finally $\log x = ay/x + a^2 \log (y - ax) - a^2 \log x + b.$

73. By proceeding as in the last article, we can reduce any homogeneous equation of the second order.

Any such equation can be brought to the form

$$f(y/x, y_1, xy_2) = 0.$$

For example, the equation of Art. 71 when divided by x becomes

$$\left(\frac{y}{x}\right)xy_2 + y_1{}^2 = 3\left(\frac{y}{x}\right)y_1,$$

while that of Art. 72 divided by x^2 becomes

$$\left(1 + \frac{y^2}{x^2}\right)\left(\frac{y}{x} - y_1\right) + \left(\frac{y}{x}\right)^2 xy_2 = 0.$$

The substitutions $y = vx$ and $x = e^t$ transform

$$f(y/x, y_1, xy_2) = 0 \quad \text{to} \quad f(v, xv_1 + v, x^2v_2 + 2xv_1) = 0,$$

and then to

$$f\left(v, \frac{dv}{dt} + v, \frac{d^2v}{dt^2} + \frac{dv}{dt}\right) = 0,$$

an equation with t absent, and therefore reducible to the first order.

Examples for solution

(1) $x^2y_2 - xy_1 + y = 0.$ (2) $x^2y_2 - xy_1 + 5y = 0.$

(3) $2x^2yy_2 + y^2 = x^2y_1{}^2.$

(4) Simplify by means of the substitution $y = z^2$, and hence solve

$$2x^2yy_2 + 4y^2 = x^2y_1{}^2 + 2xyy_1.$$

74. An equation occurring in Dynamics. The form $y_2 = f(y)$ occurs frequently in Dynamics, especially in problems on motion under a force directed to a fixed point and of magnitude depending solely on the distance from that fixed point.

Multiply each side of the equation by $2y_1$. We get

$$2y_1y_2 = 2f(y)y_1.$$

Integrating, $y_1{}^2 = 2\displaystyle\int f(y)\frac{dy}{dx}dx = 2\displaystyle\int f(y)\,dy.$

This is really the equation of energy.

Applying the method to $\dfrac{d^2x}{dt^2} = -p^2x$, (the equation of simple harmonic motion), we get

$$2\frac{dx}{dt}\frac{d^2x}{dt^2} = -2p^2x\frac{dx}{dt}.$$

Integrating with respect to t,

$$\left(\frac{dx}{dt}\right)^2 = -p^2x^2 + \text{const.} = p^2(a^2 - x^2), \text{ say.}$$

Hence
$$\pm \frac{dt}{dx} = \frac{1}{p} \frac{1}{\sqrt{(a^2 - x^2)}},$$

$$\pm t = \frac{1}{p} \sin^{-1} \frac{x}{a} + \text{const.}$$

$$x = a \sin (\pm pt + \epsilon).$$

Examples for solution

(1) $y_2 = y^3 - y$, given that $y_1 = 0$ when $y = 1$.

(2) $y_2 = e^{2y}$, given that $y = 0$ and $y_1 = 1$ when $x = 0$.

(3) $y_2 = \sec^2 y \tan y$, given that $y = 0$ and $y_1 = 1$ when $x = 0$.

(4) $\dfrac{d^2x}{dt^2} = -\dfrac{ga^2}{x^2}$, given that $x = h$ and $\dfrac{dx}{dt} = 0$ when $t = 0$.

[$h - x$ is the distance fallen from rest under gravity varying inversely as the square of the distance x from the centre of the earth, neglecting air resistance, etc.]

(5) $\dfrac{d^2u}{d\theta^2} + u = \dfrac{P}{h^2u^2}$, in the two cases

$$\text{(i) } P = \mu u^2; \quad \text{(ii) } P = \mu u^3;$$

given that $\theta = \dfrac{du}{d\theta} = 0$ when $u = \dfrac{1}{c}$, where μ, h, and c are constants.

[These give the path described by a particle attracted to a fixed point with a force varying inversely as the square and cube respectively of the distance r. u is the reciprocal of r, θ has its ordinary meaning in polar co-ordinates, μ is the acceleration at unit distance, and h is twice the areal velocity.]

75. Factorisation of the operator. The linear equation

$$(x+2)y_2 - (2x+5)y_1 + 2y = (x+1)e^x$$

may be written as

$$\{(x+2)D^2 - (2x+5)D + 2\}y = (x+1)e^x,$$

where D stands for $\dfrac{d}{dx}$, as in Chapter III.

Now the operator in this particular example can be factorised, giving
$$\{(x+2)D - 1\}(D-2)y = (x+1)e^x.$$

Put
$$(D-2)y = v.$$

Then
$$\{(x+2)D - 1\}v = (x+1)e^x.$$

This is a linear equation of the first order. Solving as in Art. 20, we get
$$v = c(x+2) + e^x,$$

i.e. $(D-2)y = c(x+2) + e^x,$

another linear equation, giving finally

$$y = a(2x+5) + be^{2x} - e^x, \text{ replacing } -\tfrac{1}{4}c \text{ by } a.$$

Of course it is only in special cases that the operator can be factorised. It is important to notice that these factors must be written in the right order, as they are not commutative. Thus, on reversing the order in this example, we get

$$(D - 2)\{(x + 2) D - 1\}y = \{(x + 2) D^2 - (2x + 4) D + 2\}y.$$

Examples for solution

(1) $(x + 1)y_2 + (x - 1)y_1 - 2y = 0.$ (2) $xy_2 + (x - 1)y_1 - y = 0.$

(3) $xy_2 + (x - 1)y_1 - y = x^2.$

(4) $xy_2 + (x^2 + 1) y_1 + 2xy = 2x$, given that $y = 2$ and $y_1 = 0$ when $x = 0$.

(5) $(x^2 - 1)y_2 - (4x^2 - 3x - 5) y_1 + (4x^2 - 6x - 5) y = e^{2x}$, given that $y = 1$ and $y_1 = 2$ when $x = 0$.

76. One integral belonging to the complementary function * known.

When one integral of the equation

$$y_2 + Py_1 + Qy = 0 \quad\quad\quad\quad\quad\quad\text{.............................(1)}$$

is known, say $y = z$, then the more general equation of the second order

$$y_2 + Py_1 + Qy = R, \quad\quad\quad\quad\text{..........................(2)}$$

where P, Q, R are functions of x, can be reduced to one of the first order by the substitution $y = vz.$

Differentiating, $y_1 = v_1 z + vz_1,$

$$y_2 = v_2 z + 2v_1 z_1 + vz_2.$$

Hence (2) becomes

$$v_2 z + v_1 (2z_1 + Pz) + v (z_2 + Pz_1 + Qz) = R,$$

$$\textit{i.e.} \quad z \frac{dv_1}{dx} + v_1 (2z_1 + Pz) = R, \quad\quad\quad\text{....................(3)}$$

since by hypothesis $z_2 + Pz_1 + Qz = 0.$

(3) is a linear equation of the first order in v_1.

Similarly a linear equation of the nth order can be reduced to one of the $(n - 1)^{\text{th}}$ if one integral belonging to the complementary function is known.

77. Example

Consider again the equation

$$(x + 2) y_2 - (2x + 5) y_1 + 2y = (x + 1) e^x. \quad\quad\text{...............(4)}$$

---

* The proof of Art. 29 that *the general solution of a linear differential equation is the sum of a Particular Integral and the Complementary Function* holds good when the coefficients are functions of x as well as in the case when they are constants.

If we notice that $y = e^{2x}$ makes the left-hand side of the equation zero, we can put $\qquad y = ve^{2x}$,

giving $\qquad\qquad\qquad y_1 = (v_1 + 2v)\,e^{2x}$,

and $\qquad\qquad\qquad y_2 = (v_2 + 4v_1 + 4v)\,e^{2x}$.

Substitution in (4) gives

$$(x+2)\,v_2 e^{2x} + \{4\,(x+2) - (2x+5)\}v_1 e^{2x}$$
$$+\{4\,(x+2) - 2\,(2x+5) + 2\}ve^{2x} = (x+1)\,e^x,$$

i.e. $\quad (x+2)\dfrac{dv_1}{dx} + (2x+3)\,v_1 = (x+1)\,e^{-x}.$

Solving this in the usual way (by finding the integrating factor) we obtain $\qquad v_1 = e^{-x} + c\,(x+2)e^{-2x}.$

Integrating, $\qquad v = -e^{-x} - \tfrac{1}{4}c\,(2x+5)e^{-2x} + b,$

whence $\qquad y = ve^{2x} = -e^x - \tfrac{1}{4}c\,(2x+5) + be^{2x}.$

Examples for solution

(1) Show that $y_2 + Py_1 + Qy = 0$ is satisfied by $y = e^x$ if $1 + P + Q = 0$, and by $y = x$ if $P + Qx = 0$.

(2) $x^2 y_2 + xy_1 - y = 8x^3$.

(3) $x^2 y_2 - (x^2 + 2x)\,y_1 + (x+2)\,y = x^3 e^x$.

(4) $xy_2 - 2\,(x+1)\,y_1 + (x+2)\,y = (x-2)\,e^{2x}$.

(5) $x^2 y_2 + xy_1 - 9y = 0$, given that $y = x^3$ is a solution.

(6) $xy_2\,(x \cos x - 2 \sin x) + (x^2 + 2)\,y_1 \sin x - 2y\,(x \sin x + \cos x) = 0$, given that $y = x^2$ is a solution.

78. Variation of Parameters. We shall now explain an elegant but somewhat artificial method for finding the complete primitive of a linear equation whose complementary function is known.

Let us illustrate the method by applying it to the example already solved in two different ways, namely,

$$(x+2)\,y_2 - (2x+5)\,y_1 + 2y = (x+1)\,e^x, \quad \ldots\ldots\ldots\ldots(1)$$

of which the complementary function is $y = a\,(2x+5) + be^{2x}$.

Assume that $\qquad y = (2x+5)\,A + e^{2x}B, \qquad \ldots\ldots\ldots\ldots\ldots(2)$

where A and B are functions of x.

This assumption is similar to, but more symmetrical than, that of Art. 77, viz.:

$$y = ve^{2x}.$$

Differentiating (2),

$$y_1 = (2x+5)\,A_1 + e^{2x}B_1 + 2A + 2e^{2x}B. \qquad \ldots\ldots\ldots\ldots(3)$$

Now so far the two functions (or *parameters*) A and B are only connected by a single equation. We can make them satisfy the additional equation

$$(2x+5)\,A_1 + e^{2x}B_1 = 0. \qquad \ldots\ldots\ldots\ldots\ldots(4)$$

(3) will then reduce to
$$y_1 = 2A + 2e^{2x}B. \qquad \text{............(5)}$$
Differentiating (5),
$$y_2 = 4e^{2x}B + 2A_1 + 2e^{2x}B_1. \qquad \text{.............(6)}$$
Substitute these values of y, y_1, and y_2 from equations (2), (5), and (6) respectively in (1). The co-factors of A and B come to zero, leaving
$$2(x+2)A_1 + 2(x+2)e^{2x}B_1 = (x+1)e^x. \qquad \text{.........(7)}$$
(4) and (7) are two simultaneous equations which we can solve for A_1 and B_1, giving
$$\frac{A_1}{e^{2x}} = \frac{B_1}{-(2x+5)} = \frac{(x+1)e^x}{2e^{2x}(x+2)(1-2x-5)} = -\frac{(x+1)e^{-x}}{4(x+2)^2}.$$
Hence
$$A_1 = -\frac{(x+1)e^x}{4(x+2)^2} = -\frac{e^x}{4}\left\{\frac{1}{x+2} - \frac{1}{(x+2)^2}\right\},$$
and, by integration, $A = -\dfrac{e^x}{4(x+2)} + a$, where a is a constant.

Similarly,
$$B_1 = \frac{(2x+5)(x+1)e^{-x}}{4(x+2)^2} = \frac{e^{-x}}{4}\left\{2 - \frac{1}{x+2} - \frac{1}{(x+2)^2}\right\},$$
and
$$B = \frac{e^{-x}}{4}\left\{\frac{1}{x+2} - 2\right\} + b.$$
Substituting in (2),
$$y = (2x+5)\left\{-\frac{e^x}{4(x+2)} + a\right\} + \frac{e^x}{4}\left\{\frac{1}{x+2} - 2\right\} + be^{2x}$$
$$= a(2x+5) + be^{2x} - e^x.$$

79. Applying these processes to the general linear equation of the second order, $\qquad y_2 + Py_1 + Qy = R, \qquad \text{...............(1)}$
of which the complementary function $au + bv$ is supposed known, a and b being arbitrary constants and u and v known functions of x, we assume that $\qquad y = uA + vB, \qquad \text{.............(2)}$
giving $\qquad\qquad\qquad y_1 = u_1A + v_1B, \qquad \text{.............(3)}$
provided that $\qquad\qquad uA_1 + vB_1 = 0. \qquad \text{.............(4)}$
Differentiating (3),
$$y_2 = u_2A + v_2B + u_1A_1 + v_1B_1. \qquad \text{.............(5)}$$
Substitute for y_2, y_1 and y in (1).

The terms involving A will be $A(u_2 + Pu_1 + Qu)$, i.e. zero, as by hypothesis, $\qquad\qquad u_2 + Pu_1 + Qu = 0.$

Similarly the terms involving B vanish, and (1) reduces to
$$u_1A_1 + v_1B_1 = R. \qquad \text{.............(6)}$$

Solving (4) and (6), $\dfrac{A_1}{v} = \dfrac{B_1}{-u} = \dfrac{R}{vu_1 - uv_1}$.

We then get A and B by integration, say
$$A = f(x) + a,$$
$$B = F(x) + b,$$
where $f(x)$ and $F(x)$ are known functions of x, and a and b are arbitrary constants.

Substituting in (2), we get finally
$$y = uf(x) + vF(x) + au + bv.$$

* 80. This method can be extended to linear equations of any order. For that of the third order,
$$y_3 + Py_2 + Qy_1 + Ry = S, \quad \dots\dots\dots\dots\dots(1)$$
of which the complementary function $y = au + bv + cw$ is supposed known, the student will easily obtain the equations
$$y = uA + vB + wC, \quad \dots\dots\dots\dots\dots(2)$$
$$y_1 = u_1A + v_1B + w_1C, \quad \dots\dots\dots\dots(3)$$
provided that $\quad 0 = uA_1 + vB_1 + wC_1; \quad \dots\dots\dots\dots(4)$
hence $\quad\quad\quad y_2 = u_2A + v_2B + w_2C, \quad \dots\dots\dots\dots(5)$
provided that $\quad 0 = u_1A_1 + v_1B_1 + w_1C_1; \dots\dots\dots\dots(6)$
then $\quad\quad\quad\quad y_3 = u_3A + v_3B + w_3C$
$$+ u_2A_1 + v_2B_1 + w_2C_1; \quad \dots\dots\dots\dots(7)$$
by substitution in (1), $S = u_2A_1 + v_2B_1 + w_2C_1. \quad \dots\dots\dots\dots(8)$

A_1, B_1, and C_1 are then found from the three equations (4), (6) and (8).

Examples for solution

(1) $y_2 + y = \operatorname{cosec} x.$ $\quad\quad\quad\quad$ (2) $y_2 + 4y = 4 \tan 2x.$

(3) $y_2 - y = \dfrac{2}{1 + e^x}.$

(4) $x^2y_2 + xy_1 - y = x^2e^x$, given the complementary function $ax + bx^{-1}$.

(5) $y_3 - 6y_2 + 11y_1 - 6y = e^{2x}.$

81. Comparison of the different methods for solving linear equations.
If it is required to solve a linear equation of the second order and no special method is indicated, it is generally best to try to guess a particular integral belonging to the complementary function and proceed as in Art. 76. This method may be used to reduce a linear equation of the nth order to one of the $(n-1)^{\text{th}}$.

---

* To be omitted on a first reading.

The method of factorisation of the operator gives a neat solution in a few cases, but these are usually examples specially constructed for this purpose. In general the operator cannot be factorised.

The method of variation of parameters is inferior in practical value to that of Art. 76, as it requires a complete knowledge of the complementary function instead of only one part of it. Moreover, if applied to equations of the third or higher order, it requires too much labour to solve the simultaneous equations for A_1, B_1, C_1, etc., and to perform the integrations.

MISCELLANEOUS EXAMPLES ON CHAPTER VII

(1) $yy_2 - y_1^2 + y_1 = 0$.

(2) $xy_2 + xy_1^2 - y_1 = 0$.

(3) $y_n^2 = 4y_{n-1}$.

(4) $y_n + y_{n-2} = 8 \cos 3x$.

(5) $(x^2 \log x - x^2)y_2 - xy_1 + y = 0$.

(6) $(x^2 + 2x - 1)y_2 - (3x^2 + 8x - 1)y_1 + (2x^2 + 6x)y = 0$.

(7) Verify that $\cos nx$ and $\sin nx$ are integrating factors of
$$y_2 + n^2 y = f(x).$$
Hence obtain two first integrals of
$$y_2 + n^2 y = \sec nx,$$
and by elimination of y_1 deduce the complete primitive.

(8) Show that the linear equation
$$Ay + By_1 + Cy_2 + \dots + Sy_n = T,$$
where A, B, C, ... T are functions of x, is *exact*, *i.e.* derivable immediately by differentiation from an equation of the next lower order, if the successive differential coefficients of A, B, C, ... satisfy the relation
$$A - B_1 + C_2 - \dots + (-1)^n S_n = 0.$$
[*N.B.*—By successive integration by parts,
$$\int Sy_n dx = Sy_{n-1} - S_1 y_{n-2} + S_2 y_{n-3} + \dots + (-1)^{n-1} S_{n-1} y + \int (-1)^n S_n y \, dx.]$$

Verify that this condition is satisfied by the following equation, and hence solve it:
$$(2x^2 + 3x)y_2 + (6x + 3)y_1 + 2y = (x + 1)e^x.$$

(9) Verify that the following non-linear equations are exact, and solve them:

       (i) $yy_2 + y_1^2 = 0$.

      (ii) $xyy_2 + xy_1^2 + yy_1 = 0$.

(10) Show that the substitution $y = ve^{-\frac{1}{2}\int P \, dx}$ transforms
$$y_2 + Py_1 + Qy = R,$$
where P, Q, and R are functions of x, into the *Normal Form*
$$v_2 + Iv = S,$$

where
$$I = Q - \tfrac{1}{2}P_1 - \tfrac{1}{4}P^2,$$

and
$$S = Re^{\tfrac{1}{2}\int P\,dx}.$$

Put into its Normal Form, and hence solve
$$y_2 - 4xy_1 + (4x^2 - 1)y = -3e^{x^2}\sin 2x.$$

(11) Show that if the two equations
$$y_2 + Py_1 + Qy = 0$$

and
$$z_2 + pz_1 + qz = 0$$

reduce to the same Normal Form, they may be transformed into each other by the relation
$$ye^{\tfrac{1}{2}\int P\,dx} = ze^{\tfrac{1}{2}\int p\,dx},$$

i.e. the condition of equivalence is that the *Invariant I* should be the same.

(12) Show that the equations
$$x^2 y_2 + 2(x^3 - x)y_1 + (1 - 2x^2)y = 0$$

and
$$x^2 z_2 + 2(x^3 + x)z_1 - (1 - 2x^2)z = 0$$

have the same invariant, and find the relation that transforms one into the other. Verify by actually carrying out this transformation.

(13) If u and su are any two solutions of
$$v_2 + Iv = 0, \quad\dotfill(1)$$

prove that
$$\frac{s_2}{s_1} = -2\frac{u_1}{u}, \quad\dotfill(2)$$

and hence that
$$\frac{s_3}{s_1} - \frac{3}{2}\left(\frac{s_2}{s_1}\right)^2 = 2I. \quad\dotfill(3)$$

From (2) show that if s is any solution of (3), $s_1^{-\frac{1}{2}}$ and $ss_1^{-\frac{1}{2}}$ are solutions of (1).

[The function of the differential coefficients of s on the left-hand side of (3) is called the *Schwarzian Derivative* (after H. A. Schwarz of Berlin) and written $\{s, x\}$. It is of importance in the theory of the Hypergeometric Series.]

(14) Calculate the Invariant I of the equation
$$x^2 y_2 - (x^2 + 2x)y_1 + (x + 2)y = 0.$$

Taking s as the quotient of the two solutions xe^x and x, verify that
$$\{s, x\} = 2I,$$

and that $s_1^{-\frac{1}{2}}$ and $ss_1^{-\frac{1}{2}}$ are solutions of the Normal Form of the original equation.

(15) If u and v are two solutions of
$$y_2 + Py_1 + Qy = 0,$$

prove that
$$uv_2 - vu_2 + P(uv_1 - vu_1) = 0,$$

and hence that
$$uv_1 - vu_1 = ae^{-\int P\,dx}.$$

Verify this for the equation of the last example.

(16) Show that $yy_1 = $ const. is a first integral of the equation formed by omitting the last term of

$$y_2 + \frac{1}{y} y_1^2 + y = 0.$$

By putting $yy_1 = C$, where C is now a function of x (in fact, *varying the parameter C*), show that if y is a solution of the full equation, then

$$C_1 = -y^2,$$

and hence

$$C^2 = \text{const.} - \tfrac{1}{2}y^4,$$

giving finally

$$y^2 = a \sin (x\sqrt{2} + b).$$

[This method applies to any equation of the form

$$y_2 + y_1^2 f(y) + F(y) = 0.]$$

(17) Solve the following equations by changing the independent variable:

(i) $x \dfrac{d^2y}{dx^2} - \dfrac{dy}{dx} - 4x^3 y = 8x^3 \sin x^2;$

(ii) $(1 + x^2)^2 \dfrac{d^2y}{dx^2} + 2x(1 + x^2) \dfrac{dy}{dx} + 4y = 0.$

(18) Transform the differential equation

$$\frac{d^2y}{dx^2} \cos x + \frac{dy}{dx} \sin x - 2y \cos^3 x = 2 \cos^5 x$$

into one having z as independent variable, where

$$z = \sin x,$$

and solve the equation. [London.]

(19) Show that if z satisfies

$$\frac{d^2z}{dx^2} + P \frac{dz}{dx} = 0,$$

by changing the independent variable from x to z, we shall transform

$$\frac{d^2y}{dx^2} + P \frac{dy}{dx} + Qy = R$$

into

$$\frac{d^2y}{dz^2} + Sy = T.$$

Hence solve $\dfrac{d^2y}{dx^2} + \left(1 - \dfrac{1}{x}\right) \dfrac{dy}{dx} + 4x^2 y e^{-2x} = 4(x^2 + x^3) e^{-3x}.$

CHAPTER VIII

NUMERICAL APPROXIMATIONS TO THE SOLUTION OF DIFFERENTIAL EQUATIONS

82. The student will have noticed that the methods given in the preceding chapters for obtaining solutions in finite form only apply to certain special types of differential equations. If an equation does not belong to one of these special types, we have to use approximate methods. The graphical method of Dr. Brodetsky, given in Chapter I., gives a good general idea of the nature of the solution, but it cannot be relied upon for numerical values.

In this chapter we shall first give Picard's * method for getting successive algebraic approximations. By putting numbers in these, we generally get excellent numerical results. Unfortunately the method can only be applied to a limited class of equations, in which the successive integrations can be easily performed.

The second method, which is entirely numerical and of much more general application, is due to Runge.† With proper precautions it gives good results in most cases, although occasionally it may involve a very large amount of arithmetical calculation. We shall treat several examples by both methods to enable their merits to be compared.

Variations of Runge's method have been given by Heun, Kutta, and the present writer.

83. Picard's method of integrating successive approximations. The differential equation

$$\frac{dy}{dx} = f(x, y),$$

---

* E. Picard, Professor at the University of Paris, was one of the most distinguished mathematicians of his day. He is well known for his researches on the Theory of Functions, and his *Traité d'analyse* is a standard text-book.

† C. Runge, Professor at the University of Göttingen, was an authority on graphical methods.

where $y = b$ when $x = a$, can be written

$$y = b + \int_a^x f(x, y)\, dx.$$

For a first approximation we replace the y in $f(x, y)$ by b; for a second we replace it by the first approximation, for a third by the second, and so on.

Ex. (i). $\quad \dfrac{dy}{dx} = x + y^2$, where $y = 0$ when $x = 0$.

Here $\quad y = \displaystyle\int_0^x (x + y^2)\, dx.$

First approximation. Put $y = 0$ in $x + y^2$, giving

$$y = \int_0^x x\, dx = \tfrac{1}{2}x^2.$$

Second approximation. Put $y = \tfrac{1}{2}x^2$ in $x + y^2$, giving

$$y = \int_0^x (x + \tfrac{1}{4}x^4)\, dx = \tfrac{1}{2}x^2 + \tfrac{1}{20}x^5.$$

Third approximation. Put $y = \tfrac{1}{2}x^2 + \tfrac{1}{20}x^5$ in $x + y^2$, giving

$$y = \int_0^x (x + \tfrac{1}{4}x^4 + \tfrac{1}{20}x^7 + \tfrac{1}{400}x^{10})\, dx$$

$$= \tfrac{1}{2}x^2 + \tfrac{1}{20}x^5 + \tfrac{1}{160}x^8 + \tfrac{1}{4400}x^{11},$$

and so on indefinitely.

Ex. (ii). $\quad \begin{cases} \dfrac{dy}{dx} = z, \\[2mm] \dfrac{dz}{dx} = x^3\,(y + z), \end{cases}$

where $y = 1$ and $z = \tfrac{1}{2}$ when $x = 0$.

Here $\quad y = 1 + \displaystyle\int_0^x z\, dx \quad$ and $\quad z = \tfrac{1}{2} + \displaystyle\int_0^x x^3\,(y + z)\, dx$.

First approximation.

$$y = 1 + \int_0^x \tfrac{1}{2}\, dx = 1 + \tfrac{1}{2}x,$$

$$z = \tfrac{1}{2} + \int_0^x x^3\,(1 + \tfrac{1}{2})\, dx = \tfrac{1}{2} + \tfrac{3}{8}x^4.$$

Second approximation.

$$y = 1 + \int_0^x (\tfrac{1}{2} + \tfrac{3}{8}x^4)\, dx = 1 + \tfrac{1}{2}x + \tfrac{3}{40}x^5,$$

$$z = \tfrac{1}{2} + \int_0^x x^3\,(\tfrac{3}{2} + \tfrac{1}{2}x + \tfrac{3}{8}x^4) = \tfrac{1}{2} + \tfrac{3}{8}x^4 + \tfrac{1}{10}x^5 + \tfrac{3}{64}x^8.$$

Third approximation.

$$y = 1 + \int_0^x (\tfrac{1}{2} + \tfrac{3}{8}x^4 + \tfrac{1}{10}x^5 + \tfrac{3}{64}x^8)\, dx$$
$$= 1 + \tfrac{1}{2}x + \tfrac{3}{40}x^5 + \tfrac{1}{60}x^6 + \tfrac{1}{192}x^9,$$
$$z = \tfrac{1}{2} + \int_0^x x^3 (\tfrac{3}{2} + \tfrac{1}{2}x + \tfrac{3}{8}x^4 + \tfrac{7}{40}x^5 + \tfrac{3}{64}x^8)\, dx$$
$$= \tfrac{1}{2} + \tfrac{3}{8}x^4 + \tfrac{1}{10}x^5 + \tfrac{3}{64}x^8 + \tfrac{7}{360}x^9 + \tfrac{1}{256}x^{12},$$

and so on.

Ex. (iii). $\dfrac{d^2y}{dx^2} = x^3 \left(\dfrac{dy}{dx} + y \right)$, where $y = 1$ and $\dfrac{dy}{dx} = \tfrac{1}{2}$ when $x = 0$.

By putting $\dfrac{dy}{dx} = z$, we reduce this to Ex. (ii).

It may be remarked that Picard's method converts the differential equation into an equation involving integrals, which is called an *Integral Equation*.

Examples for solution

Find the third approximation in the following cases. For examples (1) and (2) obtain also the exact solution by the usual methods.

(1) $\dfrac{dy}{dx} = 2y - 2x^2 - 3$, where $y = 2$ when $x = 0$.

(2) $\dfrac{dy}{dx} = 2 - \dfrac{y}{x}$, where $y = 2$ when $x = 1$.

(3)
$$\begin{cases} \dfrac{dy}{dx} = 2x + z, \\[2mm] \dfrac{dz}{dx} = 3xy + x^2z, \end{cases}$$

where $y = 2$ and $z = 0$ when $x = 0$.

(4)
$$\begin{cases} \dfrac{dy}{dx} = z, \\[2mm] \dfrac{dz}{dx} = x^2z + x^4y, \end{cases}$$

where $y = 5$ and $z = 1$ when $x = 0$.

(5) $\dfrac{d^2y}{dx^2} = x^2 \dfrac{dy}{dx} + x^4y$, where $y = 5$ and $\dfrac{dy}{dx} = 1$ when $x = 0$.

84. Determination of numerical values from these approximations.

Suppose that in Ex. (i) of the last article we desire the value of y, correct to seven places of decimals, when $x = 0.3$.

Substituting $x = 0.3$, we get $\tfrac{1}{2}(0.3)^2 = 0.045$ from the first approximation.

The second adds $\tfrac{1}{20}(0.3)^5 = 0.0001215$,

while the third adds $\tfrac{1}{160}(0.3)^8 + \tfrac{1}{4400}(0.3)^{11} = 0.00000041\ldots$.

Noticing the rapid way in which these successive increments decrease, we conclude that the next one will not affect the first seven decimal places, so the required value is 0·0451219... .

Of course for larger values of x we should have to take more than three approximations to get the result to the required degree of accuracy.

We shall prove in Chap. X. that under certain conditions the approximations obtained really do tend to a limit, and that this limit gives the solution. This is called an Existence Theorem.

Example for solution

(i) Show that in Ex. (ii) of Art. 83, $x = 0·5$ gives $y = 1·252$ ... and $z = 0·526$ ... , while $x = 0·2$ gives $y = 1·100025$ ... and $z = 0·500632$ ... .

85. Numerical approximation direct from the differential equation. The method integrating successive approximations breaks down if, as is often the case, the integrations are impracticable. But there are other methods which can always be applied. Consider the problem geometrically. The differential equation

$$\frac{dy}{dx} = f(x, y)$$

determines a family of curves (the "characteristics") which do not intersect each other and of which one passes through every point

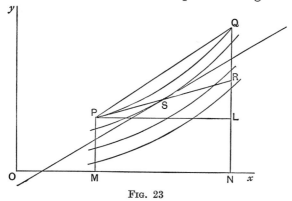

FIG. 23

in the plane.* Given a point $P(a, b)$, we know that the gradient of the characteristic through P is $f(a, b)$, and we want to determine

---

* This is on the assumption that $f(x, y)$ has a perfectly definite value for every point in the plane. If, however, $f(x, y)$ becomes indeterminate for one or more points, these points are called singular points of the equation, and the behaviour of the characteristics near such points calls for special investigation. See Art. 10.

the $y = NQ$ of any other point on the same characteristic, given that $x = ON = a + h$, say. A first approximation is given by taking the tangent PR instead of the characteristic PQ, i.e. taking

$$y = NL + LR = NL + PL \tan \angle RPL = b + hf(a, b) = b + hf_0, \text{ say.}$$

But unless h is very small indeed, the error RQ is far from negligible.

A more reasonable approximation is to take the chord PQ as parallel to the tangent to the characteristic through S, the middle point of PR.

Since S is $(a + \frac{1}{2}h, b + \frac{1}{2}hf_0)$, this gives

$$y = NL + LQ = NL + PL \tan \angle QPL = b + hf(a + \tfrac{1}{2}h, b + \tfrac{1}{2}hf_0).$$

This simple formula gives good results in some cases, as will be seen from the following examples:

Ex. (i). $\dfrac{dy}{dx} = x + y^2$; given that $y = 0$ when $x = 0$, required y when $x = 0.3$.

Here $\qquad\qquad a = b = 0, \quad h = 0.3, \quad f(x, y) = x + y^2.$

Therefore

$$f_0 = f(a, b) = 0, \quad a + \tfrac{1}{2}h = 0.15, \quad b + \tfrac{1}{2}hf_0 = 0,$$

giving $\quad b + hf(a + \tfrac{1}{2}h, b + \tfrac{1}{2}hf_0) = 0 + 0.3 \times f(0.15, 0) = 0.045.$

The value found in Art. 84 was $0.0451219...$, so the error is $0.00012...$, about $\frac{1}{4}$ per cent.

Ex. (ii). $\dfrac{dy}{dx} = 2 - \dfrac{y}{x}$; given that $y = 2$ when $x = 1$, find y when $x = 1.2$.

Here $\qquad\qquad a = 1, \quad b = 2, \quad h = 0.2, \quad f_0 = 2 - \tfrac{2}{1} = 0.$

Therefore $\quad b + hf(a + \tfrac{1}{2}h, b + \tfrac{1}{2}hf_0) = 2 + 0.2 \times f(1.1, 2)$

$$= 2 + 0.2 \times \left(2 - \frac{2}{1.1}\right) = 2.036... .$$

Now the differential equation is easily integrable, giving $y = x + \dfrac{1}{x}$, so when $x = 1.2$ the value of y is $2.033...$. The error is $0.003...$, which is rather large compared with the increment of y, namely $0.033...$.

Ex. (iii). $\qquad\qquad \dfrac{dy}{dx} = z = f(x, y, z), \text{ say,}$

$$\frac{dz}{dx} = x^3(y + z) = g(x, y, z), \text{ say;}$$

given that $y = 1$ and $z = 0.5$ when $x = 0$, find y and z when $x = 0.5$.

Here $a = 0$, $b = 1$, c (the initial value of z) $= 0.5$, $h = 0.5$.

Hence $\qquad f_0 = f(0, 1, 0.5) = 0.5; \quad g_0 = g(0, 1, 0.5) = 0.$

By an obvious extension of the method for two variables, we take
$$y = b + hf(a + \tfrac{1}{2}h, \ b + \tfrac{1}{2}hf_0, \ c + \tfrac{1}{2}hg_0) = 1 + 0.5 \times f(0.25, \ 1.125, \ 0.5) = 1.2500,$$
and
$$z = c + hg(a + \tfrac{1}{2}h, \ b + \tfrac{1}{2}hf_0, \ c + \tfrac{1}{2}hg_0)$$
$$= 0.5 + 0.5 \times g(0.25, \ 1.125, \ 0.5) = 0.5127.$$

The accurate values, found as in Art. 84, are
$$y = 1.252\ldots \quad \text{and} \quad z = 0.526\ldots.$$

Thus we have obtained a fairly good result for y, but a very bad one for z.

The uncertainty about the degree of accuracy of the result deprives the method of most of its value. However, it forms an introduction to the more elaborate method of Runge, to be explained in the next article.

Examples for solution

(1) $\dfrac{dy}{dx} = (x^2 - y)^{\frac{1}{3}} - 1$; given that $y = 4$ when $x = 2.3$, obtain the value $y = 4.122$ when $x = 2.7$. [Runge's method gives 4.118.]

(2) $\dfrac{dy}{dx} = \dfrac{1}{10}\{y^{\frac{1}{3}} - 1 + \log_e(x + y)\}$; given that $y = 2$ when $x = -1$, obtain the value $y = 2.194$ when $x = 1$. [Runge's method gives 2.192.]

(3) $\dfrac{dy}{dx} = 2x - \dfrac{y}{x}$; given that $y = 2$ when $x = 1$, obtain the value $y = 2.076$ when $x = 1.2$. Also show that $y = \dfrac{2}{3}x^2 + \dfrac{4}{3x}$, so that when $x = 1.2$, y is really $2.071\ldots$.

86. Runge's method. Suppose that the function of y defined * by
$$\frac{dy}{dx} = f(x, y), \quad y = b \text{ when } x = a,$$
is denoted by $y = F(x)$.

If this can be expanded by Taylor's theorem,
$$F(a + h) = F(a) + hF'(a) + \frac{h^2}{2!}F''(a) + \frac{h^3}{3}F'''(a) + \ldots.$$
Now
$$F'(x) = \frac{dy}{dx} = f(x, y) = f, \text{ say.}$$

We shall now take the *total* differential coefficient with respect to x (that is, taking the y in f to vary in consequence of the variation of x). Let us denote *partial* differential coefficients by
$$p = \frac{\partial f}{\partial x}, \quad q = \frac{\partial f}{\partial y}, \quad r = \frac{\partial^2 f}{\partial x^2}, \quad s = \frac{\partial^2 f}{\partial x \, \partial y}, \quad t = \frac{\partial^2 f}{\partial y^2};$$
and their values when $x = a$ and $y = b$ by p_0, q_0, etc.

* The conditions under which the differential equation and the initial condition really do define a function are discussed in Chap. X. The graphical treatment of the last article assumes that these conditions are satisfied.

Then $\qquad F''(x) = \dfrac{df}{dx} = \left(\dfrac{\partial}{\partial x} + \dfrac{dy}{dx}\dfrac{\partial}{\partial y}\right)f = p + fq.$

Similarly, $\quad F'''(x) = \left(\dfrac{\partial}{\partial x} + \dfrac{dy}{dx}\dfrac{\partial}{\partial y}\right)(p + fq)$

$$= r + pq + fs + f(s + q^2 + ft).$$

Thus

$F(a+h) - F(a)$
$$= hf_0 + \tfrac{1}{2}h^2(p_0 + f_0 q_0) + \tfrac{1}{6}h^3(r_0 + 2f_0 s_0 + f_0{}^2 t_0 + p_0 q_0 + f_0 q_0{}^2) + . \quad (1)$$

The first term represents the first approximation mentioned and rejected in Art. 85.

The second approximation of Art. 85, *i.e.*
$$y - b = hf(a + \tfrac{1}{2}h,\ b + \tfrac{1}{2}hf_0) = k_1, \text{ say,}$$
may now be expanded and compared with (1).

Now, by Taylor's theorem for two independent variables,
$f(a + \tfrac{1}{2}h,\ b + \tfrac{1}{2}hf_0)$

$$= f_0 + \tfrac{1}{2}hp_0 + \tfrac{1}{2}hf_0 q_0 + \dfrac{1}{2!}(\tfrac{1}{4}h^2 r_0 + \tfrac{1}{2}h^2 f_0 s_0 + \tfrac{1}{4}h^2 f_0{}^2 t_0) + \dots ,$$

giving $\qquad k_1 = hf_0 + \tfrac{1}{2}h^2(p_0 + f_0 q_0) + \tfrac{1}{8}h^3(r_0 + 2f_0 s_0 + f_0{}^2 t_0) + \dots \ \dots(2)$

It is obvious that k_1 is at fault in the coefficient of h^3.

Our next step is suggested by the usual methods * for the numerical integration of the simpler differential equation
$$\dfrac{dy}{dx} = f(x).$$

Our second approximation in this case reduces to the Trapezoidal Rule
$$y - b = hf(a + \tfrac{1}{2}h).$$

Now the next approximation discussed is generally Simpson's Rule, which may be written
$$y - b = \tfrac{1}{6}h\{f(a) + 4f(a + \tfrac{1}{2}h) + f(a + h)\}.$$

If we expand the corresponding formula in two variables, namely
$$\tfrac{1}{6}h\{f_0 + 4f(a + \tfrac{1}{2}h,\ b + \tfrac{1}{2}hf_0) + f(a + h,\ b + hf_0)\},$$
we easily obtain
$$hf_0 + \tfrac{1}{2}h^2(p_0 + f_0 q_0) + \tfrac{1}{6}h^3(r_0 + 2f_0 s_0 + f_0{}^2 t_0) + \dots ,\qquad \dots\dots(3)$$
which is a better approximation than k_1, but even now has not the coefficient of h^3 quite in agreement with (1).

To obtain the extra terms in h^3, Runge † replaces
$$hf(a + h,\ b + hf_0)$$

---

* See the text-books on Calculus by Gibson or Lamb.

† *Mathematische Annalen,* Vol. XLVI., pp. 167-178.

by $k''' = hf(a+h, b+k'')$, where $k'' = hf(a+h, b+hf_0)$. The modified formula may be briefly written $\frac{1}{6}\{k' + 4k_1 + k'''\}$, where $k' = hf_0$, or $\frac{2}{3}k_1 + \frac{1}{3}k_2 = k_1 + \frac{1}{3}(k_2 - k_1)$, where $k_2 = \frac{1}{2}(k' + k''')$.

The student will easily verify that the expansion of Runge's formula agrees with the right-hand side of (1) as far as the terms in h, h^2, and h^3 are concerned.

Of course this method will give bad results if the series (1) converges slowly.

If $f_0 > 1$ numerically, we rewrite our equation

$$\frac{dx}{dy} = \frac{1}{f(x, y)} = F(x, y), \text{ say,}$$

and now $F_0 < 1$ numerically, and we take y as the independent variable.

87. Method of solving examples by Runge's rule. To avoid confusion, the calculations should be formed in some definite order, such as the following:

Calculate successively $k' = hf_0$,
$$k'' = hf(a+h, b+k'),$$
$$k''' = hf(a+h, b+k''),$$
$$k_1 = hf(a+\tfrac{1}{2}h, b+\tfrac{1}{2}k'),$$
$$k_2 = \tfrac{1}{2}(k' + k'''),$$
and finally $\quad k = k_1 + \tfrac{1}{3}(k_2 - k_1).$

Moreover, as k_1 is itself an approximation to the value required, it is clear that if the difference between k and k_1, namely $\frac{1}{3}(k_2 - k_1)$, is small compared with k_1 and k, the error in k is likely to be even smaller.

Ex. (i). $\dfrac{dy}{dx} = x + y^2$; given that $y = 0$ when $x = 0$, find y when $x = 0.3$.

Here $\quad a = 0, \quad b = 0, \quad h = 0.3, \quad f(x, y) = x + y^2, \quad f_0 = 0;$

$k' = hf_0 = 0;$

$k'' = hf(a+h, b+k') = 0.3 \times f(0.3, 0) = 0.3 \times 0.3 \qquad = 0.0900;$

$k''' = hf(a+h, b+k'') = 0.3 \times f(0.3, 0.09) = 0.3 \times (0.3 + 0.0081) = 0.0924;$

$k_1 = hf(a+\tfrac{1}{2}h, b+\tfrac{1}{2}k') = 0.3 \times f(0.15, 0) = 0.3 \times 0.15 \qquad = 0.0450;$

$k_2 = \tfrac{1}{2}(k' + k''') = \tfrac{1}{2} \times 0.0924 \qquad = 0.0462;$

and

$k = k_1 + \tfrac{1}{3}(k_2 - k_1) = 0.0450 + 0.0004 \qquad = 0.0454.$

As the difference between $k = 0.0454$ and $k_1 = 0.0450$ is fairly small compared with either, it is highly probable that the error in k is less than this difference 0·0004. That is to say, we conclude that the value is 0·045, correct to the third place of decimals.

We can test this conclusion by comparing the result obtained in Art. 84, viz. 0·0451219... .

Ex. (ii). $\dfrac{dy}{dx} = \dfrac{y-x}{y+x}$; given that $y = 1$ when $x = 0$, find y when $x = 1$.

This is an example given in Runge's original paper. Divide the range into three parts, 0 to 0·2, 0·2 to 0·5, 0·5 to 1. We take a small increment for the first step because $f(x, y)$ is largest at the beginning.

First step. $a = 0,$ $b = 1,$ $h = 0·2,$ $f_0 = 1;$

$$k' = hf_0 \qquad\qquad\qquad\qquad\qquad\qquad = 0·200;$$
$$k'' = hf(a+h,\, b+k') = 0·2 \times f(0·2,\, 1·2) \quad = 0·143;$$
$$k''' = hf(a+h,\, b+k'') = 0·2 \times f(0·2,\, 1·143) = 0·140;$$
$$k_1 = hf(a+\tfrac{1}{2}h,\, b+\tfrac{1}{2}k') = 0·2 \times f(0·1,\, 1·1) = 0·167;$$
$$k_2 = \tfrac{1}{2}(k'+k'') = \tfrac{1}{2} \times 0·340 \qquad\qquad = 0·170;$$
and $$k = k_1 + \tfrac{1}{3}(k_2 - k_1) = 0·167 + 0·001 \qquad = 0·168;$$
giving $$y = 1·168 \text{ when } x = 0·2.$$

Second step.

$$a = 0·2,\quad b = 1·168,\quad h = 0·3,\quad f_0 = f(0·2,\, 1·168) = 0·708$$

Proceeding as before we get $k_1 = 0·170$, $k_2 = 0·173$ and so $k = 0·171$, giving $$y = 1·168 + 0·171 = 1·339 \text{ when } x = 0·5.$$

Third step. $a = 0·5,$ $b = 1·339,$ $h = 0·5.$

We find $k_1 = k_2 = k = 0·160$, giving $y = 1·499$ when $x = 1$.

Considering the k and k_1, the error in this result should be less than 0·001 on each of the first and second steps and negligible (to 3 decimal places) on the third, that is, less than 0·002 altogether.

As a matter of fact, the true value of y is between 1·498 and 1·499, so the error is less than 0·001. This value of y is found by integrating the equation, leading to

$$\pi - 2 \tan^{-1}\frac{y}{x} = \log_e (x^2 + y^2).$$

Examples for solution

Give numerical results to the following examples to as many places of decimals as are likely to be accurate:

(1) $\dfrac{dy}{dx} = \dfrac{1}{10}\{y^{\frac{1}{2}} - 1 + \log_e(x+y)\}$; given that $y = 2$ when $x = -1$, find y when $x = 1$, taking $h = 2$ (as f is very small).

(2) Obtain a closer approximation to the preceding question by taking two steps.

(3) $\dfrac{dy}{dx} = (x^2 - y)^{\frac{1}{2}} - 1$; given that $y = 4$ when $x = 2·3$, find y when $x = 2·7$ (*a*) in one step, (*b*) in two steps.

(4) Show that if $\dfrac{dy}{dx} = 2 - \dfrac{y}{x}$ and $y = 2$ when $x = 1$, then $y = x + \dfrac{1}{x}$.

Hence find the errors in the result given by Runge's method, taking (a) $h = 0\cdot4$, (b) $h = 0\cdot2$, (c) $h = 0\cdot1$ (a single step in each case), and compare these errors with their estimated upper limits.

(5) If $E(h)$ is the error of the result of solving a differential equation of the first order by Runge's method, prove that

$$\underset{h \to 0}{\mathrm{Lt}} \frac{E(h)}{E(nh)} = \frac{1}{n^4}.$$

Hence show that the error in a two-step solution should be about $\tfrac{1}{8}$ of that given by one step; that is to say, we get the answer correct to an extra place of decimals (roughly) by doubling the number of steps.

88. Extension * to simultaneous equations. The method is easily extended to simultaneous equations. As the proof is very similar to the work in Art. 86, though rather lengthy, we shall merely give an example. This example and those given for solution are taken, with slight modifications, from Runge's paper.

Ex.
$$\frac{dy}{dx} = 2z - \frac{y}{x} = f(x, y, z), \text{ say,}$$

$$\frac{dz}{dx} = \frac{y}{\sqrt{(1 - y^2)}} = g(x, y, z), \text{ say;}$$

given that $y = 0\cdot2027$ and $z = 1\cdot0202$ when $x = 0\cdot2$, find y and z when $x = 0\cdot4$.

Here
$a = 0\cdot2$, $b = 0\cdot2027$, $c = 1\cdot0202$, $f_0 = f(0\cdot2, 0\cdot2027, 1\cdot0202) = 1\cdot027$,
$$g_0 = 0\cdot2070, \quad h = 0\cdot2 ;$$

$\begin{aligned}
&k' = hf_0 = 0\cdot2 \times 1\cdot027 && = 0\cdot2054; \\
&l' = hg_0 = 0\cdot2 \times 0\cdot2070 && = 0\cdot0414; \\
&k'' = hf(a + h, b + k', c + l') = 0\cdot2 \times f(0\cdot4, 0\cdot4081, 1\cdot0616) && = 0\cdot2206; \\
&l'' = hg(a + h, b + k', c + l') = 0\cdot2 \times g(0\cdot4, 0\cdot4081, 1\cdot0616) && = 0\cdot0894; \\
&k''' = hf(a + h, b + k'', c + l'') = 0\cdot2 \times f(0\cdot4, 0\cdot4233, 1\cdot1096) && = 0\cdot2322; \\
&l''' = hg(a + h, b + k'', c + l'') = 0\cdot2 \times g(0\cdot4, 0\cdot4233, 1\cdot1096) && = 0\cdot0934; \\
&k_1 = hf(a + \tfrac{1}{2}h, b + \tfrac{1}{2}k', c + \tfrac{1}{2}l') = 0\cdot2 \times f(0\cdot3, 0\cdot3054, 1\cdot0409) && = 0\cdot2128; \\
&l_1 = hg(a + \tfrac{1}{2}h, b + \tfrac{1}{2}k', c + \tfrac{1}{2}l') = 0\cdot2 \times g(0\cdot3, 0\cdot3054, 1\cdot0409) && = 0\cdot0641; \\
&k_2 = \tfrac{1}{2}(k' + k''') && = 0\cdot2188; \\
&l_2 = \tfrac{1}{2}(l' + l''') && = 0\cdot0674; \\
&k = k_1 + \tfrac{1}{3}(k_2 - k_1) = 0\cdot2128 + 0\cdot0020 && = 0\cdot2148; \\
&l = l_1 + \tfrac{1}{3}(l_2 - l_1) = 0\cdot0641 + 0\cdot0011 && = 0\cdot0652;
\end{aligned}$

giving $\qquad\qquad y = 0\cdot2027 + 0\cdot2148 = 0\cdot4175$

and $\qquad\qquad z = 1\cdot0202 + 0\cdot0652 = 1\cdot0854,$

probably correct to the third place of decimals.

* The rest of this chapter may be omitted on a first reading.

Examples for solution

(1) With the equation of Art. 88, show that if $y = 0.4175$ and $z = 1.0854$ when $x = 0.4$, then $y = 0.6614$ and $z = 1.2145$ (probably correct to the third place of decimals) when $x = 0.6$.

(2) $\dfrac{dw}{dz} = -2z + \dfrac{\sqrt{(1 - w^2)}}{r}$; $\quad \dfrac{dr}{dz} = \dfrac{w}{\sqrt{(1 - w^2)}}$; $\quad$ given that $w = 0.7500$ and $r = 0.6$ when $z = 1.2145$, obtain the values $w = 0.5163$ and $r = 0.7348$ when z (which is to be taken as the independent variable) $= 1.3745$. Show that the value of r is probably correct to four decimal places, but that the third place in the value of w may be in error.

(3) By putting $w = \cos \phi$ in the last example and $y = \sin \phi$, $x = r$ in the example of Art. 88, obtain in each case the equations

$$\frac{dz}{dr} = \tan \phi; \quad 2z = \frac{\sin \phi}{r} + \cos \phi \frac{d\phi}{dr},$$

which gives the form of a drop of water resting on a horizontal plane.

89. Methods * of Heun and Kutta. These methods are very similar to those of Runge, so we shall state them very briefly. The problem is: given that $\dfrac{dy}{dx} = f(x, y)$ and $y = b$ when $x = a$, to find the increment k of y when the increment of x is h.

Heun calculates successively

$$k' = hf(a, b),$$
$$k'' = hf(a + \tfrac{1}{3}h, \, b + \tfrac{1}{3}k'),$$
$$k''' = hf(a + \tfrac{2}{3}h, \, b + \tfrac{2}{3}k''),$$

and then takes $\frac{1}{4}(k' + 3k''')$ as the approximate value of k.

Kutta calculates successively,

$$k' = hf(a, b),$$
$$k'' = hf(a + \tfrac{1}{3}h, \, b + \tfrac{1}{3}k'),$$
$$k''' = hf(a + \tfrac{2}{3}h, \, b + k'' - \tfrac{1}{3}k'),$$
$$k'''' = hf(a + h, \, b + k''' - k'' + k'),$$

and then takes $\frac{1}{8}(k' + 3k'' + 3k''' + k'''')$ as the approximate value of k.

The approximations can be verified by expansion in a Taylor's series, as in Runge's case.

Example for solution

Given that $\dfrac{dy}{dx} = \dfrac{y - x}{y + x}$ and $y = 1$ when $x = 0$, find the value of y (to 8 significant figures) when $x = 0.2$ by the methods of Runge, Heun, and Kutta, and compare them with the accurate value 1.1678417. [From Kutta's paper.]

* *Zeitschrift für Mathematik und Physik*, Vols. 45 and 46.

90. Another method, with limits for the error. The present writer has found * four formulae which give four numbers, between the greatest and least of which the required increment of y must lie. A new approximate formula can be derived from these. When applied to Runge's example, this new formula gives more accurate results than any previous method.

The method is an extension of the following well-known results concerning definite integrals.

91. Limits between which the value of a definite integral lies. Let $F(x)$ be a function which, together with its first and second differential coefficients, is continuous (and therefore finite) between $x = a$ and $x = a + h$. Let $F''(x)$ be of constant sign in the interval. In the figure this sign is taken as positive, making the curve concave upwards. LP, MQ, NR are parallel to the axis of y, M is the middle point of LN, and SQT is the tangent at Q. $OL = a$, $LN = h$.

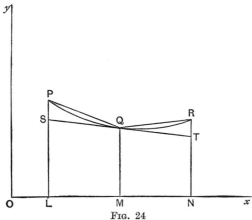

FIG. 24

Then the area $PLNR$ lies between that of the trapezium $SLNT$ and the sum of the areas of the trapezia $PLMQ$, $QMNR$.

That is, $\displaystyle\int_a^{a+h} F(x)\, dx$ lies between

$$hF(a + \tfrac{1}{2}h) = A, \text{ say,}$$

and $\qquad \tfrac{1}{4}h\{F(a) + 2F(a + \tfrac{1}{2}h) + F(a + h)\} = B, \text{ say.}$

In the figure $F''(x)$ is positive and A is the lower limit, B the upper. If $F''(x)$ were negative, A would be the upper limit and B the lower.

---

* *Phil. Mag.*, June 1919. Most of this paper is reproduced here.

As an approximation to the value of the integral it is best to take, not the arithmetic mean of A and B, but $\frac{2}{3}B + \frac{1}{3}A$, which is exact when PQR is an arc of a parabola with its axis parallel to the axis of y. It is also exact for the more general case when

$$F(x) = a + bx + cx^2 + ex^3,$$

as is proved in most treatises on the Calculus in their discussion of Simpson's Rule.

92. Extension of preceding results to functions defined by differential equations. Consider the function defined by

$$\frac{dy}{dx} = f(x, y), \quad y = b \text{ when } x = a;$$

where $f(x, y)$ is subject to the following limitations in the range of values a to $a + h$ for x and $b - h$ to $b + h$ for y. It will be seen from what follows below that the increment of y is numerically less than h, so that all values of y will fall in the above range. The limitations are:

(1) $f(x, y)$ is finite and continuous, as are also its first and second partial differential coefficients.

(2) It never numerically exceeds unity. If this condition is not satisfied, we can generally get a new equation in which it is satisfied by taking y instead of x as the independent variable.

(3) Neither d^3y/dx^3 nor $\partial f/\partial y$ changes sign.

Let m and M be any two numbers, such that

$$-1 \leq m < f < M \leq 1.$$

Then if the values of y when x is $a + \frac{1}{2}h$ and $a + h$ are denoted by $b + j$ and $b + k$ respectively,*

$$-\tfrac{1}{2}h \leq \tfrac{1}{2}mh < j < \tfrac{1}{2}Mh \leq \tfrac{1}{2}h, \quad \ldots\ldots\ldots\ldots\ldots(1$$

and

$$-h \leq mh < k < Mh \leq h. \quad \ldots\ldots\ldots\ldots\ldots(2$$

We shall now apply the formulae of the last article, taking y to be the same function as that defined by

$$y = b + \int_a^x F(x)\, dx,$$

so that

$$k = \int_a^{a+h} F(x)\, dx.$$

We have to express the formulae in terms of f instead of F.

Now, $F(a) = $ the value of dy/dx when $x = a$,

so that

$$F(a) = f(a, b).$$

---

* The following inequalities hold only if h is positive. If h is negative, they must be modified, but the final result stated at the end of this article is still true.

Similarly, $F(a + \tfrac{1}{2}h) = f(a + \tfrac{1}{2}h, \; b + j),$

and $F(a + h) = f(a + h, \; b + k).$

Now, if $\partial f/\partial y$ is positive, so that f increases with y, the inequalities (1) and (2) lead to

$$f(a + \tfrac{1}{2}h, \; b + \tfrac{1}{2}mh) < f(a + \tfrac{1}{2}h, \; b + j) < f(a + \tfrac{1}{2}h, \; b + \tfrac{1}{2}Mh), \quad \dots \dots (3)$$

and $f(a + h, \; b + mh) < f(a + h, \; b + k) < f(a + h, \; b + Mh);$ $\dots \dots \dots (4)$

while if $\partial f/\partial y$ is negative,

$$f(a + \tfrac{1}{2}h, \; b + \tfrac{1}{2}mh) > f(a + \tfrac{1}{2}h, \; b + j) > f(a + \tfrac{1}{2}h, \; b + \tfrac{1}{2}Mh), \quad \dots (5)$$

and $f(a + h, \; b + mh) > f(a + h, \; b + k) > f(a + h, \; b + Mh).$ $\dots \dots \dots (6)$

Thus if $F''(x) = d^3y/dx^3$ is positive and $\partial f/\partial y$ is also positive, the result of Art. 91,

$$A < k < B,$$

may be replaced by $p < k < Q,$ $\dots \dots \dots \dots \dots \dots \dots \dots \dots (7)$

where $p = hf(a + \tfrac{1}{2}h, \; b + \tfrac{1}{2}mh)$

and $Q = \tfrac{1}{4}h\{f(a, b) + 2f(a + \tfrac{1}{2}h, \; b + \tfrac{1}{2}Mh) + f(a + h, \; b + Mh)\};$

while if $F''(x)$ is positive, and $\partial f/\partial y$ is negative,

$$P < k < q, \dots \dots \dots \dots \dots \dots \dots \dots \dots (8)$$

where $P = hf(a + \tfrac{1}{2}h, \; b + \tfrac{1}{2}Mh)$

and $q = \tfrac{1}{4}h\{f(a, b) + 2f(a + \tfrac{1}{2}h, \; b + \tfrac{1}{2}mh) + f(a + h, \; b + mh)\}.$

Similarly, if $F''(x)$ and $\partial f/\partial y$ are both negative,

$$p > k > Q, \dots \dots \dots \dots \dots \dots \dots \dots \dots (9)$$

while if $F''(x)$ is negative and $\partial f/\partial y$ positive,

$$P > k > q. \dots \dots \dots \dots \dots \dots \dots \dots (10)$$

These results may be summed up by saying that *in every case* (subject to the limitations on f stated at the beginning of this article) k *lies between the greatest and least of the four numbers* p, P, q, *and* Q.

As an approximate formula we use $k \doteqdot \tfrac{2}{3}B + \tfrac{1}{3}A$, replacing B by Q or q, and A by p or P.

93. Application to a numerical example. Consider the example selected by Runge and Kutta to illustrate their methods,

$$\frac{dy}{dx} = \frac{y - x}{y + x}; \quad y = 1 \text{ when } x = 0.$$

It is required to find the increment k of y when x increases by $0 \cdot 2$. Here $f(x, y) = (y - x)/(y + x)$. This function satisfies the conditions laid down in the last article.*

We take $M = 1$, $m = (1 - 0 \cdot 2)/(1 \cdot 2 + 0 \cdot 2) = 4/7$.

* As $f(x, y)$ is positive, y lies between 1 and $1 \cdot 2$. When finding M and m we always take the smallest range for y that we can find. (The conditions $m < f < M$ can be replaced by $m \leq f \leq M$, without affecting the final result except to replace some $<$ signs by $\leq$ signs.)

Then
$$p = 0\cdot1654321,$$
$$P = 0\cdot1666667,$$
$$q = 0\cdot1674987,$$
$$Q = 0\cdot1690476.$$

Thus k lies between p and Q.   Errors.

$\tfrac{2}{3}Q + \tfrac{1}{3}p = 0\cdot1678424,$ 0·0000007

Kutta's value 0·1678449,   0·0000032

Runge's value 0·1678487,   0·0000070

Heun's value 0·1680250,   0·0001833

The second, third, and fourth of these were calculated by Kutta. Now this particular example admits of integration in finite terms, giving
$$\log (x^2 + y^2) - 2 \tan^{-1} (x/y) = 0.$$
Hence we may find the accurate value of k.

Accurate value = 0·1678417.

Thus in this example our result is the nearest to the accurate value, the errors being as stated above.

We may also test the method by taking a larger interval $h = 1$. Of course a more accurate way of obtaining the result would be to take several steps, say $h = 0\cdot2$, 0·3, and finally 0·5, as Runge does.

Still, it is interesting to see how far wrong the results come for the larger interval.

We take $M = 1, \quad m = (1 - 1)/(2 + 1) = 0.$

Then $\tfrac{2}{3}Q + \tfrac{1}{3}p = 0\cdot50000.$

True value = 0·49828,   Errors.

Kutta's value = 0·49914,   0·00086

Our value = 0·50000,   0·00172

Heun's value = 0·51613,   0·01785

Runge's value = 0·52381,   0·02553

This time Kutta's value is the nearest, and ours is second.

[For a systematic method of determining M and m, and for Remes' extension of the method of Arts. 90-93, see Art. 183.

For Adams' numerical method, perhaps the best of all, see Art. 182.]

CHAPTER IX

SOLUTION IN SERIES. METHOD OF FROBENIUS

94. In Chapter VII. we obtained the solution of several equation of the form

$$\frac{d^2y}{dx^2} + P\,\frac{dy}{dx} + Qy = 0,$$

where P and Q were functions of x.

In every case the solution was of the form

$$y = af(x) + bF(x),$$

where a and b were arbitrary constants.

The functions $f(x)$ and $F(x)$ were generally made up of integral or fractional powers of x, sines and cosines, exponentials, and logarithms, such as

$$(1+2x)e^x, \quad \sin x + x\cos x, \quad x^{\frac{1}{2}} + x^{-\frac{1}{2}}, \quad x + \log x, \quad e^{\frac{1}{x}}.$$

The first and second of these functions can be expanded by Maclaurin's theorem in ascending integral powers of x; the others cannot, though the last can be expanded in terms of $1/x$. .

In the present chapter, following F. G. Frobenius,* of Berlin, we shall assume as a trial solution

$$y = x^c(a_0 + a_1x + a_2x^2 + \ldots \text{ to inf.}),$$

where the a's are constants.†

The index c will be determined by a quadratic equation called the *Indicial Equation*. The roots of this equation may be equal, different and differing by an integer, or different and differing by a quantity not an integer. These cases will have to be discussed separately.

The special merit of the form of trial solution used by Frobenius is that it leads at once to another form of solution, involving $\log x$, when the differential equation has this second form of solution.

---

* Crelle, Vol. LXXVI., 1873, pp. 214-224.
† In this chapter suffixes will *not* be used to denote differentiation.

As such a function as $e^{\frac{1}{x}}$ cannot be expanded in ascending powers of x, we must expect the method to fail for differential equations having solutions of this nature. A method will be pointed out by which can be determined at once which equations have solutions of Frobenius' forms (*regular* integrals) and for what range of values of x these solutions will be convergent.

The object of the present chapter is to indicate how to deal with examples. The formal proofs of the theorems suggested will be given in the next chapter.

Among the examples will be found the important equations of Bessel,[*] Legendre, and Riccati. A sketch is also given of the Hypergeometric or Gaussian equation and its twenty-four solutions.

95. Case I. Roots of Indicial Equation unequal and differing by a quantity not an integer. Consider the equation

$$(2x + x^3)\frac{d^2y}{dx^2} - \frac{dy}{dx} - 6xy = 0. \quad\dots\dots\dots\dots\dots(1)$$

Put $z = x^c(a_0 + a_1x + a_2x^2 + \dots)$, where $a_0 \neq 0$, giving [†]

$$\frac{dz}{dx} = a_0cx^{c-1} + a_1(c+1)x^c + a_2(c+2)x^{c+1} + \dots\ ,$$

and $\quad \dfrac{d^2z}{dx^2} = a_0c(c-1)x^{c-2} + a_1(c+1)cx^{c-1} + a_2(c+2)(c+1)x^c + \dots\ .$

Substitute in (1),[‡] and equate the coefficients of the successive powers of x to zero.

The lowest power of x is x^{c-1}. Its coefficient equated to zero gives
$$a_0\{2c(c-1) - c\} = 0,$$
$$i.e. \quad c(2c - 3) = 0, \quad\dots\dots\dots\dots\dots(2)$$
as $a_0 \neq 0$.

* Friedrich Wilhelm Bessel, of Minden (1784-1846), was director of the observatory at Königsberg. He is best known by " Bessel's Functions."

Adrian Marie Legendre, of Toulouse (1752-1833), is best known by his " Zonal Harmonics " or " Legendre's Coefficients." He also did a great deal of work on Elliptic Integrals and the Theory of Numbers.

Jacopo Francesco, Count Riccati, of Venice (1676-1754), wrote on " Riccati's Equation," and also on the possibility of lowering the order of a given differential equation.

Karl Friedrich Gauss, of Brunswick (1777-1855), " the Archimedes of the nineteenth century," published researches on an extraordinary wide range of subjects, including Theory of Numbers, Determinants, Infinite Series, Theory of Errors, Astronomy, Geodesy, and Electricity and Magnetism.

† It is legitimate to differentiate a series of ascending powers of x term by term in this manner, within the region of convergence. See Bromwich, *Infinite Series,* Art. 52.

‡ Or rather in what (1) becomes when y is replaced by z.

(2) is called the *Indicial Equation*.

The coefficient of x^c equated to zero gives

$$a_1\{2(c+1)c - (c+1)\} = 0, \quad i.e. \ a_1 = 0. \ \dots\dots\dots\dots(3)$$

The coefficient of x^{c+1} has more terms in it, giving

$$a_2\{2(c+2)(c+1) - (c+2)\} + a_0\{c(c-1) - 6\} = 0,$$

$$i.e. \quad a_2(c+2)(2c+1) + a_0(c+2)(c-3) = 0,$$

$$i.e. \quad a_2(2c+1) + a_0(c-3) = 0. \ \dots\dots\dots\dots(4)$$

Similarly, $a_3(2c+3) + a_1(c-2) = 0, \ \dots\dots\dots\dots(5)$

$$a_4(2c+5) + a_2(c-1) = 0, \ \dots\dots\dots\dots(6)$$

and so on.

From (3), (5), etc., $0 = a_1 = a_3 = a_5 = \dots = a_{2n+1}$.

From (4), (6), etc.,

$$\frac{a_2}{a_0} = -\frac{c-3}{2c+1}, \qquad \frac{a_4}{a_2} = -\frac{c-1}{2c+5},$$

$$\frac{a_6}{a_4} = -\frac{c+1}{2c+9}, \qquad \frac{a_{2n}}{a_{2n-2}} = -\frac{c+2n-5}{2c+4n-3}.$$

But from (2), $c = 0$ or $\frac{3}{2}$.

Thus, if $c = 0$,

$$z = a\left\{1 + 3x^2 + \frac{3}{5}x^4 - \frac{1}{15}x^6 + \frac{1}{65}x^8\dots\right\} = au, \text{ say,}$$

replacing a_0 by a; and if $c = \frac{3}{2}$,

$$z = bx^{\frac{3}{2}}\left\{1 + \frac{3}{8}x^2 - \frac{1\cdot3}{8\cdot16}x^4 + \frac{1\cdot3\cdot5}{8\cdot16\cdot24}x^6 - \frac{1\cdot3\cdot5\cdot9}{8\cdot16\cdot24\cdot32}x^8\dots\right\}$$

$$= bv \text{ say, replacing } a_0 \text{ (which is arbitrary) by } b \text{ this time.}$$

Thus $y = au + bv$ is a solution which contains two arbitrary constants, and so may be considered the complete primitive.

In general, if the Indicial Equation has two unequal roots α and β differing by a quantity not an integer, we get two independent solutions by substituting these values of c in the series for z.

Examples for solution

(1) $4x\dfrac{d^2y}{dx^2} + 2\dfrac{dy}{dx} + y = 0.$ (2) $2x(1-x)\dfrac{d^2y}{dx^2} + (1-x)\dfrac{dy}{dx} + 3y = 0.$

(3) $9x(1-x)\dfrac{d^2y}{dx^2} - 12\dfrac{dy}{dx} + 4y = 0.$

(4) Bessel's equation of order n, taking $2n$ as non-integral,

$$x^2\dfrac{d^2y}{dx^2} + x\dfrac{dy}{dx} + (x^2 - n^2)y = 0.$$

96. Convergence of the series obtained in the last article. It is proved in nearly every treatise on Higher Algebra or Analysis that the infinite series $u_1 + u_2 + u_3 + \dots$ is convergent if

$$\underset{n \to \infty}{\text{Lt}} \left| \frac{u_{n+1}}{u_n} \right| < 1.$$

Now in the series we obtained $u_n = a_{2n-2} x^{c+2n-2}$, *i.e.*

$$\frac{u_{n+1}}{u_n} = \frac{a_{2n}}{a_{2n-2}} x^2$$

$$= -\frac{c+2n-5}{2c+4n-3} x^2,$$

and the limit when $n \to \infty$ is $-\frac{1}{2}x^2$, independent of the value of c.

Hence both series obtained are convergent for $|x| < \sqrt{2}$.

It is interesting to notice that if the differential equation is reduced to the form

$$x^2 \frac{d^2y}{dx^2} + xp(x) \frac{dy}{dx} + q(x)y = 0,$$

giving in our example $p(x) = \dfrac{-1}{2+x^2}$,

and $q(x) = \dfrac{-6x^2}{2+x^2}$,

$p(x)$ and $q(x)$ are expansible in power series which are convergent for values of x whose modulus $|x| < \sqrt{2}$.

That is, the region of convergence is identical in this example with the region for which $p(x)$ and $q(x)$ are expansible in convergent power series. We shall show in Chap. X. that this theorem is true in general.

Examples for solution

Find the region of convergence for the solutions of the last set of examples. Verify in each case that the region of convergence is identical with the region for which $p(x)$ and $q(x)$ are expansible in convergent power series.

97. Case II. Roots of Indicial Equation equal. Consider the equation

$$(x - x^2) \frac{d^2y}{dx^2} + (1 - 5x) \frac{dy}{dx} - 4y = 0.$$

Put $z = x^c(a_0 + a_1 x + a_2 x^2 + \dots),$

and after substituting in the differential equation, equate coefficients of successive powers of x to zero just as in Art. 95.

We get
$$a_0\{c(c-1)+c\}=0,$$
$$\text{i.e.}\quad c^2=0, \quad\dots\dots\dots\dots\dots\dots\dots(1)$$
$$a_1\{(c+1)c+c+1\}-a_0\{c(c-1)+5c+4\}=0,$$
$$\text{i.e.}\quad a_1(c+1)^2-a_0(c+2)^2=0, \quad\dots\dots\dots\dots(2)$$
$$a_2(c+2)^2-a_1(c+3)^2=0, \quad\dots\dots\dots\dots(3)$$
$$a_3(c+3)^2-a_2(c+4)^2=0, \quad\dots\dots\dots\dots(4)$$

and so on.

Hence
$$z=a_0x^c\left\{1+\left(\frac{c+2}{c+1}\right)^2 x+\left(\frac{c+3}{c+1}\right)^2 x^2+\left(\frac{c+4}{c+1}\right)^2 x^3+\dots\right\}$$

is a solution if $c=0$.

This gives only one series instead of two.

But if we substitute the series in the left-hand side of the differential equation (*without* putting $c=0$), we get the single term $a_0c^2x^{c-1}$. As this involves the square of c, its partial differential coefficient with respect to c, i.e. $2a_0cx^{c-1}+a_0c^2x^{c-1}\log x$, will also vanish when $c=0$.

That is,
$$\frac{\partial}{\partial c}\left[(x-x^2)\frac{d^2}{dx^2}+(1-5x)\frac{d}{dx}-4\right]z=2a_0cx^{c-1}+a_0c^2x^{c-1}\log x.$$

As the differential operators are commutative, this may be written
$$\left[(x-x^2)\frac{d^2}{dx^2}+(1-5x)\frac{d}{dx}-4\right]\frac{\partial z}{\partial c}=2a_0cx^{c-1}+a_0c^2x^{c-1}\log x.$$

Hence $\dfrac{\partial z}{\partial c}$ is a second solution of the differential equation, if c is put equal to zero after differentiation.

Differentiating,
$$\frac{\partial z}{\partial c}=z\log x+a_0x^c\left\{2\left(\frac{c+2}{c+1}\right)\cdot\frac{-1}{(c+1)^2}x+2\left(\frac{c+3}{c+1}\right)\cdot\frac{-2}{(c+1)^2}x^2\right.$$
$$\left.+2\left(\frac{c+4}{c+1}\right)\cdot\frac{-3}{(c+1)^2}x^3+\dots\right\}.$$

Putting $c=0$ and $a_0=a$ and b respectively in the two series,
$$z=a\{1^2+2^2x+3^2x^2+4^2x^3+5^2x^4+\dots\}=au,\text{ say,}$$

and
$$\frac{\partial z}{\partial c}=bu\log x-2b\{1.2x+2.3x^2+3.4x^3+\dots\}=bv,\text{ say.}$$

The complete primitive is $au+bv$.

In general, if the Indicial Equation has two equal roots $c = \alpha$, *we get two independent solutions by substituting this value of c in z and* $\frac{\partial z}{\partial c}$. The second solution will always consist of the product of the first solution (or a numerical multiple of it) and $\log x$, added to another series.

Reverting to our particular example, consideration of $p(x)$ and $q(x)$, as in Art. 96, suggests that the series will be convergent for $|x| < 1$. It may be easily shown that this is correct.

Examples for solution

(1) $(x - x^2)\dfrac{d^2y}{dx^2} + (1 - x)\dfrac{dy}{dx} - y = 0.$

(2) Bessel's equation of order zero

$$x\frac{d^2y}{dx^2} + \frac{dy}{dx} + xy = 0.$$

(3) $x\dfrac{d^2y}{dx^2} + (1 + x)\dfrac{dy}{dx} + 2y = 0.$

(4) $4(x^4 - x^2)\dfrac{d^2y}{dx^2} + 8x^3\dfrac{dy}{dx} - y = 0.$

98. Case III. Roots of Indicial Equation differing by an integer, making a coefficient of z infinite. Consider Bessel's equation of order unity,

$$x^2\frac{d^2y}{dx^2} + x\frac{dy}{dx} + (x^2 - 1)y = 0.$$

If we proceed as in Art. 95, we find

$$a_0\{c(c - 1) + c - 1\} = 0,$$

$$i.e. \quad c^2 - 1 = 0, \quad\ldots\ldots\ldots\ldots\ldots\ldots(1)$$

$$a_1\{(c + 1)^2 - 1\} = 0,$$

$$i.e. \quad a_1 = 0, \quad\ldots\ldots\ldots\ldots\ldots\ldots(2)$$

$$a_2\{(c + 2)^2 - 1\} + a_0 = 0, \quad\ldots\ldots\ldots\ldots\ldots\ldots(3)$$

and $\quad\quad a_n\{(c + n)^2 - 1\} + a_{n-2} = 0, \quad\ldots\ldots\ldots\ldots\ldots\ldots(4)$

giving

$$z = a_0 x^c \left\{ 1 - \frac{1}{(c + 1)(c + 3)} x^2 + \frac{1}{(c + 1)(c + 3)^2(c + 5)} x^4 \right.$$
$$\left. - \frac{1}{(c + 1)(c + 3)^2(c + 5)^2(c + 7)} x^6 + \ldots \right\}.$$

The roots of the indicial equation (1) are $c = 1$ or -1.

But if we put $c = -1$ in this series for z, the coefficients become infinite, owing to the factor $(c + 1)$ in the denominator.

To obviate this difficulty replace * a_0 by $(c+1)k$, giving

$$z = kx^c \left\{ (c+1) - \frac{1}{(c+3)} x^2 + \frac{1}{(c+3)^2 (c+5)} x^4 \right.$$
$$\left. - \frac{1}{(c+3)^2 (c+5)^2 (c+7)} x^6 + \ldots \right\}, \quad \ldots\ldots(5)$$

and $x^2 \dfrac{d^2z}{dx^2} + x \dfrac{dz}{dx} + (x^2 - 1)z = kx^c(c+1)(c^2-1) = kx^c(c+1)^2(c-1)$.

Just as in Case II. the occurrence of the squared factor $(c+1)^2$ shows that $\dfrac{\partial z}{\partial c}$, as well as z, satisfies the differential equation when $c = -1$. Also putting $c = 1$ in z gives a solution. So apparently we have found *three* solutions to this differential equation of only the *second* order.

On working them out, we get respectively

$$kx^{-1} \left\{ -\frac{1}{2} x^2 + \frac{1}{2^2 \cdot 4} x^4 - \frac{1}{2^2 \cdot 4^2 \cdot 6} x^6 + \ldots \right\} = ku, \text{ say,}$$

$$ku \log x + kx^{-1} \left\{ 1 + \frac{1}{2^2} x^2 - \frac{1}{2^2 \cdot 4} \left(\frac{2}{2} + \frac{1}{4} \right) x^4 \right.$$
$$\left. + \frac{1}{2^2 \cdot 4^2 \cdot 6} \left(\frac{2}{2} + \frac{2}{4} + \frac{1}{6} \right) x^6 + \ldots \right\} = kv, \text{ say,}$$

and $kx \left\{ 2 - \dfrac{1}{4} x^2 + \dfrac{1}{4^2 \cdot 6} x^4 - \dfrac{1}{4^2 \cdot 6^2 \cdot 8} x^6 + \ldots \right\} = kw, \text{ say.}$

It is obvious that $w = -4u$, so we have only found two linearly independent solutions after all, and the complete primitive is $au + bv$. The series are easily proved to be convergent for all values of x.

The identity (except for a constant multiple) of the series obtained by substituting $c = -1$ and $c = 1$ respectively in the expression for z is not an accident. It could have been seen at once from relation (4),

$$a_n\{(c+n)^2 - 1\} + a_{n-2} = 0.$$

If $c = 1$, this gives $a_n\{(1+n)^2 - 1\} + a_{n-2} = 0$. $\quad\ldots\ldots\ldots\ldots\ldots\ldots(6)$

If $c = -1$, $\qquad a_n\{(-1+n)^2 - 1\} + a_{n-2} = 0$;

hence replacing n by $n+2$,

$$a_{n+2}\{(1+n)^2 - 1\} + a_n = 0. \quad\ldots\ldots\ldots\ldots\ldots\ldots(7)$$

Thus $\qquad \left[\dfrac{a_{n+2}}{a_n} \right]_{c=-1} = \left[\dfrac{a_n}{a_{n-2}} \right]_{c=1}. \quad\ldots\ldots\ldots\ldots\ldots\ldots(8)$

As $[z]_{c=-1}$ has x^{-1} as a factor outside the bracket, while $[z]_{c=1}$ has x, relation (8) really means that the coefficients of corresponding

---

* Of course the condition $a_0 \neq 0$ is thus violated; we assume in its place that $k \neq 0$.

powers of x in the two series are in a constant ratio. The first series apparently has an extra term, namely that involving x^{-1}, but this conveniently vanishes owing to the factor $(c+1)$.

In general, if the Indicial Equation has two roots α and β (say $\alpha > \beta$) differing by an integer, and if some of the coefficients of z become infinite when $c = \beta$, we modify the form of z by replacing a_0 by $k(c - \beta)$. We then get two independent solutions by putting $c = \beta$ in the modified form of z and $\dfrac{\partial z}{\partial c}$. The result of putting $c = \alpha$ in z merely gives a numerical multiple of that obtained by putting $c = \beta$.

Examples for solution

(1) Bessel's equation of order 2,
$$x^2 \frac{d^2y}{dx^2} + x\frac{dy}{dx} + (x^2 - 4)y = 0.$$

(2) $x(1-x)\dfrac{d^2y}{dx^2} - 3x\dfrac{dy}{dx} - y = 0.$

(3) $x(1-x)\dfrac{d^2y}{dx^2} - (1+3x)\dfrac{dy}{dx} - y = 0.$

(4) $(x + x^2 + x^3)\dfrac{d^2y}{dx^2} + 3x^2\dfrac{dy}{dx} - 2y = 0.$

99. Case IV. Roots of Indicial Equation differing by an integer, making a coefficient of z indeterminate. Consider the equation
$$(1 - x^2)\frac{d^2y}{dx^2} + 2x\frac{dy}{dx} + y = 0.$$

Proceeding as usual, we get
$$c(c-1) = 0, \quad\ldots\ldots\ldots\ldots(1)$$
$$a_1(c+1)c = 0, \quad\ldots\ldots\ldots\ldots(2)$$
$$a_2(c+2)(c+1) + a_0\{-c(c-1) + 2c + 1\} = 0, \quad\ldots\ldots\ldots\ldots(3)$$
$$a_3(c+3)(c+2) + a_1\{-(c+1)c + 2(c+1) + 1\} = 0, \quad\ldots\ldots\ldots\ldots(4)$$
and so on.

(1) Gives $c = 0$ or 1.

The coefficient of a_1 in (2) vanishes when $c = 0$, but as there is no other term in the equation this makes a_1 *indeterminate* instead of *infinite*.

If $c = 1$, $a_1 = 0$.

Thus, if $c = 0$, from equations (3), (4), etc.,
$$2a_2 + a_0 = 0,$$
$$6a_3 + 3a_1 = 0,$$
$$12a_4 + 3a_2 = 0,$$
$$\text{etc.,}$$

giving $\quad [z]_{c=0} = a_0 \left\{ 1 - \dfrac{1}{2}x^2 + \dfrac{1}{8}x^4 + \dfrac{1}{80}x^6 \ldots \right\}$

$$+ a_1 \left\{ x - \dfrac{1}{2}x^3 + \dfrac{1}{40}x^5 + \dfrac{3}{560}x^7 \ldots \right\}.$$

This contains two arbitrary constants, so it may be taken as the complete primitive. The series may be proved convergent for $|x| < 1$.

But we have the other solution given by $c = 1$. Working out the coefficients,

$$[z]_{c=1} = a_0 x \left\{ 1 - \dfrac{1}{2}x^2 + \dfrac{1}{40}x^4 + \dfrac{3}{560}x^6 \ldots \right\},$$

that is, a constant multiple of the second series in the first solution.

This could have been foreseen from reasoning similar to that in Case III.

In general, if the Indicial Equation has two roots α and β (say $\alpha > \beta$) differing by an integer, and if one of the coefficients of z becomes indeterminate when $c = \beta$, the complete primitive is given by putting $c = \beta$ in z, which then contains two arbitrary constants. The result of putting $c = \alpha$ in z merely gives a numerical multiple of one of the series contained in the first solution.

Examples for solution

(1) Legendre's equation of order unity,

$$(1 - x^2)\frac{d^2y}{dx^2} - 2x\frac{dy}{dx} + 2y = 0.$$

(2) Legendre's equation of order n,

$$(1 - x^2)\frac{d^2y}{dx^2} - 2x\frac{dy}{dx} + n(n+1)y = 0.$$

(3) $\dfrac{d^2y}{dx^2} + x^2 y = 0.$ $\qquad$ (4) $(2 + x^2)\dfrac{d^2y}{dx^2} + x\dfrac{dy}{dx} + (1 + x)y = 0.$

100. Some cases where the method fails. As $e^{\frac{1}{x}}$ cannot be expanded in ascending powers of x, we must expect the method to fail in some way when the differential equation has such a solution. To construct an example, take the equation $\dfrac{d^2y}{dz^2} - y = 0$, of which e^z and e^{-z} are solutions, and transform it by putting $z = \dfrac{1}{x}$.

We have $\qquad \dfrac{dy}{dz} = \dfrac{dx}{dz} \cdot \dfrac{dy}{dx} = -\dfrac{1}{z^2}\dfrac{dy}{dx} = -x^2\dfrac{dy}{dx}$

and $\qquad \dfrac{d^2y}{dz^2} = \dfrac{dx}{dz}\dfrac{d}{dx}\left(\dfrac{dy}{dz}\right) = -x^2\dfrac{d}{dx}\left(-x^2\dfrac{dy}{dx}\right) = x^4\dfrac{d^2y}{dx^2} + 2x^3\dfrac{dy}{dx}.$

Hence the new equation is

$$x^4 \frac{d^2y}{dx^2} + 2x^3 \frac{dy}{dx} - y = 0.$$

If we try to apply the usual method, we get for the indicial equation, $-a_0 = 0$, which has no roots,* as by hypothesis $a_0 \neq 0$.

Such a differential equation is said to have *no regular integrals in ascending powers of x.* Of course $e^{\frac{1}{x}}$ and $e^{-\frac{1}{x}}$ can be expanded in powers of $\frac{1}{x}$.

The examples given below illustrate other possibilities, such as the indicial equation having one root, which may or may not give a convergent series.

It will be noticed that, writing the equation in the form

$$x^2 \frac{d^2y}{dx^2} + xp(x) \frac{dy}{dx} + q(x)y = 0,$$

in every case where the method has succeeded $p(x)$ and $q(x)$ have been finite for $x = 0$, while in all cases of failure this condition is violated.

For instance, in the above example,

$$p(x) = 2,$$

$$q(x) = -\frac{1}{x^2}, \text{ which is infinite if } x = 0.$$

Examples for solution

(1) Transform Bessel's equation by the substitution $x = 1/z$.

Hence show that it has no integrals that are regular in *descending* powers of x.

(2) Show that the following equation has only one integral that is regular in ascending powers of x, and determine it:

$$x^3 \frac{d^2y}{dx^2} + x(1 - 2x) \frac{dy}{dx} - 2y = 0.$$

(3) By putting $y = vx^2(1 + 2x)$ determine the complete primitive of the previous example.

(4) Show that the following equation has no integral that is regular in ascending powers of x, as the one series obtainable diverges for all values of x:

$$x^2 \frac{d^2y}{dx^2} - (1 - 3x) \frac{dy}{dx} + y = 0.$$

(5) Obtain two integrals of the last example regular in descending powers of x.

---

* Or we may say that it has two infinite roots.

(6) Show that the following equation has no integrals that are regular in either ascending or descending powers of x:

$$x^4(1-x^2)\frac{d^2y}{dx^2} + 2x^3\frac{dy}{dx} - (1-x^2)^3 y = 0.$$

[This is the equation whose primitive is $ae^{x+x^{-1}} + be^{-x-x^{-1}}$.]

MISCELLANEOUS EXAMPLES ON CHAPTER IX

(1) Obtain three independent solutions of

$$9x^2\frac{d^3y}{dx^3} + 27x\frac{d^2y}{dx^2} + 8\frac{dy}{dx} - y = 0.$$

(2) Obtain three independent solutions, of the form

$$z, \quad \frac{\partial z}{\partial c}, \quad \text{and} \quad \frac{\partial^2 z}{\partial c^2},$$

of the equation $x^2\dfrac{d^3y}{dx^3} + 3x\dfrac{d^2y}{dx^2} + (1-x)\dfrac{dy}{dx} - y = 0.$

(3) Show that the transformation $y = \dfrac{1}{bv}\dfrac{dv}{dx}$ reduces *Riccati's equation*

$$\frac{dy}{dx} + by^2 = cx^m$$

to the linear form $\dfrac{d^2v}{dx^2} - bcvx^m = 0.$

(4) Show that if γ is neither zero nor an integer, the *Hypergeometric Equation*

$$x(1-x)\frac{d^2y}{dx^2} + \{\gamma - (\alpha+\beta+1)x\}\frac{dy}{dx} - \alpha\beta y = 0$$

has the solutions (convergent if $|x| < 1$)

$$F(\alpha, \beta, \gamma, x) \quad \text{and} \quad x^{1-\gamma}F(\alpha-\gamma+1, \beta-\gamma+1, 2-\gamma, x),$$

where $F(\alpha, \beta, \gamma, x)$ denotes the *Hypergeometric Series*

$$1 + \frac{\alpha\beta}{1.\gamma}x + \frac{\alpha(\alpha+1)\beta(\beta+1)}{1.2.\gamma(\gamma+1)}x^2 + \frac{\alpha(\alpha+1)(\alpha+2)\beta(\beta+1)(\beta+2)}{1.2.3.\gamma(\gamma+1)(\gamma+2)}x^3 + \ldots .$$

(5) Show that the substitutions $x = 1-z$ and $x = 1/z$ transform the hypergeometric equation into

$$z(1-z)\frac{d^2y}{dz^2} + \{\alpha+\beta+1-\gamma - (\alpha+\beta+1)z\}\frac{dy}{dz} - \alpha\beta y = 0$$

and $z^2(1-z)\dfrac{d^2y}{dz^2} + z\{(1-\alpha-\beta) - (2-\gamma)z\}\dfrac{dy}{dz} + \alpha\beta y = 0$

respectively, of which the first is also of hypergeometric form.

Hence, from the last example, deduce that the original equation has the additional four solutions:

$$F(\alpha, \beta, \alpha+\beta+1-\gamma, 1-x),$$
$$(1-x)^{\gamma-\alpha-\beta}F(\gamma-\beta, \gamma-\alpha, 1+\gamma-\alpha-\beta, 1-x),$$
$$x^{-\alpha}F(\alpha, \alpha+1-\gamma, \alpha+1-\beta, x^{-1}),$$

and

$$x^{-\beta}F(\beta, \beta+1-\gamma, \beta+1-\alpha, x^{-1}).$$

(6) Show that the substitution $y=(1-x)^n Y$ transforms the hypergeometric equation into another hypergeometric equation if

$$n=\gamma-\alpha-\beta.$$

Hence show that the original equation has the additional two solutions:

$$(1-x)^{\gamma-\alpha-\beta}F(\gamma-\alpha, \gamma-\beta, \gamma, x)$$

and

$$x^{1-\gamma}(1-x)^{\gamma-\alpha-\beta}F(1-\alpha, 1-\beta, 2-\gamma, x).$$

[*Note.* Ex. 5 showed how from the original two solutions of the hypergeometric equation two others could be deduced by each of the transformations $x=1-z$ and $x=1/z$. Similarly each of the three transformations $x=\dfrac{1}{1-z}$, $x=\dfrac{z}{z-1}$, $x=\dfrac{z-1}{z}$, gives two more, thus making twelve. By proceeding as in Ex. 6 the number can be doubled, giving a total of *twenty-four*. These five transformations, together with the *identical transformation* $x=z$, form a *group*; that is, by performing two such transformations in succession we shall always get a transformation of the original set.]

(7) Show that, unless $2n$ is an odd integer (positive or negative), Legendre's equation

$$(1-x^2)\frac{d^2y}{dx^2} - 2x\frac{dy}{dx} + n(n+1)y = 0$$

has the solutions, regular in *descending* powers of x,

$$x^{-n-1}F(\tfrac{1}{2}n+\tfrac{1}{2}, \tfrac{1}{2}n+1, n+\tfrac{3}{2}, x^{-2}),$$
$$x^n F(-\tfrac{1}{2}n, \tfrac{1}{2}-\tfrac{1}{2}n, \tfrac{1}{2}-n, x^{-2}).$$

[The solution for the case $2n=-1$ can be got by changing x into x^{-1} in the result of Ex. 4 of the set following Art. 97.]

(8) Show that the form of the solution of Bessel's equation of order n depends upon whether n is zero, integral, or non-integral, although the difference of the roots of the indicial equation is not n but $2n$.

*CHAPTER X

EXISTENCE THEOREMS OF PICARD, CAUCHY,† AND FROBENIUS

101. Nature of the problem. In the preceding chapters we have studied a great many devices for obtaining solutions of differential equations of certain special forms. At one time mathematicians hoped that they would discover a method for expressing the solution of any differential equation in terms of a finite number of known functions or their integrals. When it was realised that this was impossible, the question arose as to whether a differential equation in general had a solution at all, and, if it had, of what kind.

There are two distinct methods of discussing this question. One, due to Picard, has already been illustrated by examples (Arts. 83 and 84). We obtained successive approximations, which apparently tended to a limit. We shall now prove that these approximations really do tend to a limit and that this limit gives the solution. Thus we shall prove the existence of a solution of a differential equation of a fairly general type. A theorem of this kind is called an *Existence Theorem*. Picard's method is not difficult, so we will proceed to it at once before saying anything about the second method. It must be borne in mind that the object of the present chapter is *not* to obtain practically useful solutions of particular equations. Our aim now is to prove that the assumptions made in obtaining these solutions were correct, and to state exactly the conditions that are sufficient to ensure correctness in equations similar to those treated before, but generalised as far as possible.

---

* This chapter should be omitted on a first reading.

† Augustin Louis Cauchy, of Paris (1789-1857), may be looked upon as the creator of the Theory of Functions and of the modern Theory of Differential Equations. He devised the method of determining definite integrals by Contour Integration.

102. Picard's method of successive approximation. If $\dfrac{dy}{dx} = f(x, y)$ and $y = b$ and $x = a$, the successive approximations for the value of y as a function of x are

$$b + \int_a^x f(x, b)\, dx = y_1, \text{ say,}$$

$$b + \int_a^x f(x, y_1)\, dx = y_2, \text{ say,}$$

$$b + \int_a^x f(x, y_2)\, dx = y_3, \text{ say, and so on.}$$

We have already (Arts. 83 and 84) explained the application of this method to examples. We took the case where $f(x, y) = x + y^2$: $b = a = 0$, and found

$$y_1 = \tfrac{1}{2}x^2,$$
$$y_2 = \tfrac{1}{2}x^2 + \tfrac{1}{20}x^5,$$
$$y_3 = \tfrac{1}{2}x^2 + \tfrac{1}{20}x^5 + \tfrac{1}{160}x^8 + \tfrac{1}{4400}x^{11}.$$

These functions appear to be tending to a limit, at any rate for sufficiently small values of x. It is the purpose of the present article to prove that this is the case, not merely in this particular example, but whenever $f(x, y)$ obeys certain conditions to be specified.

These conditions are that, after suitable choice of the positive numbers h and k, we can assert that, for all values of x between $a - h$ and $a + h$, and for all values of y between $b - k$ and $b + k$, we can find positive numbers M and A so that

(i) $|f(x, y)| < M$,

(ii) $|f(x, y) - f(x, y')| < A\,|\,y - y'\,|$, y and y' being any two values of y in the range considered.

In our example $f(x, y) = x + y^2$, condition (i) is obviously satisfied, taking for M any positive number greater than $|\,a\,| + h + \{|\,b\,| + k\}^2$.

Also $|\,(x + y^2) - (x + y'^2)\,| = |\,y + y'\,|\,|\,y - y'\,| < 2(|\,b\,| + k)\,|\,y - y'\,|$, so condition (ii) is also satisfied, taking $A = 2(|\,b\,| + k)$.

Returning to the general case, we consider the differences between the successive approximations.

$$y_1 - b = \int_a^x f(x, b)\, dx, \text{ by definition,}$$

but $|f(x, b)| < M$, by condition (i),

so $|\,y_1 - b\,| < \left|\int_a^x M\, dx\right|$,   i.e. $< M\,|\,x - a\, < Mh.$ ....(1)

Also $y_2 - y_1 = b + \int_a^x f(x, y_1)\, dx - b - \int_a^x f(x, b)\, dx$, by definition.

$$= \int_a^x \{f(x, y_1) - f(x, b)\}\, dx;$$

but $\quad |f(x, y_1) - f(x, b)| < A\, |y_1 - b|$, by condition (ii),

$$< AM\,|x - a|\,, \text{ from } (1),$$

so $\quad |y_2 - y_1| < \left| \int_a^x AM(x - a)\, dx \right|,\quad i.e. < \tfrac{1}{2}AM(x-a)^2 < \tfrac{1}{2}AMh^2 \ldots\ldots(2)$

Similarly, $\qquad |y_n - y_{n-1}| < \dfrac{1}{n!}\, MA^{n-1}h^n. \quad \ldots\ldots\ldots\ldots\ldots(3)$

Now the infinite series

$$b + Mh + \tfrac{1}{2}MAh^2 + \ldots + \frac{1}{n!}\, MA^{n-1}h^n \ldots = \frac{M}{A}\,(e^{Ah} - 1) + b$$

is convergent for all values of h, A, and M.

Therefore the infinite series

$$b + (y_1 - b) + (y_2 - y_1) + \ldots + (y_n - y_{n-1}) + \ldots ,$$

each term of which is equal or less in absolute value than the corresponding term of the preceding, is still more convergent.

That is to say that the sequence

$$y_1 = b + (y_1 - b),$$
$$y_2 = b + (y_1 - b) + (y_2 - y_1),$$

and so on, tends to a definite limit, say $Y(x)$, which is what we wanted to prove.

We must now prove that Y satisfies the differential equation.

At first sight this seems obvious, but it is not so really, for we must not assume without proof that

$$\operatorname*{Lt}_{n \to \infty} \int_a^x f(x, y_{n-1})\, dx = \int_a^x f(x, \operatorname*{Lt}_{n \to \infty} y_{n-1})\, dx.$$

The student who understands the idea of uniform convergence will notice that the inequalities (1), (2), (3) that we have used to prove the convergence of our series really prove its uniform convergence also. If, then, $f(x, y)$ is continuous, y_1, y_2, etc., are continuous also, and Y is a uniformly convergent series of continuous functions; that is, Y is itself continuous,* and $Y - y_{n-1}$ tends uniformly to zero as n increases.

Hence, from condition (ii), $f(x, Y) - f(x, y_{n-1})$ tends uniformly to zero.

* See Bromwich's *Infinite Series*, Art. 45.

From this we deduce that

$$\int_a^x \{f(x, \ Y) - f(x, \ y_{n-1})\} \, dx \text{ tends to zero.}$$

Thus the limit of the relation

$$y_n = b + \int_a^x f(x, \ y_{n-1}) \, dx$$

is

$$Y = b + \int_a^x f(x, \ Y) \, dx;$$

therefore* $\dfrac{dY}{dx} = f(x, \ Y)$, and $Y = b$ when $x = a$.

This completes the proof.

103. Cauchy's method. Theorems on infinite series required.
Cauchy's method is to obtain an infinite series from the differential
equation, and then prove it convergent by comparing it with another
infinite series. The second infinite series is *not* a solution of the
equation, but the relation between its coefficients is simpler than
that between those of the original series. Our first example of this
method will be for the simple case of the linear equation of the first
order

$$\frac{dy}{dx} = p(x) \cdot y.$$

Of course this equation can be solved at once by separation of
the variable, giving

$$\log y = c + \int p(x) \, dx.$$

However, we give the discussion by infinite series because it is
almost exactly similar to the slightly more difficult discussion of

$$\frac{d^2y}{dx^2} = p(x) \cdot \frac{dy}{dx} + q(x) \cdot y,$$

and other equations of higher order.

We shall need the following theorems relating to power series.
The variable x is supposed to be complex. For brevity we shall
denote absolute values by capital letters, *e.g.* A_n for $|a_n|$.

(A) A power series $\sum\limits_0^\infty a_n x^n$ is absolutely convergent at all
points within its circle of convergence $|x| = R$.

(B) The radius R of this circle is given by

$$\frac{1}{R} = \operatorname*{Lt}_{n \to \infty} \frac{A_{n+1}}{A_n},$$

provided that this limit exists.

(C) $\dfrac{d}{dx}\left(\sum_0^\infty a_n x^n\right) = \sum_0^\infty n a_n x^{n-1}$, within $\mid x \mid = R$.

(D) If we have two power series, then for points within the circle that is common to their circles of convergence,

$$\left(\sum_0^\infty a_n x^n\right)\left(\sum_0^\infty b_n x^n\right) = \sum_0^\infty (a_n b_0 + a_{n-1}b_1 + \ldots + a_0 b_n)x^n.$$

(E) If $\sum_0^\infty a_n x^n = \sum_0^\infty b_n x^n$ for all values of x within the circle $\mid x \mid = R$, then $a_n = b_n$.

(F) $A_n < MR^{-n}$, where M exceeds the absolute value of the sum of the series at points on a circle $\mid x \mid = R$ on which the series is convergent.

Proofs of these theorems will be found in Bromwich's *Infinite Series*:

A in Art. 82   [Art. 84 in 2nd ed.],
B is an obvious deduction from D'Alembert's ratio test, Art. 12,
C in Art. 52   [Art. 12 becomes Art. 12·2 in 2nd ed.],
D „   54,
E „   52,
F „   82   [Art. 84 in 2nd ed.].

Two theorems on uniform convergence will be required later on, but we will defer these until they are needed.

104. * Convergence of the solution in series of $\dfrac{dy}{dx} = y\,p(x)$. Let $p(x)$ be capable of expansion in a power series $\sum_0^\infty p_n x^n$ which is convergent everywhere within and on the circle $\mid x \mid = R$. We shall prove that a solution $y = \sum_0^\infty a_n x^n$ can be obtained which is convergent within this circle.

Substituting in the differential equation, we obtain

$$\sum_0^\infty n a_n x^{n-1} = \sum_0^\infty a_n x^n \sum_0^\infty p_n x^n \qquad \text{(Theorem C)}$$

$$= \sum_0^\infty (a_n p_0 + a_{n-1}p_1 + a_{n-2}p_2 + \ldots + a_0 p_n)\, x^n. \quad \text{(Theorem D)}.$$

Equating the coefficients of x^{n-1},   (Theorem E)

$$n a_n = a_{n-1}p_0 + a_{n-2}p_1 + a_{n-3}p_2 + \ldots + a_0 p_{n-1}. \quad \ldots\ldots\ldots\ldots(1)$$

* Revise Art. 7 before reading the following.

Hence for the absolute values of the a's and p's, denoted by the corresponding capital letters, we get

$$nA_n \leqslant A_{n-1}P_0 + A_{n-2}P_1 + A_{n-3}P_2 + \ldots + A_0P_{n-1}. \quad \ldots\ldots(2)$$

Let M be a positive number exceeding the absolute value of $p(x)$ on the circle $|x| = R$,

then $$P_n < MR^{-n}; \quad \ldots\ldots\ldots(3) \qquad \text{(Theorem F)}$$

therefore, from (2) and (3),

$$A_n < \frac{M}{n}(A_{n-1} + A_{n-2}R^{-1} + A_{n-3}R^{-2} + \ldots + A_0R^{-n+1}). \quad \ldots\ldots(4)$$

Define $B_n(n>0)$ as the right-hand side of (4), and define B_0 as any positive number greater than A_0; then $A_n < B_n$.

But $$\frac{M}{n}(A_{n-1} + A_{n-2}R^{-1} + A_{n-3}R^{-2} + \ldots + A_0R^{-n+1})$$

$$= \frac{M}{n}A_{n-1} + \frac{n-1}{nR}\frac{M}{n-1}(A_{n-2} + A_{n-3}R^{-1} + \ldots + A_0R^{-n+2}).$$

Hence, defining B_n as above,

$$B_n = \frac{M}{n}A_{n-1} + \frac{(n-1)}{n}\frac{B_{n-1}}{R},$$

whence, dividing by B_{n-1} and using k for $\dfrac{A_{n-1}}{B_{n-1}}$, so that $0 \leqslant k < 1$,

$$\frac{B_n}{B_{n-1}} = \frac{Mk}{n} + \frac{1}{R} - \frac{1}{nR},$$

whence $$\underset{n\to\infty}{\text{Lt}} \frac{B_n}{B_{n-1}} = \frac{1}{R}.$$

Therefore the series $\sum\limits_{0}^{\infty} B_n x^n$ is convergent within the circle $|x| = R$.   (Theorem B)

Still more therefore is the series $\sum\limits_{0}^{\infty} a_n x^n$ convergent within the same circle, since $$A_n < B_n.$$

The coefficients $a_1, a_2, \ldots$ can all be found from (1) in terms of the p's, which are supposed known, and the arbitrary constant a_0.

105. Remarks on this proof. The student will probably have found the last article very difficult to follow. It is important not to get confused by the details of the work. The main point is this. We should like to prove that $\underset{n\to\infty}{\text{Lt}} \dfrac{A_n}{A_{n-1}} = \dfrac{1}{R}.$ Unfortunately the relation defining the A's is rather complicated. We first simplify it by getting rid of the n quantities $P_0, P_1, \ldots, P_{n-1}$. Still the relation is too complicated, as it involves n A's. We need a simple relation involving only two. By taking a suitable definition of B_n we get such a relation between B_n and B_{n-1}, leading to

$$\operatorname*{Lt}_{n \to \infty} \frac{B_n}{B_{n-1}} = \frac{1}{R}.$$

We repeat that the object of giving such a complicated discussion of a very simple equation is to provide a model which the student can imitate in other cases.

Examples for solution

(1) Prove that, if $p(x)$ and $q(x)$ can be expanded in power series convergent at all points within and on the circle $X = R$, then a power series convergent within the same circle can be found in terms of the first two coefficients (the arbitrary constants) to satisfy

$$\frac{d^2y}{dx^2} = p(x) \cdot \frac{dy}{dx} + q(x) y.$$

[Here $n(n-1)a_n = (n-1)a_{n-1}p_0 + (n-2)a_{n-2}p_1 + \ldots + a_1 p_{n-2}$
$$\qquad\qquad + a_{n-2}q_0 + a_{n-3}q_1 + \ldots + a_0 q_{n-2}.$$

Hence, if M is any number exceeding the absolute values of both $p(x)$ and $q(x)$ at all points on the circle $X = R$,

$$A_n < \frac{M}{n} \{ (A_{n-1} + A_{n-2}R^{-1} + \ldots + A_1 R^{-n+2})$$

$$\qquad\qquad + (A_{n-2} + A_{n-3}R^{-1} + \ldots + A_0 R^{-n+2}) \}$$

$$\qquad < \frac{M}{n} (1 + R)(A_{n-1} + A_{n-2}R^{-1} + \ldots + A_0 R^{-n+1}).$$

Define the right-hand side of this inequality as B_n and then proceed as before.]

(2) Prove similar results for the equation

$$\frac{d^3y}{dx^3} = p(x) \cdot \frac{d^2y}{dx^2} + q(x) \cdot \frac{dy}{dx} + r(x) \cdot y.$$

106. Frobenius' method. Preliminary discussion. When the student has mastered the last article, he will be ready for the more difficult problem of investigating the convergence of the series given by the method of Frobenius. In the preceding chapter (which should be thoroughly known before proceeding further), we saw that in some cases we obtained two series involving only powers of x, while in others logarithms were present.

The procedure in the first case is very similar to that of the last article. But in the second case a new difficulty arises. The series with logarithms were obtained by differentiating series with respect to a parameter c. Now differentiation is a process of taking a limit and the summation of an infinite series is another process of taking a limit. It is by no means obvious that the result will be the same whichever of these two processes is performed first, even if the series of differential coefficients be convergent.

However, we shall prove that in our case the differentiation is legitimate, but this proof that our series satisfy conditions sufficient to justify term-by-term differentiation is rather long and bewildering.

To appreciate the following work the student should at first ignore all the details of the algebra, concentrating his attention on the general trend of the argument. When this has been grasped, he can go back and verify the less important steps taken for granted on a first reading.

107. Obtaining the coefficients in Frobenius' series when the roots of the indicial equation do not differ by an integer or zero. Consider the expression

$$x^2 \frac{d^2y}{dx^2} - x\, p(x) \cdot \frac{dy}{dx} - q(x) \cdot y = \phi\left(x, y, \frac{dy}{dx}, \frac{d^2y}{dx^2}\right), \text{ say,}$$

where $p(x)$ and $q(x)$ are both expansible in power series $\sum_0^\infty p_n x$ and $\sum_0^\infty q_n x^n$ which are convergent within and on the circle $\mid x \mid =$

We are trying to obtain a solution of the differential equation

$$\phi\left(x, y, \frac{dy}{dx}, \frac{d^2y}{dx^2}\right) = 0. \quad \ldots\ldots\ldots\ldots\ldots\ldots(1)$$

If y is replaced by $x^c \sum_0^\infty a_n x^n$ (with $a_0 \neq 0$), $\phi\left(x, y, \frac{dy}{dx}, \frac{d^2y}{dx^2}\right)$ becomes

$$\sum_0^\infty a_n x^{c+n}\{(c+n)(c+n-1) - (c+n)p(x) - q(x)\}$$

$$= \sum_0^\infty g_n x^{c+n}, \text{ say,}$$

where $\quad g_0 = a_0\{c(c-1) - p_0 c - q_0\}$

and $\quad g_n = a_n\{(c+n)(c+n-1) - p_0(c+n) - q_0\}$
$$- a_{n-1}\{p_1(c+n-1) + q_1\} - a_{n-2}\{p_2(c+n-2) + q_2\}$$
$$\ldots - a_0(p_n c + q_n).$$

For brevity, denote

$$c(c-1) - p_0 c - q_0 \text{ by } f(c),$$

so that $\quad (c+n)(c+n-1) - p_0(c+n) - q_0 = f(c+n).$

Then $g_n = 0$ if

$$a_n f(c+n) = a_{n-1}\{p_1(c+n-1)+q_1\} + a_{n-2}\{p_2(c+n-2)+q_2\}$$
$$+ \ldots + a_0(p_n c + q_n). \quad \ldots\ldots\ldots(2)$$

If we can choose the a's so that all the g's vanish, and if the series $\sum\limits_0^\infty a_n x^n$ so obtained is convergent, a solution of (1) will have been obtained.

Now as $a_0 \neq 0$, $g_0 = 0$ gives

$$c(c-1) - p_0 c - q_0 = 0. \quad \ldots\ldots\ldots\ldots\ldots(3)$$

This is a quadratic equation in c, and is called the Indicial Equation.

Let its roots be α and β.

If either of these values is substituted for c in the equations $g_1 = 0$, $g_2 = 0$, $g_3 = 0$, ..., values for a_1, a_2, a_3, ... are found in the form

$$a_n = a_0 h_n(c)/[f(c+n)f(c+n-1) \ldots f(c+1)], \quad \ldots\ldots\ldots(4)$$

where $h_n(c)$ is a polynomial in c. The student should work out the values of a_1 and a_2 in full if he finds any difficulty at this point.

The process by which a_n is obtained from (2) involves division by $f(c+n)$. This is legitimate only when $f(c+n) \neq 0$.

Now as
$$f(c) = (c-\alpha)(c-\beta),$$
$$f(c+n) = (c+n-\alpha)(c+n-\beta),$$
so
$$f(\alpha+n) = n(\alpha+n-\beta), \quad \ldots\ldots\ldots\ldots(5)$$
and
$$f(\beta+n) = n(\beta+n-\alpha). \quad \ldots\ldots\ldots\ldots(6)$$

Thus, if α and β do not differ by an integer, the divisors cannot vanish, so the above process for obtaining the a's is satisfactory. If $\alpha = \beta$, only one series is obtained.

108. Convergence of the series so obtained. Let M be a positive number exceeding the absolute values of $p(x)$ and $q(x)$ at all points on the circle $|x| = R$.

Then
$$P_s < MR^{-s}$$
and
$$Q_s < MR^{-s},$$
so that
$$|p_s(c+n-s) + q_s| < M(C+n-s+1)R^{-s}.$$

From these inequalities and from (2),

$$A_n < M\{A_{n-1}(C+n)R^{-1} + \ldots + A_0(C+1)R^{-n}\}/F(c+n), \quad \ldots(7)$$

say $A_n < B_n$, denoting the right-hand side of (7) by B_n. This defines B_n if $n > 0$. Define B_0 as any positive number greater than A_0. This definition of B_n gives

$$B_{n+1}F(c+n+1) - B_n F(c+n)R^{-1} = A_n M(C+n+1)R^{-1}$$
$$= kB_n M(C+n+1)R^{-1}, \quad \text{where } 0 \leqslant k < 1,$$

so that $\dfrac{B_{n+1}}{B_n} = \dfrac{F(c+n) + kM(C+n+1)}{RF(c+n+1)}$,

i.e. $= \dfrac{\mid (c+n)(c+n-1) - p_0(c+n) - q_0 \mid + kM(C+n+1)}{R \mid (c+n+1)(c+n) - p_0(c+n+1) - q_0 \mid}$.

Now for large values of n the expression on the right approaches the value

$$\frac{n^2}{Rn^2} = \frac{1}{R}.$$

Thus

$$\operatorname*{Lt}_{n\to\infty} \frac{B_{n+1}}{B_n} = \frac{1}{R}.$$

Therefore the series $\sum\limits_0^\infty B_n x^n$ and still more the series $\sum\limits_0^\infty a_n x^n$ converges within the circle $\mid x \mid = R$.

Thus, when α and β do not differ by an integer, we get two convergent infinite series satisfying the differential equation.

109. Modification required when the roots of the indicial equation differ by zero or an integer. When α and β are equal, we get one series by this method.

When α and β differ by an integer, this method holds good for the larger one, but not for the smaller, for if $\alpha - \beta = r$ (a positive integer), then from (5) and (6)

$$f(\alpha+n) = n(\alpha+n-\beta) = n(n+r),$$

but

$$f(\beta+n) = n(\beta+n-\alpha) = n(n-r),$$

which vanishes when $n = r$, giving a zero factor in the denominator of a_r when $c = \beta$. As exemplified in Arts. 98 and 99 of the preceding chapter, this may give either an infinite or indeterminate value for some of the a's. This difficulty is removed by modifying the form assumed for y, replacing a_0 by $k(c-\beta)$. This will make $a_0, a_1, \ldots ,$ a_{r-1} all zero and $a_r, a_{r+1}, \ldots$ all finite when c is put equal to β. This change in the form assumed for y will not alter the relation between the a's, and so will not affect the above investigation of convergence.

110. Differentiation of an infinite series with respect to a parameter c, the roots of the indicial equation differing by an integer. In Art. 107 we obtained an infinite series $x^c \sum\limits_0^\infty a_n x^n$ where the a's are functions of c. As in the preceding chapter, we have to consider the differentiation of this series with respect to c, c being put equal to the smaller root β after the differentiation.

Now while this differentiation is being performed we may consider x as a constant. The series can then be considered as a series of functions of the variable c, say $\sum\limits_{0}^{\infty} \psi_n(c)$, where

$$\psi_n(c) = x^{c+n} a_n$$
$$= x^{c+n} a_0 h_n(c) / [f(c+n)f(c+n-1) \dots f(c+1)], \text{ from (4),}$$

where $a_0 = k(c - \beta)$ and the factor $(c - \beta)$ is to be divided out if it occurs in the denominator.

Now Goursat (*Cours d'Analyse*, Vol. II. 2nd. ed.* p. 98) proves that if (i) all the ψ's are functions which are analytic and holomorphic within a certain region bounded by a closed contour and continuous on this contour, and if (ii) the series of ψ's is uniformly convergent on this contour, then the differentiation term by term gives a convergent series whose sum is the differential coefficient of the sum of the original series.

For the definitions of holomorphic and analytic, see the beginning of Vol. II. of Goursat. It will be seen that the ψ's satisfy these definitions and are continuous as long as we keep away from values of c that make them infinite. These values are $\alpha - 1$, $\beta - 1$, $\alpha - 2$, $\beta - 2$, etc. To avoid these take the region inside a circle of centre $c = \beta$ and of any radius less than unity.

We shall now prove that the series is uniformly convergent everywhere inside this region. This will prove it is uniformly convergent *on the contour* of a similar but slightly smaller region inside the first.

Let s be a positive integer exceeding the largest value of C within the larger region.

Then for *all* values of c within this region, for values of n exceeding s,

$$F(c+n) = |(c+n)(c+n-1) - p_0(c+n) - q_0|, \text{ by definition of } F,$$
$$\geqslant (C+n)^2 - (P_0+1)(C+n) - Q_0, \text{ as } |u-v| \geqslant |u| - |v|,$$
$$> (n-s)^2 - (M+1)(s+n) - M, \text{ as } P_0 < M \text{ and } Q_0 < M,$$
$$> n^2 + In + J, \text{ say, where } I \text{ and } J \text{ are independent of}$$
$$n, x, \text{ or } c. \quad \dots\dots\dots\dots\dots\dots\dots\dots\dots(8)$$

For sufficiently great values of n, say $n > m$, the last expression is always positive.

Let H denote the maximum value of

$$M[A_{m-1}(C+m)R^{-1} + A_{m-2}(C+m-1)R^{-2} + \dots + A_0(C+1)R^{-m}] (9)$$

for *all* the values of c in the region.

* p. 96 in 4th ed.

Then if E_m be any positive number greater than B_m, and, if, for values of $n>m$, E_n be defined by

$$E_n = \frac{M\{E_{n-1}(s+n)R^{-1}+\ldots E_m(s+m+1)R^{-n+m}\}+HR^{-n+m}}{n^2+In+J}, \quad (10)$$

so that

$$E_{m+1} = \frac{ME_m(s+m+1)R^{-1}+HR^{-1}}{(m+1)^2+I(m+1)+J},$$

which has a numerator greater than and a denominator less than those of B_{m+1}, from (8), (9), and the definition of B_n as the right-hand side of (7), we see that

$$E_{m+1}>B_{m+1}.$$

Similarly $E_n>B_n$ for all values of $n>m$.

From (10) we prove $\underset{n\to\infty}{\mathrm{Lt}}\,\dfrac{E_{n+1}}{E_n}=\dfrac{1}{R}$. This piece of work is so similar to the corresponding work at the end of Art. 108 that we leave it as an exercise for the student.

Hence $\sum\limits_{m}^{\infty} E_n R_1^n$ is convergent if $R_1< R$.

Therefore within the circle $|x|=R_1$ and within the region specified for c,

$$|a_n x^{c+n}|<A_n R_1^{s+n}<B_n R_1^{s+n}<E_n R_1^{s+n}.$$

This shows that $\Sigma a_n x^{c+n}$ satisfies Weierstrass's M-test for uniform convergence (Bromwich, Art. 44), as R_1, s, and the E's are all independent of c.

This completes the proof that $\Sigma\psi_n=\Sigma a_n x^{c+n}$ satisfies all the conditions specified, so the differentiation with respect to c is now justified. This holds within the circle $|x|=R_1$. We can take R_1 great enough to include any point within the circle $|x|=R$.

If the roots of the indicial equation are equal instead of differing by an integer, the only difference in the above work is that a_0 is *not* to be replaced by $k(c-\beta)$, as no $(c-\beta)$ can now occur in the denominator of a_n.

[For a supplement to Chaps. IX. and X. see Arts. 171-177. They deal with regular integrals, Fuchs' theorem, ordinary and singular points, equations of Fuchsian type, characteristic index, normal and subnormal integrals. For a fuller version of Art. 102, including a discussion of the uniqueness of the integral, see Ince's *Ordinary Differential Equations*, Arts. 3·2 and 3·21.]

CHAPTER XI

ORDINARY DIFFERENTIAL EQUATIONS WITH THREE VARIABLES, AND THE CORRESPONDING CURVES AND SURFACES

111. We shall now consider some simple differential equations expressing properties of curves in space and of surfaces on which these curves lie, or which they cut orthogonally (as in Electrostatics the Equipotential Surfaces cut the Lines of Force orthogonally). The ordinary * differential equations of this chapter are closely connected with the partial differential equations of the next.

Before proceeding further the student should revise his solid geometry. We need in particular the fact that the direction-cosines of the tangent to a curve are

$$\left(\frac{dx}{ds}, \quad \frac{dy}{ds}, \quad \frac{dz}{ds}\right),$$

i.e. are in the ratio $dx : dy : dz$.

Simultaneous linear equations with constant coefficients have already been discussed in Chapter III.

112. The simultaneous equations $\dfrac{dx}{P} = \dfrac{dy}{Q} = \dfrac{dz}{R}$. These equations express that the tangent to a certain curve at any point (x, y, z) has direction-cosines proportional to (P, Q, R). If P, Q, and R are constants, we thus get a straight line, or rather a doubly infinite system of straight lines, as one such line goes through any point of space. If, however, P, Q, and R are functions of x, y, and z, we get a similar system of curves, any one of which may be considered as generated by a moving point which continuously alters its direction

---

* *i.e.* not involving partial differential coefficients.

of motion. The Lines of Force of Electrostatics form such a system.*

Ex. (i).
$$\frac{dx}{1} = \frac{dy}{1} = \frac{dz}{1}. \quad\ldots\ldots\ldots\ldots\ldots\ldots(1)$$

Obvious integrals are
$$x - z = a, \quad\ldots\ldots\ldots\ldots\ldots\ldots\ldots\ldots(2)$$
$$y - z = b, \quad\ldots\ldots\ldots\ldots\ldots\ldots\ldots\ldots(3)$$

the equations of two planes, intersecting in the line
$$\frac{x-a}{1} = \frac{y-b}{1} = \frac{z}{1}, \quad\ldots\ldots\ldots\ldots\ldots\ldots(4)$$

which by suitable choice of the arbitrary constants a and b can be made to go through any given point, e.g. through (f, g, h) if $a = f - h$ and $b = g - h$.

Instead of picking out the single line of the system that goes through one given point, we may take the infinity of such lines that intersect a given curve, e.g. the circle $x^2 + y^2 = 4$, $z = 0$.

The equations of this circle, taken together with (2) and (3), give
$$x = a,$$
$$y = b,$$
and hence
$$a^2 + b^2 = 4. \quad\ldots\ldots\ldots\ldots\ldots\ldots\ldots(5)$$

This is the relation that holds between a and b if the line is to intersect the circle. Eliminating a and b from (2), (3), and (5), we get
$$(x - z)^2 + (y - z)^2 = 4,$$

the elliptic cylinder formed by those lines of the system which meet the circle.

Similarly the lines of the system which meet the curve
$$\phi(x, y) = 0, \quad z = 0$$
form the surface
$$\phi(x - z, y - z) = 0.$$

Ex. (ii).
$$\frac{dx}{z} = \frac{dy}{0} = \frac{dz}{-x}. \quad\ldots\ldots\ldots\ldots\ldots\ldots(6)$$

Obvious integrals are
$$x^2 + z^2 = a, \quad\ldots\ldots\ldots\ldots\ldots\ldots\ldots(7)$$
$$y = b, \quad\ldots\ldots\ldots\ldots\ldots\ldots\ldots\ldots(8)$$

a right circular cylinder and a plane that cuts it in a circle.

The differential equations therefore represent a system of circles, whose centres all lie on the axis of y and whose planes are all perpendicular to this axis.

One such circle goes through any point of space. That through (f, g, h) is
$$x^2 + z^2 = f^2 + h^2, \quad y = g.$$

A surface is formed by the circles of the system that intersect a given curve.

* The equations of the lines of force are $dx \left/ \dfrac{\partial V}{\partial x} \right. = dy \left/ \dfrac{\partial V}{\partial y} \right. = dz \left/ \dfrac{\partial V}{\partial z} \right.$, where V is the potential function.

If the given curve is the hyperbola

$$\frac{x^2}{A^2} - \frac{y^2}{B^2} = 1, \quad z = 0,$$

(7) and (8) give, for a circle intersecting this hyperbola,

$$x^2 = a, \quad y = b,$$

and hence

$$\frac{a}{A^2} - \frac{b^2}{B^2} = 1 \dots\dots\dots\dots\dots\dots\dots\dots\dots(9)$$

Eliminating a and b from (7), (8), and (9), we get the hyperboloid of one sheet,

$$\frac{x^2 + z^2}{A^2} - \frac{y^2}{B^2} = 1,$$

formed by those circles of the system that intersect the hyperbola.

Similarly, starting from the curve $\phi(x^2, y) = 0$, $z = 0$, we get the surface of revolution $\phi(x^2 + z^2, y) = 0$.

113. Solution of such equations by multipliers. If

$$\frac{dx}{P} = \frac{dy}{Q} = \frac{dz}{R},$$

each of these fractions is equal to

$$\frac{l\,dx + m\,dy + n\,dz}{lP + mQ + nR}.$$

This method may be used with advantage in some examples to obtain a zero denominator and a numerator that is an exact differential, or a non-zero denominator of which the numerator is the differential.

Ex. (i).
$$\frac{dx}{z(x+y)} = \frac{dy}{z(x-y)} = \frac{dz}{x^2+y^2}.$$

Each fraction $= \dfrac{x\,dx - y\,dy - z\,dz}{xz(x+y) - yz(x-y) - z(x^2+y^2)} = \dfrac{x\,dx - y\,dy - z\,dz}{0};$

therefore
$$x\,dx - y\,dy - z\,dz = 0,$$
$$i.e. \quad x^2 - y^2 - z^2 = a.$$

Similarly
$$y\,dx + x\,dy - z\,dz = 0,$$
$$i.e. \quad 2xy - z^2 = b.$$

Ex. (ii).
$$\frac{dx}{1+y} = \frac{dy}{1+x} = \frac{dz}{z}.$$

Here
$$\frac{dz}{z} = \frac{dx + dy}{2 + x + y} = \frac{dx - dy}{y - x},$$

giving
$$\log z = \log(2 + x + y) + \log a = -\log(x - y) + \log b,$$
$$i.e. \quad z = a(2 + x + y) = b/(x - y).$$

Examples for solution

Obtain the system of curves, defined by two equations with an arbitrary constant in each, satisfying the following simultaneous differential equations. Interpret geometrically whenever possible.

(1) $\dfrac{dx}{x} = \dfrac{dy}{y} = \dfrac{dz}{z}$.

(2) $\dfrac{dx}{mz - ny} = \dfrac{dy}{nx - lz} = \dfrac{dz}{ly - mx}$.

(3) $\dfrac{dx}{y^2 + z^2 - x^2} = \dfrac{dy}{-2xy} = \dfrac{dz}{-2xz}$.

(4) $\dfrac{dx}{yz} = \dfrac{dy}{zx} = \dfrac{dz}{xy}$.

(5) $\dfrac{dx}{y+z} = \dfrac{dy}{z+x} = \dfrac{dz}{x+y}$.

(6) $\dfrac{x\,dx}{z^2 - 2yz - y^2} = \dfrac{dy}{y+z} = \dfrac{dz}{y-z}$.

(7) Find the radius of the circle of Ex. 2 that goes through the point $(0, -n, m)$.

(8) Find the surface generated by the curves of Ex. 4 that intersect the circle $y^2 + z^2 = 1$, $x = 0$.

(9) Find the surface generated by the lines of Ex. 1 that intersect the helix $x^2 + y^2 = r^2$, $z = k \tan^{-1} \dfrac{y}{x}$.

(10) Find the curve which passes through the point $(1, 2, -1)$ and is such that at any point the direction-cosines of its tangent are in the ratio of the squares of the co-ordinates of that point.

114. A second integral found by the help of the first. Consider the equations

$$\frac{dx}{1} = \frac{dy}{-2} = \frac{dz}{3x^2 \sin (y + 2x)} \quad \dots\dots\dots\dots(1)$$

An obvious integral is $y + 2x = a$. ....(2)
Using this relation, we get

$$\frac{dx}{1} = \frac{dz}{3x^2 \sin a},$$

giving $z - x^3 \sin a = b$.

Substituting for a, $z - x^3 \sin (y + 2x) = b$. ....(3)
Is (3) really an integral of (1)?
Differentiating (3),

$$\{dz - 3x^2\,dx \sin (y + 2x)\} - x^3 \cos (y + 2x) \cdot \{dy + 2\,dx\} = 0,$$

which is true in virtue of (1).   So (3) *is* an integral.

Examples for solution

(1) $\dfrac{dx}{1} = \dfrac{dy}{3} = \dfrac{dz}{5z + \tan (y - 3x)}$.

(2) $\dfrac{dx}{z} = \dfrac{dy}{-z} = \dfrac{dz}{z^2 + (y + x)^2}$.

(3) $\dfrac{dx}{xz(z^2 + xy)} = \dfrac{dy}{-yz(z^2 + xy)} = \dfrac{dz}{x^4}$.

(4) $\dfrac{dx}{xy} = \dfrac{dy}{y^2} = \dfrac{dz}{zxy - 2x^2}$.

115. General and special integrals of simultaneous equations. If $u = a$ and $v = b$ are two independent integrals of the simultaneous equations
$$\frac{dx}{P} = \frac{dy}{Q} = \frac{dz}{R},$$
then $\phi(u, v) = 0$ represents a surface passing through curves of the system, and should therefore give another integral, whatever the form of the function ϕ.

An analytical proof of this is reserved for the next chapter, as its importance belongs chiefly to partial differential equations.

$\phi(u, v) = 0$ is called the *General Integral*. Some simultaneous equations possess integrals called *Special*, which are not included in the General Integral.

Examples for solution

(1) In the Ex. of Art. 113 $u = x^2 - y^2 - z^2$ and $v = 2xy - z^2$, so the General Integral is $\phi(x^2 - y^2 - z^2, 2xy - z^2) = 0$. The student should verify this in the simple cases where
$$\phi(u, v) = u - v \quad \text{or} \quad \phi(u, v) = \frac{v + 1}{u - 2}.$$

(2) Verify that for the equation
$$\frac{dx}{1 + \sqrt{(z - x - y)}} = \frac{dy}{1} = \frac{dz}{2},$$
the General Integral may be taken as
$$\phi\{2y - z, \, y + 2\sqrt{(z - x - y)}\} = 0,$$
while $z = x + y$ is a Special Integral.

116. Geometrical interpretation of the equation
$$\mathbf{P} \, dx + \mathbf{Q} \, dy + \mathbf{R} \, dz = 0.$$
This differential equation expresses that the tangent to a curve is perpendicular to a certain line, the direction-cosines of this tangent and line being proportional to (dx, dy, dz) and (P, Q, R) respectively.

But we saw that the simultaneous equations
$$\frac{dx}{P} = \frac{dy}{Q} = \frac{dz}{R}$$
expressed that the tangent to a curve was *parallel* to the line (P, Q, R). We thus get two sets of curves. If two curves, one of each set, intersect, they must intersect at right angles.

Now two cases arise. It may happen that the equation
$$P \, dx + Q \, dy + R \, dz = 0$$
is integrable. This means that a family of surfaces can be found, all curves on which are perpendicular to the curves represented by

the simultaneous equations at all points where these curves cut the surface. In fact, this is the case where an infinite number of surfaces can be drawn to cut orthogonally a doubly infinite set of curves, as equipotential surfaces cut lines of force in electrostatics. On the other hand, the curves represented by the simultaneous equations may not admit of such a family of orthogonal surfaces. In this case the single equation is non-integrable.

Ex. (i). The equation $\quad dx + dy + dz = 0$

integrates to $\qquad\qquad x + y + z = c,$

a family of parallel planes.

We saw in Ex. (i) of Art. 112 that the simultaneous equations

$$\frac{dx}{1} = \frac{dy}{1} = \frac{dz}{1}$$

represented the family of parallel lines

$$\frac{x - a}{1} = \frac{y - b}{1} = \frac{z}{1}.$$

The planes are the orthogonal trajectories of the lines.

Ex. (ii). $\qquad\qquad z\, dx - x\, dz = 0,$

$$i.e. \quad \frac{dx}{x} - \frac{dz}{z} = 0$$

integrates to $\qquad\qquad z = cx,$

a family of planes passing through the axis of y.

We saw in Ex. (ii) of Art. 112 that the corresponding simultaneous equations $\qquad \dfrac{dx}{z} = \dfrac{dy}{0} = \dfrac{dz}{-x}$

represented a system of circles whose axes all lie along the axis of y, so the planes are the orthogonal trajectories of the circles.

Examples for solution

Integrate the following equations, and whenever possible interpret the results geometrically and verify that the surfaces are the orthogonal trajectories of the curves represented by the corresponding simultaneous equations:

(1) $x\, dx + y\, dy + z\, dz = 0.$

(2) $(y^2 + z^2 - x^2)\, dx - 2xy\, dy - 2xz\, dz = 0.$ [Divide by x^2.]

(3) $yz\, dx + zx\, dy + xy\, dz = 0.$ (4) $(y + z)\, dx + (z + x)\, dy + (x + y)\, dz = 0.$

(5) $z(y\, dx - x\, dy) = y^2\, dz.$ (6) $x\, dx + z\, dy + (y + 2z)\, dz = 0.$

117. Method of integration when the solution is not obvious. When an integrable equation of the form

$$P\, dx + Q\, dy + R\, dz = 0$$

cannot be solved by inspection, we seek for a solution by considering first the simpler case where z is constant and so $dz = 0$.

For example, $yz\,dx + 2zx\,dy - 3xy\,dz = 0$ becomes, if z is constant,

$$y\,dx + 2x\,dy = 0,$$

giving $\qquad\qquad\qquad\qquad xy^2 = a.$

As this was obtained by supposing the variable z to be constant it is probable that the solution of the original equation can be obtained by replacing the constant a by some function of z, giving

$$xy^2 = f(z)$$

leading to $\qquad\qquad y^2\,dx + 2xy\,dy - \dfrac{df}{dz}\,dz = 0.$

This is identical with the original equation if

$$\frac{y^2}{yz} = \frac{2xy}{2zx} = \frac{-\dfrac{df}{dz}}{-3xy},$$

$$i.e. \quad \frac{df}{dz} = \frac{3xy^2}{z} = \frac{3f(z)}{z},$$

$$\frac{df}{f} = \frac{3dz}{z},$$

$$f(z) = cz^3,$$

giving the final solution $xy^2 = cz^3$.

For a proof that this method holds good for all integrable equations, see Art. 119.

Examples for solution

(1) $yz \log z\,dx - zx \log z\,dy + xy\,dz = 0$.

(2) $2yz\,dx + zx\,dy - xy(1+z)\,dz = 0$.

(3) $(2x^2 + 2xy + 2xz^2 + 1)\,dx + dy + 2z\,dz = 0$. [*N.B.*—Assume x constant at first.]

(4) $(y^2 + yz)\,dx + (zx + z^2)\,dy + (y^2 - xy)\,dz = 0$.

(5) $(x^2y - y^3 - y^2z)\,dx + (xy^2 - x^2z - x^3)\,dy + (xy^2 + x^2y)\,dz = 0$.

(6) Show that the integral of the following equation represents a family of planes with a common line of intersection, and that these planes are the orthogonal trajectories of the circles of Ex. 2 of the set following Art. 113:

$$(mz - ny)\,dx + (nx - lz)\,dy + (ly - mx)\,dz = 0.$$

118. Condition necessary for an equation to be integrable. If

$$P\,dx + Q\,dy + R\,dz = 0 \qquad\qquad\qquad (1)$$

has an integral $\phi(x, y, z) = c$, which on differentiation gives

$$\frac{\partial \phi}{\partial x}\,dx + \frac{\partial \phi}{\partial y}\,dy + \frac{\partial \phi}{\partial z}\,dz = 0,$$

then
$$\frac{\partial \phi}{\partial x}=\lambda P; \quad \frac{\partial \phi}{\partial y}=\lambda Q; \quad \frac{\partial \phi}{\partial z}=\lambda R.$$

Hence
$$\frac{\partial}{\partial y}(\lambda R)=\frac{\partial^2 \phi}{\partial y\,\partial z}=\frac{\partial^2 \phi}{\partial z\,\partial y}=\frac{\partial}{\partial z}(\lambda Q),$$

i.e. $\lambda\left(\dfrac{\partial Q}{\partial z}-\dfrac{\partial R}{\partial y}\right)+Q\dfrac{\partial \lambda}{\partial z}-R\dfrac{\partial \lambda}{\partial y}=0.$ ....(2)

Similarly $\lambda\left(\dfrac{\partial R}{\partial x}-\dfrac{\partial P}{\partial z}\right)+R\dfrac{\partial \lambda}{\partial x}-P\dfrac{\partial \lambda}{\partial z}=0,$ ....(3)

$\lambda\left(\dfrac{\partial P}{\partial y}-\dfrac{\partial Q}{\partial x}\right)+P\dfrac{\partial \lambda}{\partial y}-Q\dfrac{\partial \lambda}{\partial x}=0.$ ....(4)

Multiply equations (2), (3), and (4) by P, Q, and R respectively, and add. We get

$$P\left(\frac{\partial Q}{\partial z}-\frac{\partial R}{\partial y}\right)+Q\left(\frac{\partial R}{\partial x}-\frac{\partial P}{\partial z}\right)+R\left(\frac{\partial P}{\partial y}-\frac{\partial Q}{\partial x}\right)=0.$$

If the equation (1) is integrable, this condition must be satisfied.

The student familiar with vector analysis will see that if P, Q, R are the components of a vector $\mathbf{A}$, the condition may be written
$$\mathbf{A}.\operatorname{curl}\mathbf{A}=0.$$

Ex. In the worked example of the last article,
$$yz\,dx+2zx\,dy-3xy\,dz=0,$$
$$P=yz, \quad Q=2zx, \quad R=-3xy.$$

The condition gives
$$yz(2x+3x)+2zx(-3y-y)-3xy(z-2z)=0,$$
$$i.e. \quad 5xyz-8xyz+3xyz=0,$$

which is true.

Examples for solution

(1) Show that the equations in the last two sets of examples satisfy this condition.

(2) Show that there is no set of surfaces orthogonal to the curves given by
$$\frac{dx}{z}=\frac{dy}{x+y}=\frac{dz}{1}.$$

* 119. The condition of integrability is sufficient as well as necessary. We shall prove that the condition is sufficient by showing that when it is satisfied the method of Art. 117 will always be successful in giving a solution.

We require as a lemma the fact that if P, Q, R satisfy the condition, so also do $P_1=\lambda P$, $Q_1=\lambda Q$, $R_1=\lambda R$, where λ is any function of x, y, and z. We leave this as an exercise to the student.

* To be omitted on a first reading.

In Art. 117 we supposed a solution of

$$P\,dx + Q\,dy = 0$$

obtained by considering z as constant.

Let this solution be $\quad F(x, y, z) = a$,

which gives $\quad\dfrac{\partial F}{\partial x}\,dx + \dfrac{\partial F}{\partial y}\,dy = 0$,

so $\quad\dfrac{\partial F}{\partial x}\Big/ P = \dfrac{\partial F}{\partial y}\Big/ Q = \lambda$, say.

Put $\lambda P = P_1,\ \lambda Q = Q_1,\ \lambda R = R_1$.

The next step was to replace a by $f(z)$, giving

$$F(x, y, z) = f(z), \dotfill (1)$$

and thence $\quad\dfrac{\partial F}{\partial x}dx + \dfrac{\partial F}{\partial y}\,dy + \left\{\dfrac{\partial F}{\partial z} - \dfrac{df}{dz}\right\}dz = 0,$

i.e. $\quad P_1\,dx + Q_1\,dy + \left\{\dfrac{\partial F}{\partial z} - \dfrac{df}{dz}\right\}dz = 0.\dotfill (2)$

This is identical with

$$P\,dx + Q\,dy + R\,dz = 0,$$

if $\quad\dfrac{\partial F}{\partial z} - \dfrac{df}{dz} = \lambda R = R_1,$

i.e. if $\quad\dfrac{df}{dz} = \dfrac{\partial F}{\partial z} - R_1. \dotfill (3)$

In the example of Art. 117 we **got**

$$\frac{df}{dz} = \frac{3xy^2}{z} = \frac{3f(z)}{z},$$

the x and y being got rid of by virtue of the equation $xy^2 = f(z)$.

What we have to prove is that the x and y can always be got rid of from the right-hand side of equation (3) in virtue of equation (1).

In other words, we must show that $\dfrac{\partial F}{\partial z} - R_1$ involves x and y only as a function of F.

Now this will be the case if *

$$\frac{\partial F}{\partial x}\frac{\partial}{\partial y}\left\{\frac{\partial F}{\partial z} - R_1\right\} - \frac{\partial F}{\partial y}\frac{\partial}{\partial x}\left\{\frac{\partial F}{\partial z} - R_1\right\} = 0 \text{ identically.} \dotfill (4)$$

Now, by the lemma, the relation between P, Q, R leads to the similar relation

$$P_1\left\{\frac{\partial Q_1}{\partial z} - \frac{\partial R_1}{\partial y}\right\} + Q_1\left\{\frac{\partial R_1}{\partial x} - \frac{\partial P_1}{\partial z}\right\} + R_1\left\{\frac{\partial P_1}{\partial y} - \frac{\partial Q_1}{\partial x}\right\} = 0;$$

* Edwards' *Differential Calculus*, Art. 510.

also, since equation (2) is integrable,

$$P_1\left\{\frac{\partial Q_1}{\partial z}-\frac{\partial}{\partial y}\left(\frac{\partial F}{\partial z}-\frac{df}{dz}\right)\right\}+Q_1\left\{\frac{\partial}{\partial x}\left(\frac{\partial F}{\partial z}-\frac{df}{dz}\right)-\frac{\partial P_1}{\partial z}\right\}$$
$$+\left(\frac{\partial F}{\partial z}-\frac{df}{dz}\right)\left\{\frac{\partial P_1}{\partial y}-\frac{\partial Q_1}{\partial x}\right\}=0.$$

By subtraction of these last two equations we get

$$P_1\frac{\partial}{\partial y}\left(\frac{\partial F}{\partial z}-\frac{df}{dz}-R_1\right)-Q_1\frac{\partial}{\partial x}\left\{\frac{\partial F}{\partial z}-\frac{df}{dz}-R_1\right\}$$
$$-\left\{\frac{\partial F}{\partial z}-\frac{df}{dz}-R_1\right\}\left\{\frac{\partial P_1}{\partial y}-\frac{\partial Q_1}{\partial x}\right\}=0\ldots\ldots(5)$$

But $\quad P_1=\dfrac{\partial F}{\partial x},\quad Q_1=\dfrac{\partial F}{\partial y},\quad$ and $\quad\dfrac{\partial}{\partial x}\left(\dfrac{df}{dz}\right)=\dfrac{\partial}{\partial y}\left(\dfrac{df}{dz}\right)=0,$

as f is a function of z alone.

Hence (5) reduces to (4).

That is, $\dfrac{\partial F}{\partial z}-R_1$ *can* be expressed as a function of F and z, say $\psi(F,z)$. Hence from (1) and (3),

$$\frac{df}{dz}=\psi(f,z).$$

If the solution of this is $f=\chi(z)$, then $F(x,\ y,\ z)=\chi(z)$ is a solution of $\qquad P\ dx+Q\ dy+R\ dz=0,$

which is thus proved to be integrable whenever P, Q, R satisfy the condition of Art. 118.

120. The non-integrable single equation. When the condition of integrability is not satisfied, the equation

$$P\ dx+Q\ dy+R\ dz=0\ldots\ldots\ldots\ldots\ldots\ldots\ldots(1$$

represents a family of curves orthogonal to the family represented by the simultaneous equations

$$\frac{dx}{P}=\frac{dy}{Q}=\frac{dz}{R},$$

but in this case there is no family of *surfaces* orthogonal to the second family of curves.

However, we can find an infinite number of curves that lie on any given surface and satisfy (1), whether that equation is integrable or not.

Ex. Find the curves represented by the solution of

$$y\ dx+(z-y)\ dy+x\ dz=0,\qquad\ldots\ldots\ldots\ldots\ldots(1$$

which lie in the plane $\qquad 2x-y-z=1.\ldots\ldots\ldots\ldots\ldots\ldots\ldots(2$

(It is easily verified that the condition of integrability is not satisfied

The method of procedure is to eliminate one of the variables and its differential, say z and dz, from these two equations and the differential of the second of them.

Differentiating (2), $\qquad 2dx - dy - dz = 0.$

Multiplying by x and adding to (1),

$$(y + 2x)\, dx + (z - x - y)\, dy = 0,$$

or using (2), $\qquad (y + 2x)\, dx + (x - 2y - 1)\, dy = 0,$

which gives $\qquad\qquad xy + x^2 - y^2 - y = c^2. \quad\dots\dots\dots\dots\dots\dots\dots(3)$

Thus the curves of the family that lie in the plane (2) are the sections by that plane of the infinite set of rectangular hyperbolic cylinders (3).

The result of this example could have been expressed by saying that the projections on the plane of xy of curves which lie in the plane (2) and satisfy equation (1) are a family of concentric, similar and similarly situated rectangular hyperbolas.

Examples for solution

(1) Show that there is no single integral of $dz = 2y\, dx + x\, dy.$
Prove that the curves of this equation that lie in the plane $z = x + y$ lie also on surfaces of the family $(x - 1)^2(2y - 1) = c.$

(2) Show that the curves of

$$x\, dx + y\, dy + c\sqrt{\left(1 - \frac{x^2}{a^2} - \frac{y^2}{b^2}\right)}\, dz = 0$$

that lie on the ellipsoid

$$\frac{x^2}{a^2} + \frac{y^2}{b^2} + \frac{z^2}{c^2} = 1$$

lie also on the family of concentric spheres

$$x^2 + y^2 + z^2 = k^2.$$

(3) Find the orthogonal projection on the plane of xz of curves which lie on the paraboloid $3z = x^2 + y^2$ and satisfy the equation

$$2dz = (x + z)\, dx + y\, dy.$$

(4) Find the equation of the cylinder, with generators parallel to the axis of y, passing through the point $(2, 1, -1)$, and also through a curve that lies on the sphere $x^2 + y^2 + z^2 = 4$ and satisfies the equation

$$(xy + 2xz)\, dx + y^2\, dy + (x^2 + yz)\, dz = 0.$$

MISCELLANEOUS EXAMPLES ON CHAPTER XI

(1) $\dfrac{dx}{xz} = \dfrac{dy}{yz} = \dfrac{dz}{xy}.$ $\qquad$ (2) $\dfrac{dx}{y^3x - 2x^4} = \dfrac{dy}{2y^4 - x^3y} = \dfrac{dz}{9z\,(x^3 - y^3)}.$

(3) $\dfrac{dy}{dx} = z;\quad \dfrac{dz}{dx} = y.$

(4) $(z + z^3)\cos x\, \dfrac{dx}{dt} - (z + z^3)\,\dfrac{dy}{dt} + (1 - z^2)\,(y - \sin x)\,\dfrac{dz}{dt} = 0.$

(5) $(2x + y^2 + 2xz) \dfrac{dx}{dt} + 2xy \dfrac{dy}{dt} + x^2 \dfrac{dz}{dt} = 1$.

(6) Find $f(y)$ if $f(y) \, dx - zx \, dy - xy \log y \, dz = 0$ is integrable. Find the corresponding integral.

(7) Show that the following equation is not integrable:
$$3y \, dx + (z - 3y) \, dy + x \, dz = 0.$$
Prove that the projection on the plane of xy of the curves that satisfy the equation and lie in the plane $2x + y - z = a$ are the rectangular hyperbolas $\qquad x^2 + 3xy - y^2 - ay = b$.

(8) Find the differential equations of the family of twisted cubic curves $y = ax^2$; $y^2 = bzx$. Show that all these curves cut orthogonally the family of ellipsoids $\qquad x^2 + 2y^2 + 3z^2 = c^2$.

(9) Find the equations of the curve that passes through the point $(3, 2, 1)$ and cuts orthogonally the family of surfaces $x + yz = c$.

(10) Solve the following *homogeneous* equations by putting $x = uz$, $y = vz$:

(i) $(x^2 - y^2 - z^2 + 2xy + 2xz) \, dx + (y^2 - z^2 - x^2 + 2yz + 2yx) \, dy$
$$+ (z^2 - x^2 - y^2 + 2zx + 2zy) \, dz = 0;$$

(ii) $(2xz - yz) \, dx + (2yz - xz) \, dy - (x^2 - xy + y^2) \, dz = 0;$

(iii) $z^2 \, dx + (z^2 - 2yz) \, dy + (2y^2 - yz - xz) \, dz = 0$.

(11) Prove that if the equation
$$P_1 \, dx_1 + P_2 \, dx_2 + P_3 \, dx_3 + P_4 \, dx_4 = 0$$
is integrable, then
$$P_r \left(\frac{\partial P_s}{\partial x_t} - \frac{\partial P_t}{\partial x_s} \right) + P_s \left(\frac{\partial P_t}{\partial x_r} - \frac{\partial P_r}{\partial x_t} \right) + P_t \left(\frac{\partial P_r}{\partial x_s} - \frac{\partial P_s}{\partial x_r} \right) = 0,$$
where r, s, t are any three of the four suffixes 1, 2, 3, 4.

Denoting this relation by $C_{rst} = 0$, verify that
$$P_1 C_{234} - P_2 C_{134} + P_3 C_{124} - P_4 C_{123} = 0 \text{ identically,}$$
showing that only three of these four relations are independent.

Verify that these conditions are satisfied for the equation
$$(x_1^3 - x_2 x_3 x_4) \, dx_1 + (x_2^3 - x_1 x_3 x_4) \, dx_2$$
$$+ (x_3^3 - x_1 x_2 x_4) \, dx_3 + (x_4^3 - x_1 x_2 x_3) \, dx_4 = 0.$$

(12) Integrate the equation of Ex. 11 by the following process:

(i) Suppose x_3 and x_4 constant, and thus obtain
$$x_1^4 + x_2^4 - 4x_1 x_2 x_3 x_4 = a.$$

(ii) Replace a by $f(x_3, x_4)$. By differentiation and comparison with the original equation obtain $\dfrac{\partial f}{\partial x_3}$, $\dfrac{\partial f}{\partial x_4}$, and hence f and the solution
$$x_1^4 + x_2^4 + x_3^4 + x_4^4 - 4x_1 x_2 x_3 x_4 = c.$$

(13) Integrate the equation of Ex. 11 by putting $x_1 = ux_4$, $x_2 = vx_4$, $x_3 = wx_4$.

(14) Show that the following equation satisfies the conditions of integrability and obtain its integral:

$$y \sin w \, dx + x \sin w \, dy - xy \sin w \, dz - xy \cos w \, dw = 0.$$

(15) Show that the equation

$$a \, dx^2 + b \, dy^2 + c \, dz^2 + 2f \, dy \, dz + 2g \, dz \, dx + 2h \, dx \, dy = 0$$

reduces to two equations of the form

$$P \, dx + Q \, dy + R \, dz = 0$$

if $\qquad abc + 2fgh - af^2 - bg^2 - ch^2 = 0.$ (Cf. a result in Conics.)

Hence show that the solution of

$$xyz \, (dx^2 + dy^2 + dz^2) + x \, (y^2 + z^2) \, dy \, dz + y \, (z^2 + x^2) \, dz \, dx$$
$$+ z \, (x^2 + y^2) \, dx \, dy = 0$$

is $\qquad (x^2 + y^2 + z^2 - c) \, (xyz - c) = 0.$ (Cf. Art. 52.)

(16) Show that the condition of integrability of

$$P \, dx + Q \, dy + R \, dz = 0 \quad\dots\dots\dots\dots\dots\dots\dots\dots\dots(1$$

implies the orthogonality of any pair of intersecting curves of the families

$$dx/P = dy/Q = dz/R \quad\dots\dots\dots\dots\dots\dots\dots\dots(2)$$

and $\qquad dx \left/ \left(\dfrac{\partial Q}{\partial z} - \dfrac{\partial R}{\partial y} \right) \right. = dy \left/ \left(\dfrac{\partial R}{\partial x} - \dfrac{\partial P}{\partial z} \right) \right. = dz \left/ \left(\dfrac{\partial P}{\partial y} - \dfrac{\partial Q}{\partial x} \right) \right. \dots\dots\dots\dots(3)$

Hence show that the curves of (3) all lie on the surfaces of (1).

Verify this conclusion for $P = ny - mz$, $Q = lz - nx$, $R = mx - ly$.

(For the solutions of the corresponding equations, see earlier examples in this chapter.)

(17) The preceding example suggests that if $\alpha = $ const., $\beta = $ const. are two integrals of equations (3), the integral of equation (1) should be expressible in the form $f(\alpha, \beta) = $ const., and hence that

$$P \, dx + Q \, dy + R \, dz$$

should be expressible as $A \, d\alpha + B \, d\beta$, where A and B are functions of α and β.

Verify that for the case

$$P = yz \log z, \quad Q = -zx \log z, \quad R = xy,$$
$$\alpha = yz^{\frac{1}{z}}, \quad \beta = xz^{\frac{1}{z}} \log z, \quad A = -\beta, \quad \text{and} \quad B = \alpha.$$

Hence obtain an integral of (1) in the form $\alpha = c\beta$,

$$i.e. \quad y = cx \log z.$$

[For a supplement to the chapter see Arts. 168–170. They deal with an integrating factor for homogeneous equations, and with Mayer's method. A development of the method indicated by Ex. 17 is given in my paper. " A simplification of Bertrand's method for integrating a total differential equation," which appeared in the *Mathematical Gazette*, XXXVII., 1953, p. 59.]

CHAPTER XII

PARTIAL DIFFERENTIAL EQUATIONS OF THE FIRST ORDER. PARTICULAR METHODS

121. We have already (in Chap. IV.) discussed the formation of partial differential equations by elimination of arbitrary functions or of arbitrary constants. We also showed how in certain equations, of great importance in mathematical physics, simple particular solutions could be found by the aid of which more complex solutions could be built up to satisfy such initial and boundary conditions as usually occur in physical problems.

In the present chapter we shall be concerned chiefly with equations of geometrical interest, and seek for integrals of various forms, "general," "complete," and "singular," and their geometrical interpretations. Exceptional equations will be found to possess integrals of another form called "special."

122. Geometrical theorems required. The student should revise the following theorems in any treatise on solid geometry:

(i) The direction-cosines of the normal to a surface $f(x, y, z) = 0$ at the point (x, y, z) are in the ratio

$$\frac{\partial f}{\partial x} : \frac{\partial f}{\partial y} : \frac{\partial f}{\partial z}.$$

Since

$$-\frac{\partial f}{\partial x} \bigg/ \frac{\partial f}{\partial z} = \frac{\partial z}{\partial x} = p, \text{ say}, \quad \text{and} \quad -\frac{\partial f}{\partial y} \bigg/ \frac{\partial f}{\partial z} = \frac{\partial z}{\partial y} = q, \text{ say},$$

this ratio can also be written $p : q : -1$.

The symbols p and q are to be understood as here defined all through this chapter.

(ii) The envelope of the system of surfaces

$$f(x, y, z, a, b) = 0,$$

146

where a and b are variable parameters, is found by eliminating a and b from the given equation and

$$\frac{\partial f}{\partial a} = 0, \quad \frac{\partial f}{\partial b} = 0.$$

The result may contain either loci besides the envelope (cf. Chap. VI.).

123. Lagrange's linear equation and its geometrical interpretation. This is the name applied to the equation

$$Pp + Qq = R, \quad \dots\dots\dots\dots\dots\dots\dots\dots(1)$$

where P, Q, R are functions of x, y, z.

The geometrical interpretation is that the normal to a certain surface is perpendicular to a line whose direction-cosines are in the ratio $P : Q : R$. But in the last chapter we saw that the simultaneous equations

$$\frac{dx}{P} = \frac{dy}{Q} = \frac{dz}{R} \quad \dots\dots\dots\dots\dots\dots\dots\dots(2)$$

represented a family of curves such that the tangent at any point had direction-cosines in the ratio $P : Q : R$, and that $\phi(u, v) = 0$ where $u = \text{const.}$ and $v = \text{const.}$ were two particular integrals of the simultaneous equations) represented a surface through such curves.

Through every point of such a surface passes a curve of the family, lying wholly on the surface. Hence the normal to the surface must be perpendicular to the tangent to this curve, *i.e.* perpendicular to a line whose direction-cosines are in the ratio $P : Q : R$. This is just what is required by the partial differential equation.

Thus the surfaces of equation (1) are those which, taken in pairs, give the curves of equations (2). Equations (2) are called the subsidiary equations.

Thus $\phi(u, v) = 0$ is an integral of (1), if $u = \text{const.}$ and $v = \text{const.}$ are any two independent solutions of the subsidiary equations (2) and ϕ is any arbitrary function. This is called the General Integral of Lagrange's Linear Equation.

Ex. (i). $p + q = 1.$

The subsidiary equations are those discussed in Ex. (i) of Art. 112, viz.

$$\frac{dx}{1} = \frac{dy}{1} = \frac{dz}{1},$$

representing a family of parallel straight lines.

Two independent integrals are
$$x - z = a,$$
$$y - z = b,$$
representing two families of planes containing these straight lines.

The general integral is $\phi(x - z, y - z) = 0$, representing the surface formed by lines of the family passing through the curve
$$\phi(x, y) = 0, \quad z = 0.$$
If we are given a definite curve, such as the circle
$$x^2 + y^2 = 4, \quad z = 0,$$
we can construct a corresponding particular integral
$$(x - z)^2 + (y - z)^2 = 4,$$
the elliptic cylinder formed by lines of the family meeting the given circle.

Ex. (ii). $zp = -x$. [Cf. Ex. (ii) of Art. 112.]

The subsidiary equations are
$$\frac{dx}{z} = \frac{dy}{0} = -\frac{dz}{x},$$
of which two integrals are $x^2 + z^2 = a$, $y = b$.

The general integral $\phi(x^2 + z^2, y) = 0$ represents the surface of revolution formed by curves (circles in this case) of the family intersecting the curve $\phi(x^2, y) = 0$, $z = 0$.

Ex. (iii). Find the surface whose tangent planes cut off an intercept of constant length k from the axis of z.

The tangent plane at (x, y, z) is
$$Z - z = p(X - x) + q(Y - y).$$
Putting $X = Y = 0$, $\qquad Z = z - px - qy = k$.

The subsidiary equations are
$$\frac{dx}{x} = \frac{dy}{y} = \frac{dz}{z - k},$$
of which $y = ax$, $z - k = bx$, are integrals.

The general integral $\phi\left(\dfrac{y}{x}, \dfrac{z - k}{x}\right) = 0$ represents any cone with its vertex at $(0, 0, k)$, and these surfaces clearly possess the desired property.

Examples for solution

Obtain general integrals of the following equations. [Cf. the first set of examples in Chap. XI.]

(1) $xp + yq = z$.

(2) $(mz - ny)p + (nx - lz)q = ly - mx$.

(3) $(y^2 + z^2 - x^2)p - 2xyq + 2xz = 0$.

(4) $yzp + zxq = xy$.

(5) $(y + z)p + (z + x)q = x + y$.

(6) $(z^2 - 2yz - y^2)p + (xy + xz)q = xy - xz.$

(7) $p + 3q = 5z + \tan(y - 3x).$

(8) $zp - zq = z^2 + (y + x)^2.$

(9) Find a solution of Ex. (1) representing a surface meeting the parabola $y^2 = 4x$, $z = 1$.

(10) Find the most general solution of Ex. (4) representing a conicoid.

(11) Show that if the solution of Ex. (6) represents a sphere, the centre is at the origin.

(12) Find the surfaces all of whose normals intersect the axis of z.

124. Analytical investigation of the general integral. Let $u(x,y,z) = a$, $v(x, y, z) = b$, where a and b are arbitrary constants, be any two independent integrals of the equations

$$\frac{dx}{P(x, y, z)} = \frac{dy}{Q(x, y, z)} = \frac{dz}{R(x, y, z)}. \quad\quad\ldots\ldots\ldots\ldots(1)$$

By differentiation, $\quad \dfrac{\partial u}{\partial x}\,dx + \dfrac{\partial u}{\partial y}\,dy + \dfrac{\partial u}{\partial z}\,dz = 0,$

so from (1), $\quad\quad P\dfrac{\partial u}{\partial x} + Q\dfrac{\partial u}{\partial y} + R\dfrac{\partial u}{\partial z} = 0. \quad\quad\ldots\ldots\ldots\ldots\ldots(2)$

The left-hand side of (2) does not contain the constant a, so cannot vanish merely in consequence of the relation $u = a$. Hence (2) is satisfied identically.

Similarly $\quad\quad P\dfrac{\partial v}{\partial x} + Q\dfrac{\partial v}{\partial y} + R\dfrac{\partial v}{\partial z} = 0$ identically. $\ldots\ldots\ldots\ldots(3)$

Now let $w(x, y, z) = c$ be any integral containing an arbitrary constant c of the partial differential equation

$$Pp + Qq = R. \quad\quad\ldots\ldots\ldots\ldots\ldots\ldots\ldots(4)$$

Differentiating partially with respect to x and then with respect to y, regarding z as a function of x and y, we get

$$\frac{\partial w}{\partial x} + p\frac{\partial w}{\partial z} = 0 = \frac{\partial w}{\partial y} + q\frac{\partial w}{\partial z},$$

from which (4), after multiplication by $-\dfrac{\partial w}{\partial z}$, gives

$$P\frac{\partial w}{\partial x} + Q\frac{\partial w}{\partial y} + R\frac{\partial w}{\partial x} = 0 \text{ identically.} \quad\ldots\ldots\ldots\ldots(5)$$

Elimination of P, Q, R from (2), (3) and (5) gives the Jacobian

$$\begin{vmatrix} \dfrac{\partial u}{\partial x} & \dfrac{\partial u}{\partial y} & \dfrac{\partial u}{\partial z} \\[2mm] \dfrac{\partial v}{\partial x} & \dfrac{\partial v}{\partial y} & \dfrac{\partial v}{\partial z} \\[2mm] \dfrac{\partial w}{\partial x} & \dfrac{\partial w}{\partial y} & \dfrac{\partial w}{\partial z} \end{vmatrix} = 0 \text{ identically.}$$

Hence w is a function of u and v, and so also is $w - c$, say $w - c = \phi(u, v)$,

i.e. any integral of (4) *containing an arbitrary constant is included in the General Integral* $\phi(u, v) = 0$.

Sufficient conditions for the validity of this reasoning are that the nine partial derivates concerned are all continuous, and that P, Q, R do not all vanish simultaneously.

125. Special integrals. The reasoning of Art. 124 may not hold for an integral, say $w = 0$, *not* containing an arbitrary constant, for then equation (5) may not be an identity, but merely a consequence of $w = 0$. For example, for

$$xp + yq = z,$$

we may take $u = z/x$ and $v = z/y$. Then $w = z^2 - xy$ reduces the left-hand side of (5) to $2(z^2 - xy) = 2w$, which vanishes when $w = 0$. The Jacobian comes to $2z(z^2 - xy)/x^2y^2$, which also vanishes when $w = 0$. There is no relation of the form $z^2 - xy = \phi(z/x, z/y)$, but $z^2 - xy = 0$ is included in the general integral, as it is *one* way of satisfying $uv - 1 = 0$. Some call such an integral *special*, but it seems better to call it *ordinary of the second kind*. Similarly for the integral $z = 0$.

A more serious failure of the reasoning may occur when, in addition to equation (5) being satisfied merely in consequence of $w = 0$, some of the nine partial derivates become infinite when $w = 0$. For example, for

$$p - q = 2\sqrt{z},$$

we may take $u = x + y$ and $v = x - \sqrt{z}$. Then $w = z$ reduces the left-hand side of (5) to $2\sqrt{z}$. This vanishes when $z = 0$, but $\partial v/\partial z$ then becomes infinite, making the left-hand side of (3) indeterminate, so we can no longer assert that (3) and (5) are true simultaneously. The Jacobian comes to -1, which is definitely *not* zero. The integral $z = 0$ is definitely *not* included in the general integral, and is called *special*. Such integrals, discovered by Goursat and Chrystal, were discussed by Forsyth and M. J. M. Hill. A general theorem connecting a special integral with the nature of the coefficients P, Q, R, and methods of finding such integrals and their geometrical representation, are given in my papers (*J. Lond. Math. Soc.* 1938 and 1939). For a special integral to occur, it is necessary, but not sufficient, that P, Q, or R should *not* be expansible in a series of integral powers of x, y, z.

A third way in which the reasoning may fail is that P, Q, R may all vanish when $w = 0$. If, as usually happens, this can be obviated by dividing P, Q, R by a suitable factor, such an integral (which some call *singular*) seems best called *trivial*. For example, $z = 0$ is a trivial integral of $zp + z^2q = z$, but not of $p + zq = 1$.

126. The linear equation with n independent variables. The general integral of the equation

$$P_1p_1 + P_2p_2 + P_3p_3 + \ldots + P_np_n = R,$$

where $p_1 = \dfrac{\partial z}{\partial x_1}$, $p_2 = \dfrac{\partial z}{\partial x_2}$, ... etc., and the P's and R are functions of the x's and z, is $\quad \phi(u_1, u_2, u_3, \ldots u_n) = 0$,

where $u_1 = $ const., $u_2 = $ const., ... etc., are any n independent integrals of the subsidiary equations

$$\frac{dx_1}{P_1} = \frac{dx_2}{P_2} = \frac{dx_3}{P_3} = \ldots = \frac{dz}{R}.$$

This may be verified as in Art. 124. The student should write out the proof for the case of three independent variables.

Besides this general integral, special integrals exist for exceptional equations, just as in the case of two independent variables.

Examples for solution

(1) $p_2 + p_3 = 1 + p_1$.

(2) $x_1p_1 + 2x_2p_2 + 3x_3p_3 + 4x_4p_4 = 0$.

(3) $(x_3 - x_2)p_1 + x_2p_2 - x_3p_3 = x_2(x_1 + x_3) - x_2^2$.

(4) $x_2x_3p_1 + x_3x_1p_2 + x_1x_2p_3 + x_1x_2x_3 = 0$.

(5) $p_1 + x_1p_2 + x_1x_2p_3 = x_1x_2x_3\sqrt{z}$.

(6) $p_1 + p_2 + p_3\{1 + \sqrt{(z - x_1 - x_2 - x_3)}\} = 3$.

127. The equation $P \dfrac{\partial f}{\partial x} + Q \dfrac{\partial f}{\partial y} + R \dfrac{\partial f}{\partial z} = 0$. If P, Q, R are functions of x, y, z but *not* of f, the equation can be viewed from two different aspects.

Consider, for example,

$$\frac{\partial f}{\partial x} - \frac{\partial f}{\partial y} + 2\sqrt{z}\,\frac{\partial f}{\partial z} = 0. \quad \ldots\ldots\ldots\ldots\ldots\ldots(1)$$

We may regard this as equivalent to the three-dimensional equation $\qquad p - q = 2\sqrt{z}, \quad \ldots\ldots\ldots\ldots\ldots\ldots\ldots\ldots(2)$
of which $\phi(x+y, x - \sqrt{z}) = 0$ is the general integral and $z = 0$ a special integral.

On the other hand, regarding (1) as an equation in four variables, we get the general integral

$$\phi(f, x+y, x - \sqrt{z}) = 0,$$

which is equivalent to $f = \psi(x+y, x - \sqrt{z})$, where ψ is an arbitrary function, but if

$$f = z, \quad \frac{\partial f}{\partial x} - \frac{\partial f}{\partial y} + 2\sqrt{z}\,\frac{\partial f}{\partial z} = 2\sqrt{z} = 2\sqrt{f}.$$

Thus $f = z$ is *not* an integral of (1), although $f = z = 0$ certainly gives a solution.

In general it may be proved that

$$P \frac{\partial f}{\partial x} + Q \frac{\partial f}{\partial y} + R \frac{\partial f}{\partial z} = 0,$$

regarded as four-dimensional, where P, Q, R do not contain f, has no special integrals.* A similar theorem is true for any number of independent variables.

Examples for solution

(1) Verify that if $f = x$, $f = 0$ is a surface satisfying

$$\sqrt{x}\,\frac{\partial f}{\partial x} + \sqrt{y}\,\frac{\partial f}{\partial y} + \sqrt{z}\,\frac{\partial f}{\partial z} = 0,$$

and hence that this differential equation, interpreted three-dimensionally, admits the three special integrals $x = 0$, $y = 0$, $z = 0$ and the general integral $\phi(\sqrt{z} - \sqrt{x}, \sqrt{z} - \sqrt{y}) = 0$.

(2) Show that the general integral of the last example represents surfaces through curves which, if they do not go through the origin, either touch the co-ordinate planes or lie wholly in one of them.

[*Hint.* Prove that $\dfrac{dx}{ds} = \sqrt{\left(\dfrac{x}{x + y + z}\right)}$, and that $dx/ds = 0$ if $x = 0$, unless x, y, z are all zero.]

* See Appendix B.

(3) Show that $\sqrt{x}\,\dfrac{\partial z}{\partial x}+\sqrt{y}\,\dfrac{\partial z}{\partial y}=0$, regarded two-dimensionally, represents a family of parabolas $\sqrt{y}=\sqrt{x}+c$, and their envelope, the co-ordinate axes $x=0$, $y=0$; while regarded three-dimensionally it represents the surfaces $z=\phi(y^{\frac{1}{2}}-x^{\frac{1}{2}})$.

128. Non-linear equations. We shall now consider equations in which p and q occur other than in the first degree. Before giving the general method we shall discuss four simple standard forms, for which a " complete integral " (*i.e.* one involving two arbitrary constants) can be obtained by inspection or by other simple means. In Arts. 133-135 we shall show how to deduce general and singular integrals from the complete integrals.

129. Standard I. Only p and q present. Consider, for example, this equation $q=3p^2$.

The most obvious solution is to take p and q as *constants* satisfying the equation, say $p=a$, $q=3a^2$.

Then, since $\qquad dz=p\,dx+q\,dy=a\,dx+3a^2\,dy,$
$$z=ax+3a^2y+c.$$

This is the complete integral, containing two arbitrary constants a and c.

In general, the complete integral of $f(p, q)=0$ is
$$z=ax+by+c,$$
where a and b are connected by the relation $f(a, b)=0$.

Examples for solution

Find complete integrals of the following:

(1) $p=2q^2+1$.   (2) $p^2+q^2=1$.
(3) $p=e^q$.   (4) $p^2q^3=1$.
(5) $p^2-q^2=4$.   (6) $pq=p+q$.

130. Standard II. Only p, q, and z present. Consider the equation
$$z^2(p^2z^2+q^2)=1. \qquad\qquad\ldots\ldots\ldots\ldots\ldots\ldots\ldots(1)$$

As a trial solution assume that z is a function of $x+ay$ ($=u$, say), where a is an arbitrary constant.

Then $\qquad p=\dfrac{\partial z}{\partial x}=\dfrac{dz}{du}\cdot\dfrac{\partial u}{\partial x}=\dfrac{dz}{du}; \quad q=\dfrac{\partial z}{\partial y}=\dfrac{dz}{du}\cdot\dfrac{\partial u}{\partial y}=a\dfrac{dz}{du}.$

Substituting in (1), $\qquad z^2\left(\dfrac{dz}{du}\right)^2(z^2+a^2)=1,$

$$i.e.\quad \dfrac{du}{dz}=\pm z(z^2+a^2)^{\frac{1}{2}},$$
$$i.e.\quad u+b=\pm\tfrac{1}{3}(z^2+a^2)^{\frac{3}{2}},$$
$$i.e.\quad 9(x+ay+b)^2=(z^2+a^2)^3.$$

In general, this method reduces $f(z, p, q) = 0$ to the *ordinary* differential equation

$$f\left(z, \frac{dz}{du}, a\frac{dz}{du}\right) = 0.$$

Examples for solution

Find complete integrals of the following:

(1) $4z = pq$.
(2) $z^2 = 1 + p^2 + q^2$.
(3) $q^2 = z^2 p^2 (1 - p^2)$.
(4) $p^3 + q^3 = 27z$.
(5) $p(z + p) + q = 0$.
(6) $p^2 = zq$.

131. Standard III. $f(x, p) = F(y, q)$. Consider the equation

$$p - 3x^2 = q^2 - y.$$

As a trial solution put each side of this equation equal to an arbitrary constant a, giving

$$p = 3x^2 + a; \quad q = \pm\sqrt{(y + a)}.$$

But $dz = p \, dx + q \, dy$

$$= (3x^2 + a) \, dx \pm \sqrt{(y + a)} \, dy;$$

therefore $z = x^3 + ax \pm \tfrac{2}{3}(y + a)^{\frac{3}{2}} + b,$

which is the complete integral required.

Examples for solution

Find complete integrals of the following:

(1) $p^2 = q + x$.
(2) $pq = xy$.
(3) $yp = 2yx + \log q$.
(4) $q = xyp^2$.
(5) $pe^y = qe^x$.
(6) $q(p - \cos x) = \cos y$.

132. Standard IV. Partial differential equations analogous to Clairaut's form. In Chap. VI. we showed that the complete primitive of

$$y = px + f(p)$$

was $y = cx + f(c)$, a family of straight lines.

Similarly the complete integral of the partial differential equation

$$z = px + qy + f(p, q)$$

is $z = ax + by + f(a, b)$, a family of planes.

For example, the complete integral of

$$z = px + qy + p^2 + q^2$$

is $z = ax + by + a^2 + b^2.$

Corresponding to the singular solution of Clairaut's form, giving the envelope of the family of straight lines, we shall find in the next article a " singular integral " of the partial differential equation, giving the envelope of the family of planes.

Examples for solution

(1) Prove that the complete integral of $z = px + qy - 2p - 3q$ represents all possible planes through the point $(2, 3, 0)$.

(2) Prove that the complete integral of $z = px + qy + \sqrt{(p^2 + q^2 + 1)}$ represents all planes at unit distance from the origin.

(3) Prove that the complete integral of $z = px + qy + pq/(pq - p - q)$ represents all planes such that the algebraic sum of the intercepts on the three co-ordinate axes is unity.

133. Singular Integrals. In Chap. VI. we showed that if the family of curves represented by the complete primitive of an ordinary differential equation of the first order had an envelope, the equation of this envelope was a singular solution of the differential equation. A similar theorem is true concerning the family of surfaces represented by the complete integral of a partial differential equation of the first order. If they have an envelope, its equation is called a " singular integral." To see that this is really an integral we have merely to notice that at any point of the envelope there is a surface of the family touching it. Therefore the normals to the envelope and this surface coincide, so the values of p and q at any point of the envelope are the same as that of some surface of the family, and therefore satisfy the same equation.

We gave two methods of finding singular solutions, namely from the c-discriminant and from the p-discriminant, and we showed that these methods gave also node-loci, cusp-loci, and tac-loci, whose equations did not satisfy the differential equations. The geometrical reasoning of Chap. VI. can be extended to surfaces, but the discussion of the extraneous loci which do not furnish singular integrals is more complicated.* As far as the envelope is concerned, the student who has understood Chap. VI. will have no difficulty in understanding that this surface is included among those found by eliminating a and b from the complete integral and the two derived equations

$$f(x, y, z, a, b) = 0,$$

$$\frac{\partial f}{\partial a} = 0,$$

$$\frac{\partial f}{\partial b} = 0;$$

* See a paper by M. J. M. Hill, *Phil. Trans.* (A), 1892.

or by eliminating p and q from the differential equation and the two derived equations

$$F(x, y, z, p, q) = 0,$$

$$\frac{\partial F}{\partial p} = 0,$$

$$\frac{\partial F}{\partial q} = 0.$$

In any actual example one should test whether what is apparently a singular integral really satisfies the differential equation.

Ex. (i). The complete integral of the equation of Art. 132 was
$$z = ax + by + a^2 + b^2.$$
Differentiating with respect to a, $\quad 0 = x \qquad + 2a.$
Similarly $\qquad\qquad\qquad 0 = \qquad y \qquad + 2b.$
Eliminating a and b, $\qquad\qquad 4z = -(x^2 + y^2).$
It is easily verified that this satisfies the differential equation
$$z = px + qy + p^2 + q^2$$
and represents a paraboloid of revolution, the envelope of all the planes represented by the complete integral.

Ex. (ii). The complete integral of the equation of Art. 130 was
$$9(x + ay + b)^2 = (z^2 + a^2)^3. \quad\dots\dots\dots\dots\dots\dots\dots(1)$$
Differentiating with respect to a,
$$18y(x + ay + b) = 6a(z^2 + a^2)^2. \quad\dots\dots\dots\dots\dots(2)$$
Similarly $\qquad 18(x + ay + b) = 0. \quad\dots\dots\dots\dots\dots\dots\dots\dots(3)$
Hence from (2), $\qquad\qquad a = 0. \quad\dots\dots\dots\dots\dots\dots\dots\dots(4)$
Substituting from (3) and (4) in (1), $z = 0$.

But $z = 0$ gives $p = q = 0$, and these values do *not* satisfy the differential equation $\qquad\qquad z^2(p^2z^2 + q^2) = 1.$
Hence $z = 0$ is *not* a singular integral.

Ex. (iii). Consider the equation $\quad p^2 = zq$.
Differentiating with respect to p, $2p = 0$.
Similarly $\qquad\qquad\qquad\qquad 0 = z$.
Eliminating p and q from these three equations, we get
$$z = 0.$$
This satisfies the differential equation, so it really is a singular integral.

But it is derivable by putting $b = 0$ in
$$z = be^{ax + a^2y},$$
which is easily found to be a complete integral.

So $z = 0$ is both a singular integral and a particular case of the complete integral.

Examples for solution

Find the singular integrals of the following:

(1) $z = px + qy + \log pq$.   (2) $z = px + qy + p^2 + pq + q^2$.

(3) $z = px + qy + \frac{1}{2}p^2q^2$.   (4) $z = px + qy + p/q$.

(5) $4z = pq$.   (6) $z^2 = 1 + p^2 + q^2$.   (7) $p^3 + q^3 = 27z$.

(8) Show that no equation belonging to Standard I. or III. has a singular integral. [The usual process leads to the equation $0 = 1$.]

(9) Show that $z = 0$ is both a singular integral and a particular case of a complete integral of $q^2 = z^2p^2(1 - p^2)$.

134. General Integrals. We have seen, in Ex. (i) of the last article, that *all* the planes represented by the complete integral

$$z = ax + by + a^2 + b^2 \quad\quad\quad\quad\quad (1)$$

touch the paraboloid of revolution represented by the singular integral

$$4z = -(x^2 + y^2). \quad\quad\quad\quad\quad (2)$$

Now consider, not *all* these planes, but merely those perpendicular to the plane $y = 0$. These are found by putting $b = 0$ in (1), giving

$$z = ax + a^2,$$

of which the envelope is the parabolic cylinder

$$4z = -x^2. \quad\quad\quad\quad\quad (3)$$

Take another set, those which pass through the point $(0, 0, 1)$.

From (1), $1 = a^2 + b^2$,

so (1) becomes $z = ax \pm y\sqrt{(1 - a^2)} + 1$,

of which the envelope is easily found to be the right circular cone

$$(z - 1)^2 = x^2 + y^2. \quad\quad\quad\quad\quad (4)$$

In general, we may put $b = f(a)$, where f is any function of a,

giving $z = ax + yf(a) + a^2 + \{f(a)\}^2. \quad\quad\quad\quad (5)$

The envelope of (5) is found by eliminating a between it and the equation found by differentiating it partially with respect to a,

i.e. $0 = x + yf'(a) + 2a + 2f(a)f'(a). \quad\quad\quad\quad (6)$

If f is left as a perfectly arbitrary function, the eliminant is called the " general integral " of the original differential equation. Equations (3) and (4) are particular integrals derived from the general integral.

We may define the general integral of a partial differential equation of the first order as the equation representing the aggregate of the envelopes of every possible *singly-infinite* set of surfaces that can be chosen out of the *doubly-infinite* set represented by the complete integral. These sets are defined by putting $b = f(a)$ in the complete integral.

It is usually impossible to actually perform the elimination of a between the two equations giving the envelope, on account of the arbitrary function f and its differential coefficient. The geometrical interest lies chiefly in particular cases formed by taking f as some definite (and preferably simple) function of a.

135. Characteristics.   The curve of intersection of two consecutive surfaces belonging to any singly-infinite set chosen from those represented by the complete integral is called a *characteristic*.

Now such a curve is found from the equation of the family of surfaces by the same two equations that give the envelope. For instance, equations (5) and (6) of the last article, for any definite numerical values of a, $f(a)$, and $f'(a)$, define a straight line (as the intersection of two planes), and this straight line is a characteristic. The characteristics in this example consist of the triply-infinite set of straight lines that touch the paraboloid of revolution (2).

The parabolic cylinder (3) is generated by one singly-infinite set of characteristics, namely those perpendicular to the plane $y = 0$, while the cone (4) is generated by another set, namely those that pass through the fixed point (0, 0, 1). Thus we see that the *general integral represents the aggregate of all such surfaces generated by the characteristics*.

If a singular integral exists, it must be touched by *all* the characteristics, and therefore by the surfaces generated by particular sets of them represented by the general integral. It is easily verified that the parabolic cylinder and right circular cone of the last article touch the paraboloid of revolution.

136. Peculiarities of the linear equation.   To discuss the linear equation
$$Pp + Qq = R \dots\dots\dots\dots\dots\dots\dots\dots(1)$$
on these lines, suppose that $u = \text{const.}$
and $$v = \text{const.}$$
are two independent integrals of the subsidiary equations.*

Then it is easily verified that an integral of (1) is

$$u + av + b = 0. \dots\dots\dots\dots\dots\dots\dots(2)$$

---

* Since u and v are independent, at least one of them must contain z. Let this one be u. We make this stipulation to prevent $u + av + b$ being a function of x and y alone, in which case $u + av + b = 0$ would make terms in (1) indeterminate, instead of definitely satisfying it in the ordinary way.

This may be taken as the complete integral. The general integral is found from

$$u + av + f(a) = 0, \quad \dots\dots\dots\dots\dots\dots\dots(3)$$

$$v + f'(a) = 0. \quad \dots\dots\dots\dots\dots\dots\dots(4)$$

From (4), a is a function of v alone,

say $a = F(v)$.

Substituting in (3), $u = $ a function of v,

say $u = \psi(v)$,

which is equivalent to the general integral $\phi(u, v) = 0$ found at the beginning of the chapter.

The linear equation is exceptional in that its complete integral (2) is a particular case of the general integral. Another peculiarity is that the characteristics, which are here the curves represented by the subsidiary equations, are only doubly-infinite in number instead of triply-infinite. Only one passes through a given point (in general), whereas in the non-linear case, exemplified in the last article, an infinite number may do so, forming a surface.

Examples for solution

(1) Find the surface generated by characteristics of

$$z = px + qy + p^2 + pq + q^2$$

that are parallel to the axis of x. Verify that it really satisfies the differential equation and touches the surface represented by the singular integral.

(2) Prove that $z^2 = 4xy$ is an integral of

$$z = px + qy + \log pq$$

representing the envelope of planes included in the complete integral and passing through the origin.

(3) Prove that the characteristics of $q = 3p^2$ that pass through the point $(-1, 0, 0)$ generate the cone $(x + 1)^2 + 12yz = 0$.

(4) What is the nature of the integral $(y + 1)^2 + 4xz = 0$ of the equation

$$z = px + qy + p/q?$$

(5) Show that either of the equations

$$z = (x + y)^2 + ax + by,$$

$$z = (x + y)^2 + \frac{mx^2 + ny^2}{x + y}$$

may be taken as the complete integral of a certain differential equation, and that the other may be deduced from it as a particular case of the general integral.   [London.]

(6) Show that $z = (x + a)^2 e^{by}$ is a complete integral of the differential equation $p^2 = 4ze^{qy/z}$.

Show that $y^2 z = 4 \left(\dfrac{xy}{2 - y} \right)^{2-y}$ is part of the general integral of the same equation, and deduce it from the above given complete integral.

[London.]

MISCELLANEOUS EXAMPLES ON CHAPTER XII

(1) $z = px + qy - p^2 q$.

(2) $0 = px + qy - (px + z)^2 q$.

(3) $z(z^2 + xy)(px - qy) = x^4$.

(4) $p^{\frac{1}{2}} - q^{\frac{1}{3}} = 3x - 2y$.

(5) $p_1^2 + 2x_2 p_2 + x_3^2 p_3 = 0$.

(6) $x_3 p_1 + x_2 p_2 + x_1 p_3 = 0$.

(7) $p^3 + q^3 - 3pqz = 0$.

(8) $p_1^2 + p_2^2 + p_3^2 = 4z$.

(9) $p_1 + p_2 + p_3 = 4z$.

(10) $p^2 + 6p + 2q + 4 = 0$.

(11) $z^2 p^2 y + 6zpxy + 2zqx^2 + 4x^2 y = 0$.

(12) $zpy^2 = x(y^2 + z^2 q^2)$.

(13) $p^2 z^2 + q^2 = p^2 q$.

(14) $(z - px - qy)x^3 y^2 = q^2 z x^3 - 3p^3 z^2 y^2$.

(15) Find the particular case of the general integral of $p + q = pq$ that represents the envelope of planes included in the complete integral and passing through the point $(1, 1, 1)$.

(16) Prove that if the equation $P\,dx + Q\,dy + R\,dz = 0$ is integrable, it represents a family of surfaces orthogonal to the family represented by
$$Pp + Qq = R.$$
Hence find the family orthogonal to
$$\phi\{z(x + y)^2,\ x^2 - y^2\} = 0.$$

(17) Find the surfaces whose tangent planes all pass through the origin.

(18) Find the surfaces whose normals all intersect the circle
$$x^2 + y^2 = 1, \quad z = 0.$$

(19) Find the surfaces whose tangent planes form with the co-ordinate planes a tetrahedron of constant volume.

(20) Prove that there is no non-developable surface such that every tangent plane cuts off intercepts from the axes whose algebraic sum is zero.

(21) Show that if two surfaces are polar reciprocals with respect to the quadric $x^2 + y^2 = 2z$, and (x, y, z), (X, Y, Z) are two corresponding points (one on each surface) such that the tangent plane at either point is the polar plane of the other, then
$$X = p; \quad Y = q; \quad Z = px + qy - z; \quad x = P; \quad y = Q.$$
Hence show that if one surface satisfies
$$f(x, y, z, p, q) = 0,$$
the other satisfies $f(P, Q, PX + QY - Z, X, Y) = 0$.

(These equations are said to be derived from each other by the *Principle of Duality*.)

(22) Show that the equation dual to

$$z = px + qy + pq$$

is

$$0 = Z + XY,$$

giving

$$x = P = \frac{\partial Z}{\partial X} = -Y, \quad y = Q = -X,$$

$$z = PX + QY - Z = -XY.$$

Hence derive (as an integral of the first equation) $z = -xy$.

(23) By means of a partial differential equation eliminate the arbitrary function from the equation

$$x + y + z = f(x^2 + y^2 + z^2).$$

[Differentiating partially with respect to x and y, we get

$$1 + p = \{f'(x^2 + y^2 + z^2)\}(2x + 2zp),$$

and

$$1 + q = \{f'(x^2 + y^2 + z^2)\}(2y + 2zq).$$

Hence

$$(1 + p)(y + zq) = (1 + q)(x + zp),$$

or

$$(y - z)p + (z - x)q = x - y.]$$

(24) Use the method of Ex. 23 to verify the solutions of the examples on p. 148.

(25) Find particular integrals of the following partial differential equations to represent surfaces passing through the given curves.

(i) $p + q = 1$; $x = 0$, $y^2 = z$.   (ii) $xp + yq = z$; $x + y = 1$, $yz = 1$.

(iii) $(y - z)p + (z - x)q = x - y$; $z = 0$, $y = 2x$.

(iv) $x(y - z)p + y(z - x)q = z(x - y)$; $x = y = z$.

(v) $yp - 2xyq = 2xz$; $x = t$, $y = t^2$, $z = t^3$.

(vi) $(y - z)\{2xyp + (x^2 - y^2)q\} + z(x^2 - y^2) = 0$; $x = t^2$, $y = 0$, $z = t^3$.

[Eliminate x, y, z from the two equations of the curve and two independent integrals $u(x, y, z) = a$, $v(x, y, z) = b$ of the subsidiary equations. This gives a relation between a and b. Replace a by $u(x, y, z)$, b by $v(x, y, z)$, and we get the integral required.

Thus for (i) $u(x, y, z) \equiv x - z = a$, $v(x, y, z) \equiv y - z = b$. (Cf. p. 148.)

From these and the curve equations $x = 0$, $y^2 = z$, we get $a = -y^2$, $b = y - y^2$, so $(b - a)^2 = -a$.

Replace a by $x - z$, b by $y - z$, and we get the integral

$$(y - x)^2 = z - x.$$

Similarly for (ii), (iii), and (iv). In (v) and (vi) we eliminate x, y, z, t from five equations.]

Answers.   (ii) $yz = (x + y)^2$.   (iii) $5(x + y + z)^2 = 9(x^2 + y^2 + z^2)$.

(iv) $(x + y + z)^3 = 27xyz$.   (v) $(x^2 + y)^5 = 32y^2z^2$.

(vi) $x^3 - 3xy^2 = z^2 - 2yz$.

CHAPTER XIII

PARTIAL DIFFERENTIAL EQUATIONS OF THE FIRST
ORDER. GENERAL METHODS

137. We shall now explain Charpit's method of dealing with
equations with two independent variables and Jacobi's method for
equations with any number of independent variables. Jacobi's
method leads naturally to the discussion of simultaneous partial
differential equations.

The methods of this chapter are considerably more complicated
than those of the last. We shall therefore present them in their
simplest form, and pass lightly over several points which might be
considerably elaborated.

138. Charpit's † method. In Art. 131 we solved the equation

$$p - 3x^2 = q^2 - y \quad \dots\dots\dots\dots\dots\dots\dots\dots\dots(1$$

by using an *additional* differential equation

$$p - 3x^2 = a, \quad \dots\dots\dots\dots\dots\dots\dots\dots\dots(2$$

solving for p and q in terms of x and y, and substituting in

$$dz = p\,dx + q\,dy, \quad \dots\dots\dots\dots\dots\dots\dots\dots(3$$

which then becomes *integrable*, considered as an ordinary differential
equation in the three variables x, y, z.

We shall now apply a somewhat similar method to the general
partial differential equation of the first order with two independent
variables

$$F(x, y, z, p, q) = 0. \quad \dots\dots\dots\dots\dots\dots\dots(4$$

We must find another equation, say

$$f(x, y, z, p, q) = 0, \quad \dots\dots\dots\dots\dots\dots\dots(5$$

---

* To be omitted on a first reading.

† This method was partly due to Lagrange, but was perfected by Charpit.
Charpit's memoir was presented to the Paris Academy of Sciences in 1784, but
the author died soon afterwards and the memoir was never printed.

such that p and q can be found from (4) and (5) as functions of x, y, z which make (3) integrable.

The necessary and sufficient condition that (3) should be integrable is that

$$P\left(\frac{\partial Q}{\partial z} - \frac{\partial R}{\partial y}\right) + Q\left(\frac{\partial R}{\partial x} - \frac{\partial P}{\partial z}\right) + R\left(\frac{\partial P}{\partial y} - \frac{\partial Q}{\partial x}\right) = 0 \quad \text{(identically)},$$

where $$P = p, \ Q = q, \ R = -1,$$

$$\textit{i.e.} \quad p\frac{\partial q}{\partial z} - q\frac{\partial p}{\partial z} - \frac{\partial p}{\partial y} + \frac{\partial q}{\partial x} = 0. \quad \dots\dots\dots\dots(6)$$

By differentiating (4) partially with respect to x, keeping y and z constant, but regarding p and q as denoting the functions of x, y, z obtained by solving (4) and (5), we get

$$\frac{\partial F}{\partial x} + \frac{\partial F}{\partial p}\frac{\partial p}{\partial x} + \frac{\partial F}{\partial q}\frac{\partial q}{\partial x} = 0. \quad \dots\dots\dots\dots(7)$$

Similarly

$$\frac{\partial f}{\partial x} + \frac{\partial f}{\partial p}\frac{\partial p}{\partial x} + \frac{\partial f}{\partial q}\frac{\partial q}{\partial x} = 0. \quad \dots\dots\dots\dots(8)$$

From (7) and (8),

$$J\frac{\partial q}{\partial x} = \frac{\partial F}{\partial x}\frac{\partial f}{\partial p} - \frac{\partial F}{\partial p}\frac{\partial f}{\partial x}. \quad \dots\dots\dots\dots(9)$$

where J stands for

$$\frac{\partial F}{\partial p}\frac{\partial f}{\partial q} - \frac{\partial F}{\partial q}\frac{\partial f}{\partial p}.$$

Similarly

$$J\frac{\partial q}{\partial z} = \frac{\partial F}{\partial z}\frac{\partial f}{\partial p} - \frac{\partial F}{\partial p}\frac{\partial f}{\partial z}, \quad \dots\dots\dots\dots(10)$$

$$J\frac{\partial p}{\partial y} = -\frac{\partial F}{\partial y}\frac{\partial f}{\partial q} + \frac{\partial F}{\partial q}\frac{\partial f}{\partial y}, \quad \dots\dots\dots\dots(11)$$

$$J\frac{\partial p}{\partial z} = -\frac{\partial F}{\partial z}\frac{\partial f}{\partial q} + \frac{\partial F}{\partial q}\frac{\partial f}{\partial z}, \quad \dots\dots\dots\dots(12)$$

Substituting in (6) multiplied by * J, we get

$$p\left(\frac{\partial F}{\partial z}\frac{\partial f}{\partial p} - \frac{\partial F}{\partial p}\frac{\partial f}{\partial z}\right) + q\left(\frac{\partial F}{\partial z}\frac{\partial f}{\partial q} - \frac{\partial F}{\partial q}\frac{\partial f}{\partial z}\right)$$

$$+ \frac{\partial F}{\partial y}\frac{\partial f}{\partial q} - \frac{\partial F}{\partial q}\frac{\partial f}{\partial y} + \frac{\partial F}{\partial x}\frac{\partial f}{\partial p} - \frac{\partial F}{\partial p}\frac{\partial f}{\partial x} = 0,$$

$$\textit{i.e.} \quad -\frac{\partial F}{\partial p}\frac{\partial f}{\partial x} - \frac{\partial F}{\partial q}\frac{\partial f}{\partial y} - \left(p\frac{\partial F}{\partial p} + q\frac{\partial F}{\partial q}\right)\frac{\partial f}{\partial z}$$

$$+ \left(\frac{\partial F}{\partial x} + p\frac{\partial F}{\partial z}\right)\frac{\partial f}{\partial p} + \left(\frac{\partial F}{\partial y} + q\frac{\partial F}{\partial z}\right)\frac{\partial f}{\partial q} = 0. \quad \dots(13)$$

---

* J cannot vanish identically, for this would imply that F and f, regarded as functions of p and q, were not independent. This is contrary to our hypothesis that equations (4) and (5) can be solved for p and q.

This is a linear equation of the form considered in Art. 126 with x, y, z, p, q as independent variables and f as the dependent variable.

The corresponding subsidiary equations are

$$\frac{dx}{-\dfrac{\partial F}{\partial p}} = \frac{dy}{-\dfrac{\partial F}{\partial q}} = \frac{dz}{-p\dfrac{\partial F}{\partial p} - q\dfrac{\partial F}{\partial q}} = \frac{dp}{\dfrac{\partial F}{\partial x} + p\dfrac{\partial F}{\partial z}} = \frac{dq}{\dfrac{\partial F}{\partial y} + q\dfrac{\partial F}{\partial z}} = \frac{df}{0}. \quad (14)$$

If any integral of these equations can be found involving p or q or both, the integral may be taken as the additional differential equation (5), which in conjunction with (4) will give values of p and q to make (3) integrable. This will give a complete integral of (4), from which general and singular integrals can be deduced in the usual way.

139. As an example of the use of this method, consider the equation $$2xz - px^2 - 2qxy + pq = 0. \quad \dotfill (1)$$

Taking the left-hand side of this equation as F, and substituting in the simultaneous equations (14) of the last article, we get

$$\frac{dx}{x^2 - q} = \frac{dy}{2xy - p} = \frac{dz}{px^2 + 2xyq - 2pq} = \frac{dp}{2z - 2qy} = \frac{dq}{0} = \frac{df}{0},$$

of which an integral is $$q = a. \quad \dotfill (2)$$

From (1) and (2), $$p = \frac{2x(z - ay)}{x^2 - a}.$$

Hence $$dz = p\,dx + q\,dy = \frac{2x(z - ay)\,dx}{x^2 - a} + a\,dy,$$

i.e. $$\frac{dz - a\,dy}{z - ay} = \frac{2x\,dx}{x^2 - a},$$

i.e. $$z = ay + b(x^2 - a).$$

This is the complete integral. It is easy to deduce the Singular Integral $$z = x^2 y.$$

The form of the complete integral shows that (1) could have been reduced to $$z = PX + qy - Pq,$$

which is a particular case of a standard form, by the transformation

$$x^2 = X; \quad P = \frac{\partial z}{\partial X} = \frac{1}{2x}\frac{\partial z}{\partial x}.$$

Equations that can be solved by Charpit's method are often solved more easily by some such transformation.

Examples for solution

Apply Charpit's method to find complete integrals of the following:

(1) $2z + p^2 + qy + 2y^2 = 0$. (2) $yzp^2 = q$.
(3) $pxy + pq + qy = yz$. (4) $2x(z^2q^2 + 1) = pz$.
(5) $q = 3p^2$. (Cf. Art. 129.) (6) $z^2(p^2z^2 + q^2) = 1$. (Cf. Art. 130.)
(7) $p - 3x^2 = q^2 - y$. (Cf. Art. 131.)
(8) $z = px + qy + p^2 + q^2$. (Cf. Art. 132.)
(9) Solve Ex. 2 by putting $y^2 = Y$, $z^2 = Z$.
(10) Solve Ex. 4 by a suitable transformation of the variables.

140. Three or more independent variables. Jacobi's * method.

Consider the equation

$$F(x_1, x_2, x_3, p_1, p_2, p_3) = 0, \quad\quad\quad\quad\quad\quad (1)$$

where the dependent variable z does not occur except by its partial differential coefficients p_1, p_2, p_3 with respect to the three independent variables x_1, x_2, x_3. The fundamental idea of Jacobi's method is very similar to that of Charpit's.

We try to find *two* additional equations

$$F_1(x_1, x_2, x_3, p_1, p_2, p_3) = a_1, \quad\quad\quad\quad\quad (2)$$

$$F_2(x_1, x_2, x_3, p_1, p_2, p_3) = a_2 \quad\quad\quad\quad\quad (3)$$

(where a_1 and a_2 are arbitrary constants), such that p_1, p_2, p_3 can be found from (1), (2), (3) as functions of x_1, x_2, x_3 that make

$$dz = p_1\,dx_1 + p_2\,dx_2 + p_3\,dx_3 \quad\quad\quad\quad\quad (4)$$

integrable, for which the conditions are

$$\frac{\partial p_2}{\partial x_1} = \frac{\partial^2 z}{\partial x_1\,\partial x_2} = \frac{\partial p_1}{\partial x_2}; \quad \frac{\partial p_3}{\partial x_1} = \frac{\partial p_1}{\partial x_3}; \quad \frac{\partial p_3}{\partial x_2} = \frac{\partial p_2}{\partial x_3}. \quad\quad (5)$$

Now, by differentiating (1) partially with respect to x_1, keeping x_2 and x_3 constant, but regarding p_1, p_2, p_3 as denoting the functions of x_1, x_2, x_3 obtained by solving (1), (2), (3), we get

$$\frac{\partial F}{\partial x_1} + \frac{\partial F}{\partial p_1}\frac{\partial p_1}{\partial x_1} + \frac{\partial F}{\partial p_2}\frac{\partial p_2}{\partial x_1} + \frac{\partial F}{\partial p_3}\frac{\partial p_3}{\partial x_1} = 0. \quad\quad (6)$$

Similarly $\quad \dfrac{\partial F_1}{\partial x_1} + \dfrac{\partial F_1}{\partial p_1}\dfrac{\partial p_1}{\partial x_1} + \dfrac{\partial F_1}{\partial p_2}\dfrac{\partial p_2}{\partial x_1} + \dfrac{\partial F_1}{\partial p_3}\dfrac{\partial p_3}{\partial x_1} = 0. \quad\quad (7)$

* Carl Gustav Jacob Jacobi of Potsdam (1804-1851) may be considered as one of the creators of the Theory of Elliptic Functions. The "Jacobian" or "Functional Determinant" reminds us of the large part he played in bringing determinants into general use.

From (6) and (7),

$$\frac{\partial(F, F_1)}{\partial(x_1, p_1)} + \frac{\partial(F, F_1)}{\partial(p_2, p_1)}\frac{\partial p_2}{\partial x_1} + \frac{\partial(F, F_1)}{\partial(p_3, p_1)}\frac{\partial p_3}{\partial x_1} = 0, \dots\dots\dots(8)$$

where $\dfrac{\partial(F, F_1)}{\partial(x_1, p_1)}$ denotes the " Jacobian " $\dfrac{\partial F}{\partial x_1}\dfrac{\partial F_1}{\partial p_1} - \dfrac{\partial F}{\partial p_1}\dfrac{\partial F_1}{\partial x_1}$.

Similarly

$$\frac{\partial(F, F_1)}{\partial(x_2, p_2)} + \frac{\partial(F, F_1)}{\partial(p_1, p_2)}\frac{\partial p_1}{\partial x_2} + \frac{\partial(F, F_1)}{\partial(p_3, p_2)}\frac{\partial p_3}{\partial x_2} = 0, \dots\dots\dots(9)$$

and

$$\frac{\partial(F, F_1)}{\partial(x_3, p_3)} + \frac{\partial(F, F_1)}{\partial(p_1, p_3)}\frac{\partial p_1}{\partial x_3} + \frac{\partial(F, F_1)}{\partial(p_2, p_3)}\frac{\partial p_2}{\partial x_3} = 0. \dots\dots\dots(10)$$

Add equations (8), (9), (10).

Two terms are

$$\frac{\partial(F, F_1)}{\partial(p_2, p_1)}\frac{\partial p_2}{\partial x_1} + \frac{\partial(F, F_1)}{\partial(p_1, p_2)}\frac{\partial p_1}{\partial x_2} = \frac{\partial^2 z}{\partial x_1 \partial x_2}\left\{\frac{\partial(F, F_1)}{\partial(p_2, p_1)} + \frac{\partial(F, F_1)}{\partial(p_1, p_2)}\right\} = 0.$$

Similarly two other pairs of terms vanish, leaving

$$\frac{\partial(F, F_1)}{\partial(x_1, p_1)} + \frac{\partial(F, F_1)}{\partial(x_2, p_2)} + \frac{\partial(F, F_1)}{\partial(x_3, p_3)} = 0, \dots\dots\dots(11)$$

i.e.

$$\frac{\partial F}{\partial x_1}\frac{\partial F_1}{\partial p_1} - \frac{\partial F}{\partial p_1}\frac{\partial F_1}{\partial x_1} + \frac{\partial F}{\partial x_2}\frac{\partial F_1}{\partial p_2} - \frac{\partial F}{\partial p_2}\frac{\partial F_1}{\partial x_2} + \frac{\partial F}{\partial x_3}\frac{\partial F_1}{\partial p_3} - \frac{\partial F}{\partial p_3}\frac{\partial F_1}{\partial x_3} = 0. \quad (12)$$

This equation is generally written as $(F, F_1) = 0$.

Similarly $(F, F_2) = 0$ and $(F_1, F_2) = 0$.

But these are linear equations of the form of Art. 126. Hence we have the following rule:

Try to find two independent integrals, $F_1 = a_1$ and $F_2 = a_2$, of the subsidiary equations

$$\frac{dx_1}{-\dfrac{\partial F}{\partial p_1}} = \frac{dp_1}{\dfrac{\partial F}{\partial x_1}} = \frac{dx_2}{-\dfrac{\partial F}{\partial p_2}} = \frac{dp_2}{\dfrac{\partial F}{\partial x_2}} = \frac{dx_3}{-\dfrac{\partial F}{\partial p_3}} = \frac{dp_3}{\dfrac{\partial F}{\partial x_3}}.$$

If these satisfy the condition

$$(F_1, F_2) \equiv \sum\left(\frac{\partial F_1}{\partial x_r}\frac{\partial F_2}{\partial p_r} - \frac{\partial F_1}{\partial p_r}\frac{\partial F_2}{\partial x_r}\right) = 0,$$

and if the p's can be found as functions of the x's from

$$F = F_1 - a_1 = F_2 - a_2 = 0,$$

*integrate the equation * formed by substituting these functions in*

$$dz = p_1\,dx_1 + p_2\,dx_2 + p_3\,dx_3.$$

* For a proof that this equation will always be integrable, see Appendix C.

141. Examples on Jacobi's method

Ex. (i). $$2p_1x_1x_3 + 3p_2x_3{}^2 + p_2{}^2p_3 = 0. \quad \text{.........................(1)}$$

The subsidiary equations are

$$\frac{dx_1}{-2x_1x_3} = \frac{dp_1}{2p_1x_3} = \frac{dx_2}{-3x_3{}^2 - 2p_2p_3} = \frac{dp_2}{0} = \frac{dx_3}{-p_2{}^2} = \frac{dp_3}{2p_1x_1 + 6p_2x_3},$$

of which integrals are $$F_1 \equiv p_1x_1 = a_1, \quad \text{..............................(2)}$$

and $$F_2 \equiv p_2 = a_2. \quad \text{...............................(3)}$$

Now with these values (F_1, F_2) is obviously zero, so (2) and (3) can be taken as the two additional equations required.

$$p_1 = a_1x_1{}^{-1}, \quad p_2 = a_2, \quad p_3 = -a_2{}^{-2}(2a_1x_3 + 3a_2x_3{}^2).$$

Hence $$dz = a_1x_1{}^{-1}\,dx_1 + a_2\,dx_2 - a_2{}^{-2}(2a_1x_3 + 3a_2x_3{}^2)\,dx_3$$

or $$z = a_1 \log x_1 + a_2x_2 - a_2{}^{-2}(a_1x_3{}^2 + a_2x_3{}^3) + a_3,$$

the complete integral.

Ex. (ii). $$(x_2 + x_3)(p_2 + p_3)^2 + zp_1 = 0. \quad \text{.........................(4)}$$

This equation is not of the form considered in Art. 140, as it involves . But put

$$z = x_4, \quad p_1 = \frac{\partial z}{\partial x_1} = \frac{\partial x_4}{\partial x_1} = -\frac{\partial u}{\partial x_1} \Big/ \frac{\partial u}{\partial x_4} = -P_1/P_4, \text{ say,}$$

where $u = 0$ is an integral of (4).

Similarly $$p_2 = -P_2/P_4; \quad p_3 = -P_3/P_4.$$

(4) becomes $$(x_2 + x_3)(P_2 + P_3)^2 - x_4P_1P_4 = 0, \quad \text{.......................(5)}$$

an equation in four independent variables, not involving the dependent variable u.

The subsidiary equations are

$$\frac{dx_1}{x_4P_4} = \frac{dP_1}{0} = \frac{dx_2}{-2(x_2 + x_3)(P_2 + P_3)} = \frac{dP_2}{(P_2 + P_3)^2} = \frac{dx_3}{-2(x_2 + x_3)(P_2 + P_3)}$$

$$= \frac{dP_3}{(P_2 + P_3)^2} = \frac{dx_4}{x_4P_1} = \frac{dP_4}{-P_1P_4},$$

of which integrals are $$F_1 \equiv P_1 = a_1, \quad \text{..................................(6)}$$

$$F_2 \equiv P_2 - P_3 = a_2, \quad \text{...........................(7)}$$

$$F_3 \equiv x_4P_4 = a_3. \quad \text{.................................(8)}$$

We have to make sure that $(F_r, F_s) = 0$, where r and s are *any* two of the indices 1, 2, 3. This is easily seen to be true.

Solving (5), (6), (7), (8), we get

$$P_1 = a_1; \quad P_4 = a_3x_4{}^{-1}; \quad 2P_2 = a_2 \pm \sqrt{\{a_1a_3/(x_2 + x_3)\}}; \quad P_3 = P_2 - a_2;$$

so $$du = a_1\,dx_1 + a_3x_4{}^{-1}\,dx_4 + \tfrac{1}{2}a_2\,(dx_2 - dx_3)$$
$$\pm \tfrac{1}{2}\sqrt{\{a_1a_3/(x_2 + x_3)\}}\,(dx_2 + dx_3),$$

i.e. $$u = a_1x_1 + a_3 \log x_4 + \tfrac{1}{2}a_2(x_2 - x_3) \pm \sqrt{\{a_1a_3(x_2 + x_3)\}} + a_4.$$

So $u = 0$ gives, replacing
$$x_4 \text{ by } z, \quad a_1/a_3 \text{ by } A_1, \quad \tfrac{1}{2}a_2/a_3 \text{ by } A_2, \quad a_4/a_3 \text{ by } A_3,$$
$$\log z + A_1 x_1 + A_2(x_2 - x_3) \pm \sqrt{\{A_1(x_2 + x_3)\}} + A_3 = 0,$$
the complete integral of (4).

Examples for solution

Apply Jacobi's method to find complete integrals of the following:

(1) $p_1^3 + p_2^2 + p_3 = 1.$ (2) $x_3^2 p_1^2 p_2^2 p_3^2 + p_1^2 p_2^2 - p_3^2 = 0.$

(3) $p_1 x_1 + p_2 x_2 = p_3^2.$ (4) $p_1 p_2 p_3 + p_4^3 x_1 x_2 x_3 x_4^3 = 0.$

(5) $p_1 p_2 p_3 = z^3 x_1 x_2 x_3.$ (6) $p_3 x_3 (p_1 + p_2) + x_1 + x_2 = 0.$

(7) $p_1^2 + p_2 p_3 - z(p_2 + p_3) = 0.$

(8) $(p_1 + x_1)^2 + (p_2 + x_2)^2 + (p_3 + x_3)^2 = 3(x_1 + x_2 + x_3).$

142. Simultaneous partial differential equations.   The following examples illustrate some typical cases :

Ex. (i). $F \equiv p_1^2 + p_2 p_3 x_2 x_3^2 = 0, \dots\dots\dots\dots\dots\dots(1)$

$F_1 \equiv p_1 + p_2 x_2 = 0. \dots\dots\dots\dots\dots\dots\dots(2)$

Here
$$(F, F_1) \equiv \sum \left(\frac{\partial F}{\partial x_r} \frac{\partial F_1}{\partial p_r} - \frac{\partial F}{\partial p_r} \frac{\partial F_1}{\partial x_r} \right) = (p_2 p_3 x_3^2) x_2 - (p_3 x_2 x_3^2) p_2 = 0.$$

Thus the problem may be considered as the solution of the equation (1), with part of the work (the finding of F_1) already done.

The next step is to find F_2 such that
$$(F, F_2) = 0 = (F_1, F_2).$$

The subsidiary equations derived by Jacobi's process from F are
$$\frac{dx_1}{-2p_1} = \frac{dp_1}{0} = \frac{dx_2}{-p_3 x_2 x_3^2} = \frac{dp_2}{p_2 p_3 x_3^2} = \frac{dx_3}{-p_2 x_2 x_3^2} = \frac{dp_3}{2 p_2 p_3 x_2 x_3} .$$

An integral is $p_1 = a. \dots\dots\dots\dots\dots\dots\dots(3)$

We may take F_2 as p_1, since this satisfies $(F, F_2) = 0 = (F_1, F_2)$.

Solving (1), (2), (3) and substituting in $dz = p_1 \, dx_1 + p_2 \, dx_2 + p_3 \, dx_3,$
$$dz = a \, dx_1 - a x_2^{-1} \, dx_2 + a x_3^{-2} \, dx_3,$$

so $z = a(x_1 - \log x_2 - x_3^{-1}) + b.$

Ex. (ii). $F \equiv p_1 x_1 + p_2 x_2 - p_3^2 = 0, \dots\dots\dots\dots\dots(4)$

$F_1 \equiv p_1 - p_2 + p_3 - 1 = 0. \dots\dots\dots\dots\dots(5)$

Here $(F, F_1) = p_1 + p_2(-1) = p_1 - p_2.$

This must vanish if the expression for dz is to be integrable.

Hence we have the additional equation
$$p_1 - p_2 = 0. \dots\dots\dots\dots\dots\dots\dots\dots(6)$$

Solving (4), (5), (6) and substituting,
$$dz = \frac{dx_1 + dx_2}{x_1 + x_2} + dx_3,$$
$$z = \log(x_1 + x_2) + x_3 + a.$$

In examples of this type we do not have to use the subsidiary equations. The result has only one arbitrary constant, whereas in Ex. (i) we got two.

Ex. (iii).
$$F \equiv x_1{}^2 + x_2{}^2 + p_3 = 0, \quad\dots\dots\dots\dots\dots(7)$$

$$F_1 \equiv p_1 + p_2 + x_3{}^2 = 0. \quad\dots\dots\dots\dots\dots(8)$$

Here
$$(F, F_1) = 2x_1 + 2x_2 - 2x_3.$$

As x_1, x_2, x_3 are independent variables, this cannot be always zero.

Hence we cannot find an *integrable* expression for dz from these equations, which have no common integral.

Ex. (iv).
$$F \equiv p_1 + p_2 + p_3{}^2 - 3x_1 - 3x_2 - 4x_3{}^2 = 0, \dots\dots\dots(9)$$

$$F_1 \equiv x_1 p_1 - x_2 p_2 - 2x_1{}^2 + 2x_2{}^2 = 0, \quad\dots\dots\dots(10)$$

$$F_2 \equiv p_3 - 2x_3 = 0. \quad\dots\dots\dots\dots\dots(11)$$

Solving (9), (10), (11) and substituting in the expression for dz,
$$dz = (2x_1 + x_2)\, dx_1 + (x_1 + 2x_2)\, dx_2 + 2x_3\, dx_3,$$
so
$$z = x_1{}^2 + x_1 x_2 + x_2{}^2 + x_3{}^2 + a.$$

This time there is no need to work out (F, F_1), (F, F_2), (F_1, F_2).

Ex. (v).
$$F \equiv p_1 + p_2 - 1 - x_2 = 0, \quad\dots\dots\dots\dots\dots(12)$$

$$F_1 \equiv p_1 + p_3 - x_1 - x_2 = 0, \quad\dots\dots\dots\dots(13)$$

$$F_2 \equiv p_2 + p_3 - 1 - x_1 = 0. \quad\dots\dots\dots\dots(14)$$

These give $dz = x_2\, dx_1 + dx_2 + x_1\, dx_3$.

As this cannot be integrated, the simultaneous equations have no common integral.

Ex. (vi).
$$F \equiv x_1 p_1 - x_2 p_2 + p_3 - p_4 = 0, \quad\dots\dots\dots(15)$$

$$F_1 \equiv p_1 + p_2 - x_1 - x_2 = 0. \quad\dots\dots\dots\dots(16)$$

Here $(F, F_1) = p_1 - x_1(-1) - p_2 + x_2(-1) = p_1 - p_2 + x_1 - x_2$.

As in Ex. (ii), this gives us a new equation
$$F_2 \equiv p_1 - p_2 + x_1 - x_2 = 0. \quad\dots\dots\dots\dots(17)$$

Now
$$(F, F_2) = p_1 - x_1 - p_2(-1) + x_2(-1) = F_1 = 0,$$
and
$$(F_1, F_2) = (-1) - 1 + (-1)(-1) - (-1) = 0,$$
so we cannot get any more equations by this method.

The subsidiary equations derived from F are
$$\frac{dx_1}{-x_1} = \frac{dp_1}{p_1} = \frac{dx_2}{x_2} = \frac{dp_2}{-p_2} = \frac{dx_3}{-1} = \frac{dp_3}{0} = \frac{dx_4}{1} = \frac{dp_4}{0}.$$

A suitable integral is
$$F_3 \equiv p_3 = a, \quad\dots\dots\dots\dots\dots\dots(18)$$
for this satisfies $(F, F_3) = (F_1, F_3) = (F_2, F_3) = 0$.

We have now four equations (15), (16), (17), (18). These give
$$p_1 = x_2; \quad p_2 = x_1; \quad p_3 = a; \quad p_4 = a;$$
so
$$z = x_1 x_2 + a(x_3 + x_4) + b.$$

But in this example we can obtain a more general integral. The two given equations (15) and (16) and the derived one (17) are equivalent to the simpler set:

$$p_1 = x_2, \quad\dotfill(19$$
$$p_2 = x_1, \quad\dotfill(20$$
$$p_3 - p_4 = 0. \quad\dotfill(21$$

From (19) and (20), $z = x_1 x_2 + $ any function of x_3 and x_4.

(21) is a linear equation of Lagrange's type, of which the general integral is
$$\phi(z, x_3 + x_4) = 0,$$

i.e. z is any function of $(x_3 + x_4)$, and may of course also involve x and x_2.

Hence a general integral of all three equations, or of the two given equations, is
$$z = x_1 x_2 + \psi(x_3 + x_4),$$

involving an arbitrary function. The complete integral obtained by the other method is included as a particular case. The general integral could have been obtained from the complete, as in Art. 134.

Examples for solution

Obtain common complete integrals (if possible) of the following simultaneous equations:

(1) $p_1{}^2 + p_2{}^2 - 8(x_1 + x_2)^2 = 0,$
$(p_1 - p_2)(x_1 - x_2) + p_3 x_3 - 1 = 0.$

(2) $x_1{}^2 p_2 p_3 = x_2{}^2 p_3 p_1 = x_3{}^2 p_1 p_2 = 1.$

(3) $p_1 p_2 p_3 - 8 x_1 x_2 x_3 = 0,$
$p_2 + p_3 - 2x_2 - 2x_3 = 0.$

(4) $2x_3 p_1 p_3 - x_4 p_4 = 0,$
$2p_1 - p_2 = 0.$

(5) $p_1 x_3{}^2 + p_3 = 0,$
$p_2 x_3{}^2 + p_3 x_2{}^2 = 0.$

(6) $p_2{}^2 + p_3{}^3 + x_1 + 2x_2 + 3x_3 = 0,$
$p_1 + p_4{}^2 x_4 - 1 = 0.$

(7) $2p_1 + p_2 + p_3 + 2p_4 = 0,$
$p_1 p_3 - p_2 p_4 = 0.$

(8) Find the general integral of Ex. (5).

(9) Find the general integral of Ex. (7).

MISCELLANEOUS EXAMPLES ON CHAPTER XIII

(1) $2x_1 x_3 z p_1 p_3 + x_2 p_2 = 0.$

(2) $x_2 p_3 + x_1 p_4 = p_1 p_3 - p_2 p_4 + x_4{}^2 = 0$

(3) $9x_1 x_4 p_1 (p_2 + p_3) - 4p_4{}^2 = 0,$
$p_1 x_1 + p_2 - p_3 = 0.$

(4) $9x_1 z p_1 (p_2 + p_3) - 4 = 0,$
$p_1 x_1 + p_2 - p_3 = 0.$

(5) $x_1 p_2 p_3 = x_2 p_3 p_1 = x_3 p_1 p_2 = z^2 x_1 x_2 x_3.$

(6) $p_1 z^2 - x_1{}^2 = p_2 z^2 - x_2{}^2 = p_3 z^2 - x_3{}^2 = 0.$

(7) Find a singular integral of $z = p_1 x_1 + p_2 x_2 + p_3 x_3 + p_1{}^2 + p_2{}^2 + p_3{}^2$ representing the envelope of all the hyper-surfaces (in this case hyper-planes) included in the complete integral.

(8) Show that no equation of the form $F(x_1, x_2, x_3, p_1, p_2, p_3) = 0$ has a singular integral.

(9) Show that if z is absent from the equation $F(x, y, z, p, q) = 0$, Charpit's method coincides with Jacobi's.

(10) Show that if a system of partial differential equations is linear and homogeneous in the p's and has a common integral

$$z = a_1 u_1 + a_2 u_2 + \dots ,$$

where the u's are functions of the x's, then a more general integral is

$$z = \phi(u_1, u_2, \dots).$$

Find a general integral of the simultaneous equations

$$x_1 p_1 - x_2 p_2 + x_2 p_3 = 0,$$
$$x_4 p_3 - x_4 p_4 + x_5 p_5 = 0.$$

(11) If p_1 and p_2 are functions of the independent variables x_1, x_2 satisfying the simultaneous equations

$$F(x_1, x_2, p_1, p_2) = 0 = F_1(x_1, x_2, p_1, p_2),$$

prove that $(F, F_1) + \left(\dfrac{\partial p_1}{\partial x_2} - \dfrac{\partial p_2}{\partial x_1} \right) \dfrac{\partial(F, F_1)}{\partial(p_1, p_2)} = 0.$

Hence show that if the simultaneous equations, taken as partial differential equations, have a common integral, $(F, F_1) = 0$ is a necesary but not a sufficient condition.

Examine the following pairs of simultaneous equations:

$$\text{(i)} \quad F \equiv p_1 + 2p_2 - 2 = 0,$$
$$F_1 \equiv (p_1 + 2p_2)^2 - 1 = 0.$$

[Here $\dfrac{\partial(F, F_1)}{\partial(p_1, p_2)} = 0$ identically, and the equations cannot be solved for p_1 and p_2.]

$$\text{(ii)} \quad F \equiv p_1 - p_2{}^2 = 0,$$
$$F_1 \equiv p_1 + 2p_2 x_1 + x_1{}^2 = 0.$$

[Here (F, F_1) and $\dfrac{\partial(F, F_1)}{\partial(p_1, p_2)}$ both come to functions which vanish when the p's are replaced by their values in terms of x_1 and x_2. There is no common integral.]

$$\text{(iii)} \quad F \equiv p_1 - p_2{}^2 + x_2 = 0,$$
$$F_1 \equiv p_1 + 2p_2 x_1 + x_1{}^2 + x_2 = 0.$$

[These have a common integral, although $\dfrac{\partial(F, F_1)}{\partial(p_1, p_2)}$ comes to a function that vanishes when the p's are replaced by their values.]

Note on Charpit's Method (pp. 162-164).

Sometimes we can find an equation $f(x, y, z, p, q) = 0$ which is an integral, not of the subsidiary equations (14), but of simpler equations obtained from these by using the original differential equation (4). This will satisfy (13), not identically, but in virtue of (4), and in conjunction with (4) will still make (3) integrable. Thus in Ex. 2, Art. 139, $pz = a$ is an integral, not of $dz/(-2yzp^2 + q) = dp/yp^3$, but of $dz/(-yzp^2) = dp/yp^3$, giving finally the result on p. xvi. Similarly for Jacobi's method.

CHAPTER XIV

PARTIAL DIFFERENTIAL EQUATIONS OF THE SECOND AND HIGHER ORDERS

143. We shall first give some simple examples that can be integrated by inspection. After this we shall deal with linear partial differential equations with constant coefficients; these are treated by methods similar to those used for ordinary linear equations with constant coefficients. The rest of the chapter will be devoted to the more difficult subject of Monge's * methods. It is hoped that the treatment will be full enough to enable the student to solve examples and to make him believe in the correctness of the method, but a discussion of the theory will not be attempted.†

Several examples will deal with the determination of the arbitrary functions involved in the solutions by geometrical conditions.‡

The miscellaneous examples at the end of the chapter contain several important differential equations occurring in the theory of vibrations of strings, bars, membranes, etc.

The second partial differential coefficients $\dfrac{\partial^2 z}{\partial x^2}$, $\dfrac{\partial^2 z}{\partial x\,\partial y}$, $\dfrac{\partial^2 z}{\partial y^2}$ will be denoted by r, s, t respectively.

144. Equations that can be integrated by inspection.

Ex. (i). $$s = 2x + 2y.$$

Integrating with respect to x (keeping y constant),

$$q = x^2 + 2xy + \phi(y).$$

Similarly, integrating with respect to y,

$$z = x^2 y + xy^2 + \int \phi(y)\,dy + f(x),$$

say $$z = x^2 y + xy^2 + f(x) + F(y).$$

---

* Gaspard Monge, of Beaune (1746-1818), Professor at Paris, created Descriptive Geometry. He applied differential equations to questions in solid geometry.

† The student who desires this should consult Goursat, *Sur l'intégration des equations aux dérivées partielles du second ordre.*

‡ Frost's *Solid Geometry*, Chap. XXV., may be read with advantage.

Ex. (ii). Find a surface passing through the parabolas
$$z=0, \quad y^2=4ax \quad \text{and} \quad z=1, \quad y^2=-4ax,$$
and satisfying
$$xr+2p=0.$$
The differential equation is
$$x\frac{\partial p}{\partial x}+2p=0,$$
giving
$$x^2p=f(y),$$
$$p=\frac{1}{x^2}f(y),$$
$$z=-\frac{1}{x}f(y)+F(y).$$

The functions f and F are to be determined from the geometrical conditions.

Putting $z=0$ and $x=y^2/4a$,
$$0=-\frac{4a}{y^2}f(y)+F(y).$$

Similarly
$$1=\frac{4a}{y^2}f(y)+F(y).$$

Hence
$$F(y)=\frac{1}{2}, \quad f(y)=\frac{y^2}{8a}$$

and
$$z=\frac{1}{2}-\frac{y^2}{8ax},$$

i.e. $8axz=4ax-y^2$, a conicoid.

Examples for solution

(1) $r=6x$. (2) $xys=1$.

(3) $t=\sin xy$. (4) $xr+p=9x^2y^3$.

(5) $ys+p=\cos(x+y)-y\sin(x+y)$. (6) $t-xq=x^2$.

(7) Find a surface satisfying $s=8xy$ and passing through the circle
$$z=0=x^2+y^2-1.$$

(8) Find the most general conicoid satisfying $xs+q=4x+2y+2$.

(9) Find a surface of revolution that touches $z=0$ and satisfies
$$r=12x^2+4y^2.$$

(10) Find a surface satisfying $t=6x^3y$, containing the two lines
$$y=0=z, \quad y=1=z.$$

145. Homogeneous linear equations with constant coefficients. In Chap. III. we dealt at some length with the equation
$$(D^n+a_1D^{n-1}+a_2D^{n-2}+\ldots+a_n)y=f(x), \quad \ldots\ldots\ldots\ldots(1)$$
where $D\equiv\dfrac{d}{dx}$.

We shall now deal briefly with the corresponding equation in two independent variables,

$$(D^n + a_1 D^{n-1} D' + a_2 D^{n-2} D'^2 + \ldots + a_n D'^n) z = f(x, y), \quad \ldots\ldots(2)$$

where $D \equiv \dfrac{\partial}{\partial x}, \quad D' \equiv \dfrac{\partial}{\partial y}$.

The simplest case is $\quad (D - mD') z = 0$,

$$i.e. \quad p - mq = 0,$$

of which the solution is $\phi(z, y + mx) = 0$,

$$i.e. \quad z = F(y + mx).$$

This suggests, what is easily verified, that the solution of (2) if $f(x, y) = 0$ is

$$z = F_1(y + m_1 x) + F_2(y + m_2 x) + \ldots + F_n(y + m_n x),$$

where the $m_1, m_2, \ldots m_n$ are the roots (supposed all different) of

$$m^n + a_1 m^{n-1} + a_2 m^{n-2} + \ldots + a_n = 0.$$

Ex.
$$\frac{\partial^3 z}{\partial x^3} - 3 \frac{\partial^3 z}{\partial x^2 \partial y} + 2 \frac{\partial^3 z}{\partial x \partial y^2} = 0,$$

$$i.e. \quad (D^3 - 3D^2 D' + 2DD'^2) z = 0.$$

The roots of $m^3 - 3m^2 + 2m = 0$ are 0, 1, 2.

Hence $\quad z = F_1(y) + F_2(y + x) + F_3(y + 2x)$.

Examples for solution

(1) $(D^3 - 6D^2 D' + 11DD'^2 - 6D'^3) z = 0$.

(2) $2r + 5s + 2t = 0$.

(3) $\dfrac{\partial^2 z}{\partial x^2} - \dfrac{\partial^2 z}{\partial y^2} = 0$.

(4) Find a surface satisfying $r + s = 0$ and touching the elliptic paraboloid $z = 4x^2 + y^2$ along its section by the plane $y = 2x + 1$. [*N.B.*— The values of p (and also of q) for the two surfaces must be equal for any point on $y = 2x + 1$.]

146. Case when the auxiliary equation has equal roots.

Consider the equation $\quad (D - mD')^2 z = 0$. $\quad\ldots\ldots\ldots\ldots\ldots\ldots\ldots\ldots\ldots(1)$

Put $\quad (D - mD') z = u$.

(1) becomes $\quad (D - mD') u = 0$,

giving $\quad u = F(y + mx)$;

therefore $\quad (D - mD') z = F(y + mx)$,

or $\quad p - mq = F(y + mx)$.

The subsidiary equations are

$$\frac{dx}{1} = \frac{dy}{-m} = \frac{dz}{F(y + mx)},$$

giving $\qquad\qquad y + mx = a,$

and $\qquad\qquad dz - F(a)\,dx = 0,$

$\qquad i.e. \quad z - xF(y + mx) = b,$

so the general integral is

$\qquad \phi\{z - xF(y + mx),\, y + mx\} = 0 \quad \text{or} \quad z = xF(y + mx) + F_1(y + mx).$

Similarly we can prove that the integral of

$$(D - mD')^n z = 0$$

is $\qquad z = x^{n-1}F(y + mx) + x^{n-2}F_1(y + mx) + \ldots + F_{n-1}(y + mx).$

Examples for solution

(1) $(4D^2 + 12DD' + 9D'^2)z = 0.$ (2) $25r - 40s + 16t = 0.$

(3) $(D^3 - 4D^2D' + 4DD'^2)z = 0.$

(4) Find a surface passing through the two lines $z = x = 0,$ $z - 1 = x - y = 0,$ satisfying $r - 4s + 4t = 0.$

147. The Particular Integral. We now return to equation (2) of Art. 145, and write it for brevity as

$$F(D, D')z = f(x, y).$$

We can prove, following Chap. III. step by step, that the most general value of z is the sum of a particular integral and the complementary function (which is the value of z when the differential equation has $f(x, y)$ replaced by zero).

The particular integral may be written $\dfrac{1}{F(D, D')} \cdot f(x, y),$ and we may treat the symbolic function of D and D' as we did that of D alone, factorising it, resolving into partial fractions, or expanding in an infinite series.

$E.g. \quad \dfrac{1}{D^2 - 6DD' + 9D'^2}(12x^2 + 36xy) = \dfrac{1}{D^2}\left(1 - \dfrac{3D'}{D}\right)^{-2}(12x^2 + 36xy)$

$\qquad = \dfrac{1}{D^2}\left(1 + \dfrac{6D'}{D} + 27\dfrac{D'^2}{D^2} + \ldots\right) \cdot (12x^2 + 36xy)$

$\qquad = \dfrac{1}{D^2} \cdot (12x^2 + 36xy) + \dfrac{6}{D^3} \cdot 36x$

$\qquad = x^4 + 6x^3y + 9x^4 = 10x^4 + 6x^3y,$

so the solution of $\quad (D^2 - 6DD' + 9D'^2)z = 12x^2 + 36xy$

is $\qquad z = 10x^4 + 6x^3y + \phi(y + 3x) + x\psi(y + 3x).$

Examples for solution

(1) $(D^2 - 2DD' + D'^2)z = 12xy.$

(2) $(2D^2 - 5DD' + 2D'^2)z = 24(y - x).$

(3) Find a real function V of x and y, reducing to zero when $y =$
and satisfying $\qquad \dfrac{\partial^2 V}{\partial x^2} + \dfrac{\partial^2 V}{\partial y^2} = -4\pi(x^2 + y^2).$

148. Short methods. When $f(x, y)$ is a function of $ax + by$
shorter methods may be used.

Now $D\phi(ax + by) = a\phi'(ax + by); \quad D'\phi(ax + by) = b\phi'(ax + by).$

Hence $F(D, D')\phi(ax + by) = F(a, b)\phi^{(n)}(ax + by),$

where $\phi^{(n)}$ is the nth derived function of ϕ, n being the degree c
$F(D, D').$

Conversely

$$\frac{1}{F(D, D')}\phi^{(n)}(ax + by) = \frac{1}{F(a, b)}\phi(ax + by), \quad \dots\dots\dots(\text{A}$$

provided $F(a, b) \neq 0$, e.g.

$$\frac{1}{D^3 - 4D^2D' + 4DD'^2}\cos(2x + 3y) = \frac{-\sin(2x + 3y)}{2^3 - 4 \cdot 2^2 \cdot 3 + 4 \cdot 2 \cdot 3^2}$$

$$= -\frac{1}{32}\sin(2x + 3y),$$

since $\phi(2x + 3y)$ may be taken as $-\sin(2x + 3y)$ if

$$\phi'''(2x + 3y) = \cos(2x + 3y).$$

To deal with the case when $F(a, b) = 0$, we consider the equatio
$$(D - mD')z \equiv p - mq = x^r\psi(y + mx),$$

of which the solution is easily found to be

$$z = \frac{x^{r+1}}{r+1}\psi(y + mx) + \phi(y + mx),$$

so we may take

$$\frac{1}{D - mD'} \cdot x^r\psi(y + mx) = \frac{x^{r+1}}{r+1}\psi(y + mx).$$

Hence

$$\frac{1}{(D - mD')^n}\psi(y + mx) = \frac{1}{(D - mD')^{n-1}} \cdot x\psi(y + mx) = \dots$$

$$= \frac{x^n}{n!}\psi(y + mx), \quad \dots\dots\dots\dots\dots\dots\dots\dots(\text{B}$$

$$e.g. \quad \frac{1}{D^2 - 2DD' + D'^2}\tan(y + x) = \tfrac{1}{2}x^2\tan(y + x),$$

while

$$\frac{1}{D^2 - 5DD' + 4D'^2}\sin(4x + y) = \frac{1}{D - 4D'} \cdot \frac{1}{D - D'}\sin(4x + y)$$

$$= \frac{1}{D - 4D'} \cdot -\tfrac{1}{3}\cos(4x + y) \text{ by (A}$$

$$= -\tfrac{1}{3}x\cos(4x + y) \text{ by (B).}$$

Examples for solution

(1) $(D^2 - 2DD' + D'^2)z = e^{x+2y}$.

(2) $(D^2 - 6DD' + 9D'^2)z = 6x + 2y$.

(3) $(D^3 - 4D^2D' + 4DD'^2)z = 4 \sin(2x + y)$.

(4) $2r - s - 3t = 5e^x/e^y$.   (5) $\dfrac{\partial^2 V}{\partial x^2} + \dfrac{\partial^2 V}{\partial y^2} = 12(x + y)$.

(6) $4r - 4s + t = 16 \log(x + 2y)$.

149. General method. To find a general method of getting a particular integral, consider
$$(D - mD')z \equiv p - mq = f(x, y).$$
The subsidiary equations are
$$\frac{dx}{1} = \frac{dy}{-m} = \frac{dz}{f(x, y)},$$
of which one integral is $y + mx = c.$

Using this integral to find another,
$$dz = f(x, c - mx)\, dx,$$
$$z = \int f(x, c - mx)\, dx + \text{constant},$$
where c is to be replaced by $y + mx$ after integration.

Hence we may take $\dfrac{1}{D - mD'} \cdot f(x, y)$ *as* $\displaystyle\int f(x, c - mx)\, dx$, *where c is replaced by $y + mx$ after integration.*

Ex. $(D - 2D')(D + D')z = (y - 1)e^x.$

Here $\displaystyle\int f(x, c - 2x)\, dx = \int (c - 2x - 1)e^x\, dx = (c - 2x + 1)e^x.$

Therefore $\dfrac{1}{D - 2D'} \cdot (y - 1)e^x = (y + 1)e^x$, replacing c by $y + 2x$.

Similarly $\dfrac{1}{D + D'} \cdot (y + 1)e^x$ is found from $\displaystyle\int (c + x + 1)e^x\, dx = (c + x)e^x$ by replacing c by $y - x$, giving ye^x, which is the particular integral required.

Hence $z = ye^x + \phi(y + 2x) + \psi(y - x).$

Examples for solution

(1) $(D^2 + 2DD' + D'^2)z = 2\cos y - x \sin y$.

(2) $(D^2 - 2DD' - 15D'^2)z = 12xy$.   (3) $r + s - 6t = y \cos x$.

(4) $\dfrac{\partial^2 z}{\partial x^2} - \dfrac{\partial^2 z}{\partial x \partial y} - 2\dfrac{\partial^2 z}{\partial y^2} = (2x^2 + xy - y^2)\sin xy - \cos xy$.

(5) $r - t = \tan^3 x \tan y - \tan x \tan^3 y$.

(6) $\dfrac{\partial^2 y}{\partial x^2} - 4\dfrac{\partial^2 y}{\partial t^2} = \dfrac{4x}{t^2} - \dfrac{t}{x^2}$.

150. Non-homogeneous linear equations. The simplest case is

$$(D - mD' - a)z = 0,$$

$$i.e. \quad p - mq = az,$$

giving
$$\phi(ze^{-ax}, y + mx) = 0,$$

or
$$z = e^{ax}\psi(y + mx).$$

Similarly we can show that the integral of

$$(D - mD' - a)(D - nD' - b)z = 0$$

is
$$z = e^{ax}f(y + mx) + e^{bx}F(y + nx),$$

while that of
$$(D - mD' - a)^2z = 0$$

is
$$z = e^{ax}f(y + mx) + xe^{ax}F(y + mx).$$

But the equations where the symbolical operator cannot be resolved into factors *linear* in D and D' cannot be integrated in this manner.

Consider for example $(D^2 - D')z = 0$.

As a trial solution put $z = e^{hx+ky}$, giving

$$(D^2 - D')z = (h^2 - k)e^{hx+ky}.$$

So $z = e^{h(x+hy)}$ is a particular integral, and a more general one is $\Sigma Ae^{h(x+hy)}$, where the A and h in each term are perfectly arbitrary, and any number of terms may be taken.

This form of integral is best suited to physical problems, as was explained at some length in Chap. IV. Of course the integral of any linear partial differential equation with constant coefficients may be expressed in this manner, but the shorter forms involving arbitrary functions are generally to be preferred.

Examples for solution

(1) $DD'(D - 2D' - 3)z = 0.$ (2) $r + 2s + t + 2p + 2q + z = 0.$

(3) $\dfrac{\partial^2 V}{\partial x^2} = \dfrac{\partial V}{\partial t}.$ (4) $(D^2 - D'^2 + D - D')z = 0.$

(5) $(2D^4 - 3D^2D' + D'^2)z = 0.$ (6) $\dfrac{\partial^2 V}{\partial x^2} + \dfrac{\partial^2 V}{\partial y^2} = n^2V.$

(7) $(D - 2D' - 1)(D - 2D'^2 - 1)z = 0.$

(8) Find a solution of Ex. (4) reducing to 1 when $x = +\infty$ and to h^2 when $x = 0$.

151. Particular Integrals. The methods of obtaining particular integrals of non-homogeneous equations are very similar to those in Chap. III., so we shall merely give a few examples.

Ex. (i). $(D^3 - 3DD' + D + 1)z = e^{2x+3y},$

$$\frac{1}{D^3 - 3DD' + D + 1} \cdot e^{2x+3y} = \frac{e^{2x+3y}}{2^3 - 3 \cdot 2 \cdot 3 + 2 + 1} = -\tfrac{1}{7}e^{2x+3y}.$$

Hence $\qquad z = -\frac{1}{7}e^{2x+3y} + \Sigma Ae^{hx+ky},$

where $\qquad h^3 - 3hk + h + 1 = 0.$

Ex. (ii). $\qquad (D + D' - 1)(D + 2D' - 3)z = 4 + 3x + 6y.$

$$\frac{1}{D + D' - 1} \cdot \frac{1}{D + 2D' - 3} = \tfrac{1}{3}\{1 - (D + D')\}^{-1}\left\{1 - \frac{D + 2D'}{3}\right\}^{-1}$$

$$= \tfrac{1}{3}\{1 + D + D' + \text{terms of higher degree}\}$$

$$\times \left\{1 + \frac{D + 2D'}{3} + \text{terms of higher degree}\right\}$$

$$= \tfrac{1}{3}\left\{1 + \frac{4D + 5D'}{3} + \text{terms of higher degree}\right\}.$$

Acting on $4 + 3x + 6y$, this operator gives

$$\tfrac{1}{3}\{4 + 3x + 6y + 4 + 10\} = 6 + x + 2y.$$

Hence $\qquad z = 6 + x + 2y + e^x f(y - x) + e^{3x} F(y - 2x).$

Ex. (iii). $\qquad (D^2 - DD' - 2D)z = \sin(3x + 4y).$

$$\frac{1}{D^2 - DD' - 2D} \cdot \sin(3x + 4y) = \frac{1}{-3^2 - (-3 \cdot 4) - 2D} \cdot \sin(3x + 4y)$$

$$= \frac{1}{3 - 2D} \cdot \sin(3x + 4y)$$

$$= \frac{3 + 2D}{9 - 4D^2} \cdot \sin(3x + 4y) = \frac{3\sin(3x + 4y) + 6\cos(3x + 4y)}{9 - 4(-3^2)}$$

$$= \tfrac{1}{15}\sin(3x + 4y) + \tfrac{2}{15}\cos(3x + 4y).$$

Hence $\quad z = \tfrac{1}{15}\sin(3x + 4y) + \tfrac{2}{15}\cos(3x + 4y) + \Sigma Ae^{hx+ky},$

where $\qquad h^2 - hk - 2h = 0.$

Examples for solution

(1) $(D - D' - 1)(D - D' - 2)z = e^{2x-y}.$

(2) $s + p - q = z + xy.$ $\qquad$ (3) $(D - D'^2)z = \cos(x - 3y).$

(4) $r - s + p = 1.$ $\qquad$ (5) $\dfrac{\partial^2 y}{\partial x^2} - \dfrac{\partial^2 y}{\partial z^2} = y + e^{x+z}.$

(6) $(D - 3D' - 2)^2 z = 2e^{2x}\tan(y + 3x).$

152. Examples in elimination. We shall now consider the result of eliminating an arbitrary function from a partial differential equation of the first order.

Ex. (i). $\qquad 2px - qy = \phi(x^2y).$

Differentiating partially, first with respect to x and then to y, we get

$$2rx - sy + 2p = 2xy\phi'(x^2y),$$

and $\qquad 2sx - ty - q = x^2\phi'(x^2y),$

whence $\qquad x(2rx - sy + 2p) = 2y(2sx - ty - q)$

or $\qquad 2x^2r - 5xys + 2y^2t + 2(px + qy) = 0,$

which is of the first degree in r, s, t.

The same equation results from eliminating ψ from
$$px - 2qy = \psi(xy^2).$$

Ex. (ii). $\qquad\qquad p^2 + q = \phi(2x + y).$

This gives $\qquad\qquad 2pr + s = 2\phi'(2x + y),$

and $\qquad\qquad\qquad 2ps + t = \phi'(2x + y),$

whence $\qquad\qquad\quad 2pr + s = 4ps + 2t,$

again of the first degree in r, s, t.

Ex. (iii). $\qquad\qquad y - p = \phi(x - q).$

This gives $\qquad\qquad -r = (1 - s)\phi'(x - q),$

and $\qquad\qquad\quad 1 - s = -t\phi'(x - q),$

whence $\qquad\qquad\quad rt = (1 - s)^2$

or $\qquad\qquad\qquad 2s + (rt - s^2) = 1.$

This example differs from the other two in that p and q occur in the arbitrary function as well as elsewhere. The result contains a term in $\qquad\qquad\qquad (rt - s^2).$

Examples for solution

Eliminate the arbitrary function from the following:

(1) $py - q + 3y^2 = \phi(2x + y^2).$ $\qquad$ (2) $x - \dfrac{1}{q} = \phi(z).$

(3) $p + x - y = \phi(q - x + y).$ $\qquad$ (4) $px + qy = \phi(p^2 + q^2).$

(5) $p^2 - x = \phi(q^2 - 2y).$ $\qquad\qquad$ (6) $p + zq = \phi(z).$

153. Generalisation of the preceding results. If u and v are known functions of x, y, z, p, q, and we treat the equation $u = \phi(v)$ as before, we get

$$r\frac{\partial u}{\partial p} + s\frac{\partial u}{\partial q} + \frac{\partial u}{\partial x} + p\frac{\partial u}{\partial z} = \left(r\frac{\partial v}{\partial p} + s\frac{\partial v}{\partial q} + \frac{\partial v}{\partial x} + p\frac{\partial v}{\partial z}\right) . \phi'(v),$$

and $\quad s\dfrac{\partial u}{\partial p} + t\dfrac{\partial u}{\partial q} + \dfrac{\partial u}{\partial y} + q\dfrac{\partial u}{\partial z} = \left(s\dfrac{\partial v}{\partial p} + t\dfrac{\partial v}{\partial q} + \dfrac{\partial v}{\partial y} + q\dfrac{\partial v}{\partial z}\right) . \phi'(v).$

Eliminating $\phi'(v)$ we find that the terms in rs and st cancel out, leaving a result of the form

$$Rr + Ss + Tt + U(rt - s^2) = V,$$

where R, S, T, U and V involve p, q, and the partial differential coefficients of u and v with respect to x, y, z, p, q.

The coefficient $\qquad U = \dfrac{\partial u}{\partial p}\dfrac{\partial v}{\partial q} - \dfrac{\partial v}{\partial p}\dfrac{\partial u}{\partial q},$

which vanishes if v is a function of x, y, z only and not of p or q.

These results will show us what to expect when we start with the equations of the second order and try to obtain equations of the first order from them.

154. Monge's method of integrating $Rr + Ss + Tt = V$. We shall now consider equations of the first degree in r, s, t, whose coefficients R, S, T, V are functions of p, q, x, y, z, and try to reverse the process of Arts. 152 and 153.

Since $$dp = \frac{\partial p}{\partial x} dx + \frac{\partial p}{\partial y} dy = r\, dx + s\, dy$$

and $$dq = s\, dx + t\, dy,$$

$$Rr + Ss + Tt - V = 0$$

becomes $$R\left(\frac{dp - s\, dy}{dx}\right) + Ss + T\left(\frac{dq - s\, dx}{dy}\right) - V = 0$$

i.e. $R\, dp\, dy + T\, dq\, dx - V\, dy\, dx - s(R\, dy^2 - S\, dy\, dx + T\, dx^2) = 0.$

The chief feature of Monge's method is obtaining one or two relations between p, q, x, y, z (each relation involving an arbitrary function) to satisfy the simultaneous equations

$$R\, dy^2 - S\, dy\, dx + T\, dx^2 = 0,$$

$$R\, dp\, dy + T\, dq\, dx - V\, dy\, dx = 0.$$

These relations are called Intermediate Integrals.

The method of procedure will be best understood by studying worked examples.

Ex. (i). $2x^2 r - 5xys + 2y^2 t + 2(px + qy) = 0.$

Proceeding as above, we obtain the simultaneous equations

$$2x^2\, dy^2 + 5xy\, dy\, dx + 2y^2\, dx^2 = 0, \quad\quad\quad\quad\quad\ldots\ldots\ldots\ldots\ldots\ldots(1)$$

and $2x^2\, dp\, dy + 2y^2\, dq\, dx + 2(px + qy)\, dy\, dx = 0. \ldots\ldots\ldots\ldots(2)$

(1) gives $(x\, dy + 2y\, dx)(2x\, dy + y\, dx) = 0,$

i.e. $x^2 y = a$ or $xy^2 = b.$

If we take $x^2 y = a$ and divide each term of (2) by $x\, dy$ or its equivalent $-2y\, dx$, we get

$$2x\, dp - y\, dq + 2p\, dx - q\, dy = 0,$$

i.e. $2px - qy = c.$

This, in conjunction with $x^2 y = a$, suggests the intermediate integral

$$2px - qy = \phi(x^2 y), \quad\quad\quad\quad\quad\quad\quad\ldots\ldots\ldots\ldots\ldots\ldots\ldots\ldots(3)$$

where ϕ is an arbitrary function.   [Cf. Ex. (i) of Art. 152.]

Similarly $xy^2 = b$ and equation (2) leads to

$$px - 2qy = \psi(xy^2). \quad\quad\quad\quad\quad\quad\quad\ldots\ldots\ldots\ldots\ldots\ldots\ldots\ldots(4)$$

Solving (3) and (4),

$$3px = 2\phi(x^2 y) - \psi(xy^2),$$

$$3qy = \phi(x^2 y) - 2\psi(xy^2),$$

so $dz = p\, dx + q\, dy = \frac{1}{3}\phi\,(x^2y)\,.\,\left(\dfrac{2dx}{x} + \dfrac{dy}{y}\right) - \frac{1}{3}\psi\,(xy^2)\,.\,\left(\dfrac{dx}{x} + \dfrac{2dy}{y}\right),$

i.e. $z = \frac{1}{3}\displaystyle\int \phi\,(x^2y)\,.\,d\log\,(x^2y) - \frac{1}{3}\displaystyle\int \psi\,(xy^2)\,.\,d\log\,(xy^2),$

or $z = f(x^2y) + F(xy^2).$

Ex. (ii). $y^2r - 2ys + t = p + 6y.$

Eliminating r and t as before, we are led to the simultaneous equations

$$y^2\,dy^2 + 2y\,dy\,dx + dx^2 = 0, \quad\dots\dots\dots\dots\dots\dots(5)$$

and

$$y^2\,dp\,dy + dq\,dx - (p + 6y)\,dy\,dx = 0. \quad\dots\dots\dots\dots(6)$$

(5) gives $(y\,dy + dx)^2 = 0,$

i.e. $2x + y^2 = a.$

Using this integral and dividing each term of (6) by $y\,dy$ or its equivalent $-dx$, we get

$$y\,dp - dq + (p + 6y)\,dy = 0,$$

i.e. $py - q + 3y^2 = c.$

This suggests the intermediate integral

$$py - q + 3y^2 = \phi\,(2x + y^2).$$

As we have only one intermediate integral, we must integrate this by Lagrange's method.

The subsidiary equations are

$$\frac{dx}{y} = \frac{dy}{-1} = \frac{dz}{-3y^2 + \phi\,(2x + y^2)}.$$

One integral is $2x + y^2 = a.$ Using this to find another,

$$dz + \{-3y^2 + \phi\,(a)\}\,dy = 0,$$

i.e. $z - y^3 + y\phi\,(2x + y^2) = b.$

Hence the general integral is

$$\psi\{z - y^3 + y\phi\,(2x + y^2),\ 2x + y^2\} = 0,$$

or $z = y^3 - y\phi\,(2x + y^2) + f(2x + y^2).$

Ex. (iii). $pt - qs = q^3.$

The simultaneous equations are

$$q\,dy\,dx + p\,dx^2 = 0, \quad\dots\dots\dots\dots\dots\dots\dots(7)$$

and $p\,dq\,dx - q^3\,dy\,dx = 0. \quad\dots\dots\dots\dots\dots\dots(8)$

(7) gives $dx = 0\quad\text{or}\quad q\,dy + p\,dx\ (= dz) = 0,$

i.e. $x = a\quad\text{or}\quad z = b.$

If $dx = 0$ (8) reduces to $0 = 0.$

If $z = b$, $q\,dy = -p\,dx$ and (8) reduces to

$$p\,dq + q^2p\,dx = 0,$$

i.e. $dq/q^2 + dx = 0,$

giving $-\dfrac{1}{q} + x = c = \psi\,(z). \quad\dots\dots\dots\dots\dots(9)$

(9) may be integrated by Lagrange's method, but a shorter way is to rewrite it

$$\frac{\partial y}{\partial z} = \frac{1}{q} = x - \psi(z),$$

giving

$$y = xz - \int \psi(z)\, dz + F(x),$$

$$y = xz + f(z) + F(x).$$

Examples for solution

(1) $r - t \cos^2 x + p \tan x = 0.$

(2) $(x - y)(xr - xs - ys + yt) = (x + y)(p - q).$

(3) $(q + 1)s = (p + 1)t.$ (4) $t - r \sec^4 y = 2q \tan y.$

(5) $xy(t - r) + (x^2 - y^2)(s - 2) = py - qx.$

(6) $(1 + q)^2 r - 2(1 + p + q + pq)s + (1 + p)^2 t = 0.$

(7) Find a surface satisfying $2x^2 r - 5xys + 2y^2 t + 2(px + qy) = 0$ and touching the hyperbolic paraboloid $z = x^2 - y^2$ along its section by the plane $y = 1$.

(8) Obtain the integral of $q^2 r - 2pqs + p^2 t = 0$ in the form

$$y + xf(z) = F(z),$$

and show that this represents a surface generated by straight lines that are all parallel to a fixed plane.

* 155. Monge's method of integrating $\mathbf{Rr + Ss + Tt + U(rt - s^2) = V}$. As before, the coefficients R, S, T, U, V are functions of p, q, x, y, z.

The process of solution falls naturally into two parts:

         (i) the formation of intermediate integrals;

        (ii) the further integration of these integrals.

For the sake of clearness we shall consider these two parts separately.

156. Formation of intermediate integrals. As in Art. 154,

$$r = (dp - s\, dy)/dx$$

and

$$t = (dq - s\, dx)/dy.$$

Substitute for r and t in

$$Rr + Ss + Tt + U(rt - s^2) = V,$$

multiply up by dx and dy (to clear of fractions), and we get

$$R\, dp\, dy + T\, dq\, dx + U\, dp\, dq - V\, dx\, dy$$

$$- s(R\, dy^2 - S\, dx\, dy + T\, dx^2 + U\, dp\, dx + U\, dq\, dy) = 0,$$

say $N - sM = 0$.

---

* The remainder of this chapter should be omitted on a first reading. This extension of Monge's ideas is due to André Marie Ampère, of Lyons (1775-1836), whose name has been given to the unit of electric current.

We now try to obtain solutions of the simultaneous equations
$$M = 0,$$
$$N = 0.$$
So far we have imitated the methods employed in Art. 154, but we cannot now factorise M as we did before, on account of the presence of the terms $U\,dp\,dx + U\,dq\,dy$.

As there is no hope of factorising M or N separately, let us try to factorise $M + \lambda N$, where λ is some multiplier to be determined later.

Writing M and N in full, the expression to be factorised is
$$R\,dy^2 + T\,dx^2 - (S + \lambda V)\,dx\,dy + U\,dp\,dx + U\,dq\,dy$$
$$+ \lambda R\,dp\,dy + \lambda T\,dq\,dx + \lambda U\,dp\,dq.$$
As there are no terms in dp^2 or dq^2, dp can only appear in one factor and dq in the other.

Suppose the factors are
$$A\,dy + B\,dx + C\,dp \quad \text{and} \quad E\,dy + F\,dx + G\,dq.$$
Then equating coefficients of dy^2, dx^2, $dp\,dq$,
$$AE = R; \quad BF = T; \quad CG = \lambda U.$$
We may take
$$A = R, \ E = 1, \ B = kT, \ F = 1/k, \ C = mU, \ G = \lambda/m.$$
Equating the coefficients of the other five terms, we get
$$kT + R/k = -(S + \lambda V), \quad\quad\quad\quad\quad (1)$$
$$\lambda R/m = U, \quad\quad\quad\quad\quad\quad\quad (2)$$
$$kT\lambda/m = \lambda T, \quad\quad\quad\quad\quad\quad (3)$$
$$mU = \lambda R, \quad\quad\quad\quad\quad\quad\quad (4)$$
$$mU/k = U. \quad\quad\quad\quad\quad\quad\quad (5)$$
From (5), $m = k$, and this satisfies (3).

From (2) or (4), $m = \lambda R/U$.

Hence, from (1),
$$\lambda^2(RT + UV) + \lambda US + U^2 = 0. \quad\quad\quad\quad\quad (6)$$
So if λ is a root of (6), the factors required are
$$\left(R\,dy + \lambda\frac{RT}{U}\,dx + \lambda R\,dp \right) \left(dy + \frac{U}{\lambda R}\,dx + \frac{U}{R}\,dq \right),$$

i.e. $\dfrac{R}{U}(U\,dy + \lambda T\,dx + \lambda U\,dp) \cdot \dfrac{1}{\lambda R}(\lambda R\,dy + U\,dx + \lambda U\,dq).$

We shall therefore try to obtain integrals from the linear equations $U\,dy + \lambda T\,dx + \lambda U\,dp = 0 \quad\quad\quad\quad (7)$

and $\lambda R\,dy + U\,dx + \lambda U\,dq = 0, \quad\quad\quad\quad (8)$

where λ satisfies (6).

The rest of the procedure will be best understood from worked examples.

157. Examples

Ex. (i). $\qquad 2s + (rt - s^2) = 1.$

Substituting $R = T = 0$, $S = 2$, $U = V = 1$ in equation (6) of the last article,* we get $\qquad \lambda^2 + 2\lambda + 1 = 0,$

a quadratic with equal roots -1 and -1.

With $\lambda = -1$, equations (7) and (8) give

$$dy - dp = 0,$$
$$dx - dq = 0,$$

of which obvious integrals are

$$y - p = \text{const.}$$

and $\qquad\qquad x - q = \text{const.}$

Combining these as in Art. 154, we get the intermediate integral

$$y - p = f(x - q).$$

Ex. (ii). $\qquad r + 3s + t + (rt - s^2) = 1.$

The quadratic in λ comes to

$$2\lambda^2 + 3\lambda + 1 = 0,$$

so $\lambda = -1$ or $-\frac{1}{2}$.

With $\lambda = -1$, equations (7) and (8) give

$$dy - dx - dp = 0,$$
$$-dy + dx - dq = 0,$$

of which obvious integrals are

$$p + x - y = \text{const.} \quad\dots\dots\dots\dots\dots\dots\dots(1)$$

and $\qquad\qquad q - x + y = \text{const.} \quad\dots\dots\dots\dots\dots\dots\dots(2)$

Similarly $\lambda = -\frac{1}{2}$ leads to

$$p + x - 2y = \text{const.} \quad\dots\dots\dots\dots\dots\dots\dots(3)$$

and $\qquad\qquad q - 2x + y = \text{const.} \quad\dots\dots\dots\dots\dots\dots\dots(4)$

In what pairs shall we combine these four integrals?

Consider again the simultaneous equations denoted by $M = 0$, $N = 0$ in the last article. If these are both satisfied, then $M + \lambda_1 N = 0$ and $M + \lambda_2 N = 0$ are also *both* satisfied (where λ_1 and λ_2 are the roots of the quadratic in λ). Therefore one of the linear factors vanishes for $\lambda = \lambda_1$ and one (obviously the *other* one, or else $dy = 0$) for $\lambda = \lambda_2$.

That is, we combine integrals (1) and (4), and also (2) and (3), giving the two intermediate integrals

$$p + x - y = f(q - 2x + y)$$

and $\qquad\qquad p + x - 2y = F(q - x + y).$

---

* We quote the results of the last article to save space, but the student is advised to work each example from first principles.

Ex. (iii). $2yr + (px + qy)s + xt - xy(rt - s^2) = 2 - pq.$

The quadratic in λ comes to

$$\lambda^2 xypq - \lambda xy(px + qy) + x^2 y^2 = 0,$$

giving $\lambda = y/p \quad \text{or} \quad x/q.$

Substituting in (7) and (8) of the last article, we get, after a littl
reduction,

$$p\, dy - dx + y\, dp = 0, \quad \dotfill (5$$

$$2y\, dy - px\, dx - xy\, dq = 0, \quad \dotfill (6$$

$$-qy\, dy + x\, dx - xy\, dp = 0, \quad \dotfill (7$$

and $-2\, dy + q\, dx + x\, dq = 0.$ $\dotfill (8$

Combining the obvious integrals of (5) and (8), we get

$$yp - x = f(-2y + qx).$$

But (6) and (7) are non-integrable. This may be seen from the
way that p and q occur in them. Thus, although the quadratic in λ ha
two different roots, we get only one intermediate integral.

Examples for solution

Obtain an intermediate integral (or two if possible) of the following:

(1) $3r + 4s + t + (rt - s^2) = 1.$ (2) $r + t - (rt - s^2) = 1.$

(3) $2r + te^x - (rt - s^2) = 2e^x.$ (4) $rt - s^2 + 1 = 0.$

(5) $3s + (rt - s^2) = 2.$

(6) $qxr + (x + y)s + pyt + xy(rt - s^2) = 1 - pq.$

(7) $(q^2 - 1)zr - 2pqzs + (p^2 - 1)zt + z^2(rt - s^2) = p^2 + q^2 - 1.$

158. Further integration of intermediate integrals

Ex. (i). Consider the intermediate integral obtained in Ex. (i) o
Art. 157, $y - p = f(x - q).$

We can obtain a " complete " integral involving arbitrary constant
a, b, c by putting $x - q = a$

and $y - p = f(a) = b,$ say.

Hence $dz = p\, dx + q\, dy = (y - b)\, dx + (x - a)\, dy$

and $z = xy - bx - ay + c.$

An integral of a more general form can be obtained by supposing
the arbitrary function f occurring in the intermediate integral to be
linear, giving $y - p = m(x - q) + n.$

Integrating this by Lagrange's method, we get

$$z = xy + \phi(y + mx) - nx.$$

Ex. (ii). Consider the two intermediate integrals of Ex. (ii), Art. 157.

$$p + x - y = f(q - 2x + y)$$

and $p + x - 2y = F(q - x + y).$

If we attempt to deal with these simultaneous equations as we dealt with the single equation in Ex. (i), we get

$$q - 2x + y = \alpha,$$
$$q - x + y = \beta,$$
$$p + x - y = f(\alpha),$$
$$p + x - 2y = F(\beta).$$

If the terms on the right-hand side are constants, we get the absurd result that x, y, p, q are all constants!

But now suppose that α and β are not *constants*, but *parameters*, capable of variation.

Solving the four equations, we get

$$x = \beta - \alpha,$$
$$y = f(\alpha) - F(\beta),$$
$$p = y - x + f(\alpha),$$
$$q = x - y + \beta,$$

giving
$$dz = p\,dx + q\,dy$$
$$= (y - x)(dx - dy) + f(\alpha)\,dx + \beta\,dy$$
$$= -\tfrac{1}{2}d(x-y)^2 + f(\alpha)\,d\beta - f(\alpha)\,d\alpha + \beta f'(\alpha)\,d\alpha - \beta F'(\beta)\,d\beta;$$

i.e.
$$z = -\tfrac{1}{2}(x-y)^2 - \int f(\alpha)\,d\alpha - \int \beta F'(\beta)\,d\beta + \beta f(\alpha).$$

To obtain a result free from symbols of integration, put

$$\int f(\alpha)\,d\alpha = \phi(\alpha) \quad \text{and} \quad \int F(\beta)\,d\beta = \psi(\beta).$$

Now $\displaystyle\int \beta F'(\beta)\,d\beta = \beta F(\beta) - \int F(\beta)\,d\beta$, integrating by parts,

$$= \beta\psi'(\beta) - \psi(\beta)$$

Hence $z = -\tfrac{1}{2}(x-y)^2 - \phi(\alpha) - \beta\psi'(\beta) + \psi(\beta) + \beta\phi'(\alpha),$

or finally
$$\begin{cases} z = -\tfrac{1}{2}(x-y)^2 - \phi(\alpha) + \psi(\beta) + \beta y, \\ x = \beta - \alpha, \\ y = \phi'(\alpha) - \psi'(\beta). \end{cases}$$

These three equations constitute the parametric form of the equation of a surface. As the solution contains two arbitrary functions, it may be regarded as of the most general form possible.

Examples for solution (completing the solution of the preceding set).

Integrate by the methods explained above:

(1) $p + x - 2y = f(q - 2x + 3y).$ (2) $p - x = f(q - y).$

(3) $p - e^x = f(q - 2y).$ (4) $p - y = f(q + x),$
$\qquad\qquad\qquad\qquad\qquad\qquad p + y = F(q - x).$

(5) $p - y = f(q - 2x),$ (6) $px - y = f(qy - x).$
$\quad\; p - 2y = F(q - x).$ (7) $(zp - x) = f(zq - y).$

(8) Obtain a particular solution of (4) by putting $\phi(\alpha) = -\tfrac{1}{2}\alpha^2$ $\psi(\beta) = \tfrac{1}{2}\beta^2$ and eliminating α and β.

MISCELLANEOUS EXAMPLES ON CHAPTER XIV

(1) $r = 2y^2$.   (2) $\log s = x + y$.   (3) $2yq + y^2 t = 1$.

(4) $r - 2s + t = \sin(2x + 3y)$.   (5) $x^2 r - 2xs + t + q = 0$.

(6) $rx^2 - 3sxy + 2ty^2 + px + 2qy = x + 2y$.

(7) $y^2 r + 2xys + x^2 t + px + qy = 0$.

(8) $5r + 6s + 3t + 2(rt - s^2) + 3 = 0$.

(9) $2pr + 2qt - 4pq(rt - s^2) = 1$.

(10) $rt - s^2 - s(\sin x + \sin y) = \sin x \sin y$.

(11) $7r - 8s - 3t + (rt - s^2) = 36$.

(12) Find a surface satisfying $r = 6x + 2$ and touching $z = x^3 + y^3$ along its section by the plane $x + y + 1 = 0$.

(13) Find a surface satisfying $r - 2s + t = 6$ and touching the hyperbolic paraboloid $z = xy$ along its section by the plane $y = x$.

(14) A surface is drawn satisfying $r + t = 0$ and touching $x^2 + z^2 = 1$ along its section by $y = 0$. Obtain its equation in the form
$$z^2(x^2 + z^2 - 1) = y^2(x^2 + z^2).$$
    [London.]

(15) Show that of the four linear differential equations in x, y, p, q obtained by the application of Monge's method to
$$2r + qs + xt - x(rt - s^2) = 2,$$
two are integrable, leading to the intermediate integral
$$p - x = f(qx - 2y),$$
while the other two, although non-integrable singly, can be combined to give the integral $p + \tfrac{1}{4}q^2 - x = a$.

Hence obtain the solutions
$$z = \tfrac{1}{2}x^2 - 2mxy - \tfrac{2}{3}m^2 x^3 + nx + \phi(y + \tfrac{1}{2}mx^2)$$
and $z = (a - \tfrac{1}{4}b^2)x + \tfrac{1}{2}x^2 + by + c$,
and show that one is a particular case of the other.

(16) A surface is such that its section by any plane parallel to $x = 0$ is a circle passing through the axis of x. Prove that it satisfies the functional and differential equations
$$y^2 + z^2 + yf(x) + zF(x) = 0,$$
$$(y^2 + z^2)t + 2(z - yq)(1 + q^2) = 0.$$

(17) Obtain the solution of $x^2 r + 2xys + y^2 t = 0$ in the form
$$z = f\left(\frac{y}{x}\right) + xF\left(\frac{y}{x}\right),$$
and show that this represents a surface generated by lines that intersect the axis of z.

(18) Show that $rt - s^2 = 0$ leads to the "complete" integral
$$z = ax + by + c.$$

Show that the " general " integral derived from this (as in Art. 134) represents a developable surface (see Smith's *Solid Geometry*, Arts. 222-223).

Hence show that for any developable surface $q = f(p)$.

(19) Find the developable surfaces that satisfy
$$pq(r-t) - (p^2 - q^2)s + (py - qx)(rt - s^2) = 0.$$

[Assume $q = f(p)$. This is called Poisson's method. We get
$$q = ap \quad \text{or} \quad p^2 + q^2 = b^2,$$
giving
$$z = \phi(x + ay) \quad \text{or} \quad z = bx \cos \alpha + by \sin \alpha + c.$$

The second of these integrals represents a plane which generates the developable surface given by the corresponding "general" integral.]

(20) Show that if
$$X = p, \quad Y = q, \quad Z = px + qy - z,$$
then
$$r = T/(RT - S^2), \quad s = -S/(RT - S^2), \quad t = R/(RT - S^2),$$
where
$$R = \frac{\partial^2 Z}{\partial X^2}, \text{ etc.}$$

Hence show that the equation
$$ar + bs + ct + e(rt - s^2) = 0$$
transforms into
$$AT - BS + CR + E = 0,$$
where a, b, c, e are any functions of x, y, p, q, and A, B, C, E the corresponding functions of P, Q, X, Y.

Apply this *Principle of Duality* (cf. No. 21 of the Miscellaneous Examples at the end of Chap. XII.) to derive two intermediate integrals of
$$pq(r - t) - (p^2 - q^2)s + (py - qx)(rt - s^2) = 0.$$

(21) Prove that if x, y, u, v are real and $u + iv = f(x + iy)$, then $V = u$ and $V = v$ are both solutions of
$$\frac{\partial^2 V}{\partial x^2} + \frac{\partial^2 V}{\partial y^2} = 0,$$
and the two systems of curves $u = \text{const.}$,
$$v = \text{const.},$$
are mutually orthogonal.

Verify these properties for the particular cases
(i) $u + iv = x + iy$,
(ii) $u + iv = (x + iy)^2$,
(iii) $u + iv = 1/(x + iy)$.

[The differential equation is the two-dimensional form of Laplace's equation, which is of fundamental importance in gravitation, electrostatics and hydrodynamics. u and v are called *Conjugate Functions*. See Ramsey's *Hydromechanics*, Vol. II. Art. 41.]

(22) Obtain the solution of
$$\frac{\partial^2 y}{\partial t^2} = a^2 \frac{\partial^2 y}{\partial x^2},$$

subject to the conditions $y = f(x)$ and $\dfrac{\partial y}{\partial t} = F(x)$ when $t = 0$, in the form

$$y = \tfrac{1}{2}f(x + at) + \tfrac{1}{2}f(x - at) + \frac{1}{2a}\int_{x-at}^{x+at} F(\lambda)\,d\lambda.$$

[y is the transverse displacement of any point x of a vibrating string of infinite length, whose initial displacement and velocity are given by $f(x)$ and $F(x)$. See Ramsey's *Hydromechanics*, Vol. II. Art. 248.]

(23) If $y = f(x) \cos(nt + \alpha)$ is a solution of

$$\frac{\partial^2 y}{\partial t^2} + a^4 \frac{\partial^4 y}{\partial x^4} = 0,$$

show that $f(x) = A \sin mx + B \cos mx + H \sinh mx + K \cosh mx$, where

$$m = \sqrt{(n/a^2)}.$$

[The differential equation is that approximately satisfied by the lateral vibrations of bars, neglecting rotatory inertia. See Rayleigh's *Sound*, Art. 163.]

(24) Show that

$$w = A \sin(m\pi x/a) \sin(n\pi y/b) \cos(pct + \alpha)$$

satisfies
$$\frac{\partial^2 w}{\partial t^2} = c^2\left(\frac{\partial^2 w}{\partial x^2} + \frac{\partial^2 w}{\partial y^2}\right),$$

and vanishes when

$$x = 0, \quad y = 0, \quad x = a \quad \text{or} \quad y = b,$$

provided that m and n are positive integers satisfying

$$(p/\pi)^2 = (m/a)^2 + (n/b)^2.$$

[This gives one solution of the differential equation of a vibrating membrane with a fixed rectangular boundary. See Rayleigh's *Sound*, Arts. 194-199.]

(25) Show that $\qquad w = A J_0(nr) \cos(nct + \alpha)$

satisfies
$$\frac{\partial^2 w}{\partial t^2} = c^2\left(\frac{\partial^2 w}{\partial r^2} + \frac{1}{r}\frac{\partial w}{\partial r}\right),$$

where J_0 is Bessel's function of order zero (see Ex. 2 of the set following Art. 97).

[This refers to a vibrating membrane with a fixed circular boundary. See Rayleigh's *Sound*, Arts. 200-206.]

(26) Show that $\qquad V = (Ar^n + Br^{-n-1}) P_n(\cos\theta)$

satisfies
$$\frac{\partial^2 V}{\partial r^2} + \frac{2}{r}\frac{\partial V}{\partial r} + \frac{1}{r^2}\frac{\partial^2 V}{\partial \theta^2} + \frac{\cot\theta}{r^2}\frac{\partial V}{\partial \theta} = 0,$$

where P_n is Legendre's function of order n (for Legendre's equation, see Ex. 2 of the set following Art. 99).

[*N.B.* Take $\mu = \cos\theta$ as a new variable. This equation is the form taken by Laplace's potential equation in three dimensions, when V is known to be symmetrical about an axis. See Routh's *Analytical Statics*, Vol. II. Art. 300.]

CHAPTER XV

MISCELLANEOUS METHODS

159. This chapter consists of six sections. The first (Arts. 160-161) is supplementary to Chap. VI., and deals with some difficulties in the theory of singular solutions, especially the definition of an envelope and the way in which particular solutions may occur in the discriminants. The conception of discriminant-loci as *boundaries* appears to be very little known.

The second section (Arts. 162-167) deals with Riccati's equation, chiefly in its generalised form. The examples include a series which indicate in what cases Riccati's original equation can be integrated in finite terms.

The third section (Arts. 168-170) deals with total differential equations, and is supplementary to Chap. XI. The use of an integrating factor for homogeneous equations will appeal to the elementary student, while Mayer's method is of great interest from the point of view of theory.

The fourth section (Arts. 171-177) deals with linear differential equations of the second order and their solution by series. It is supplementary to Chaps. IX. and X. A few results concerning equations of higher order are included.

The fifth section (Arts. 178-181) deals with some equations of Mathematical Physics, particularly those concerned with wave-motion. It is supplementary to Chaps. IV. and XIV.

Finally the sixth section (Arts. 182-183) deals with numerical approximations to the solution of differential equations (supplementary to Chap. VIII.). After describing the method of Adams, perhaps the best that has yet been devised, it gives a summary of some extensions (due to E. Remes) of the author's method (*i.e.* that of Arts. 90-93).

160. Some difficulties in the theory of singular solutions.* We shall now supplement Chap. VI. by pointing out some difficulties concerning envelopes, singular solutions, and particular integrals.

The old definition of an envelope of a family of curves, as *the locus of the ultimate intersections of consecutive curves*, must be abandoned, for it has been found to lead to the ridiculous conclusion that a curve is not the envelope of its own circles of curvature.† De la Vallée Poussin's definition is *the locus of the isolated characteristic points (i.e. of ordinary points on a curve whose distances from neighbouring curves are small to an order beyond the first)*. However, it has been pointed out that this is still unsatisfactory in certain respects.‡ For our purposes the most convenient definition appears to be *a curve which touches every member of the family, and which, at each point, is touched by some member of the family*. This agrees with the definition given on p. 66; the second part of the definition was not explicitly stated there, but it was implied by the following sentence.

There are at least three different definitions of a singular solution. Our definition (p. 66) is that it is *a solution corresponding to an envelope § of the family of curves represented by the complete primitive*. However, in exceptional cases the envelope is also a particular curve of the family. Thus the parabola $y = c(x-c)^2$ touches the line $y = 0$ at the point $(c, 0)$, so $y = 0$ is the envelope of the family obtained by giving all possible non-zero values to c, as well as the particular curve given by $c = 0$. In accordance with our definition, $y = 0$ must be considered to be both a singular solution and a particular integral of the differential equation of the family (Ex. 6, p. 76). But some prefer to confine the term singular to *a solution which cannot be obtained by giving any constant value to the arbitrary constant occurring*

---

* For envelopes, see Fowler's *Elementary Differential Geometry of Plane Curves*, Chap. V. For singular solutions, see the *Encyklopädie der Mathematischen Wissenschaften* II. A 4a and III. D 8.

† C and C', the centres of curvature corresponding to two neighbouring points P and P' of a curve, lie on the evolute of that curve. The difference between the radii of curvature CP and $C'P'$ is the arc CC' of the evolute. This arc is in general greater than the chord CC', i.e. greater than the distance between the centres of curvature. Thus one circle of curvature completely encloses the other, and there are no real intersections. For other cases where the old definition fails, see Ex. 13, following Art. 161.

‡ Neville, *Proc. Camb. Phil. Soc.*, Vol. XXI. p. 97, 1922.

§ But see the end of this article for the exceptional case of envelopes parallel to the axis of y.

n the complete primitive. A third definition * of a singular solution s that it is *a solution which occurs in the p-discriminant.* It will be shown in Art. 161 that such a solution need not represent an envelope. It may be a particular solution, or its limiting form.

It is natural for the student to suppose that every family of curves depending on one parameter will possess an envelope and consequently that every differential equation of the first order and of degree higher than the first will possess a singular solution. But this is not the case. In discussing envelopes, it is implicitly assumed that the functions occurring in the equation of the family satisfy certain conditions concerning continuity. These conditions are usually satisfied for the complete primitives of the simple differential equations given in an elementary treatment of singular solutions, but this is due to the fact that in constructing such examples the complete primitives were really taken as the starting point. If we start from the most general differential equation of similar form, there is no reason to suppose that the complete primitive will satisfy the conditions required for the existence of an envelope. In fact, we may say that the existence of a singular solution must be considered as the exception rather than the rule.†

It should be noticed that the usual process for finding envelopes (Art. 56) may fail for one form of the complete primitive, and yet be effective for another. For example, it fails for $x^{\frac{1}{2}} + y^{\frac{1}{2}} = c^{\frac{1}{2}}$, or for $x + \sin^{-1} y = c$, but is effective for

$$(x + y - c)^2 = 4xy, \quad \text{or for} \quad y = \sin(c - x).$$

The equation $x^{\frac{1}{2}} + y^{\frac{1}{2}} = c^{\frac{1}{2}}$, leading to $y = xp^2$, illustrates another point. The differential equation is satisfied by $y = 0$, but hardly by $x = 0$, which, giving $p = \infty$, leaves both sides indeterminate. However, $x = 0$ and $y = 0$ are both envelopes of the family of curves (parabolas touching the axes) and both satisfy $y(dx)^2 = x(dy)^2$, a differential relation which really represents the geometrical facts more accurately than the differential equation. [Cf. Ex. 9, p. 79 and Ex. 11, p. 233. In the first $x = 0$ is a limiting form of a particular curve, and in the second an envelope and also a cusp locus.] In such cases we feel compelled to refuse $x = 0$ a place among

---

* This is the one adopted by most advanced treatises (cf. Ince's *Ordinary Differential Equations*, p. 87, and Bieberbach's *Differentialgleichungen*, p. 85). In quoting results from various sources it is necessary to give the definitions on which they are based, or much confusion may be caused.

† See Ex. 10, following Art. 161.

the solutions, but the rejection may be considered as due to the failure of the differential equation to represent fairly directions parallel to the axis of y, rather than to any peculiarity in the envelope itself.

161. Discriminants, Particular Solutions, and Boundaries. In this article we shall confine ourselves to complete primitives of the form $f(x, y, c) = 0$, where $f(x, y, c)$ is a polynomial in x, y, and c, which may also be written in the form

$$a_0(x,y)c^n + na_1(x,y)c^{n-1} + \tfrac{1}{2}n(n-1)a_2(x,y)c^{n-2} + \ldots + a_n(x,y) = 0.$$

The c-discriminant Δ_c is defined (except for a numerical factor) as the product of a_0^{2n-2} and the squares of the differences of the roots. The a_0^{2n-2} is introduced to make the result a polynomial in a_0, $a_1 \ldots a_n$. Thus for $n = 2, 3, 4$ we get respectively

$$a_0a_2 - a_1^2,$$
$$(a_0a_3 - a_1a_2)^2 - 4(a_0a_2 - a_1^2)(a_1a_3 - a_2^2),$$
$$(a_0a_4 - 4a_1a_3 + 3a_2^2)^3 - 27(a_0a_2a_4 + 2a_1a_2a_3 - a_0a_3^2 - a_1^2a_4 - a_2^3)^2.$$

As in Chap. VI. we shall sometimes use the word *discriminant* to denote, not only the *function* Δ_c, but also the *equation* $\Delta_c = 0$ and the *loci* represented by this equation.

In working examples on singular solutions it is desirable to employ a systematic method of calculating the discriminants. For quadratics, cubics and quartics, the above results may be used. If, as in Art. 56, we obtain Δ_c by elimination, there is a risk that some factors will be overlooked. It is often recommended that Sylvester's *dialytic method* should be used to perform this elimination. To apply this here, we multiply f by c^{n-2}, c^{n-3}, ..., c, 1, and $\partial f/\partial c$ by c^{n-1}, c^{n-2}, ..., c, 1, and then eliminate c^{2n-2}, c^{2n-3}, ..., c, 1 from the $(2n-1)$ equations thus formed, giving a determinant of $(2n-1)$ rows and columns. For the quadratic $a_0c^2 + 2a_1c + a_2 = 0$, this gives

$$\begin{vmatrix} a_0, & 2a_1, & a_2 \\ 2a_0, & 2a_1, & 0 \\ 0, & 2a_0, & 2a_1 \end{vmatrix} = 4a_0(a_0a_2 - a_1^2).$$

But this contains the superfluous factor a_0. It is easy to see that the same superfluous factor will occur whatever the degree of f, giving an expression of degree $(2n-1)$ instead of the proper degree $(2n-2)$. If Sylvester's method is employed for the examples at the end of this article, this factor must be removed.

---

* In using these, remember that the a's are not the actual coefficients, which have also binomial numerical factors; *e.g.* for a quartic the coefficient of c^2 is not a_2, but $6a_2$.

The primary purpose of these examples is to illustrate some ways in which particular solutions or their limiting forms may be given by the c- and p-discriminants. In some cases the solutions occur as merely one part of a particular curve (Ex. 1). Their geometrical significance takes various forms. They may be envelopes and so also singular solutions (Ex. 2), or node-loci (Ex. 3), or cusp-loci (Ex. 4), or tac-loci (Ex. 5), or asymptotes (Ex. 6), or tangents touching all the curves of a family at the same point (Ex. 8). They may be merely lines (not tangents) through a common point of a family (Ex. 7). In connection with Clairaut's form they are furnished (Ex. 9) by the inflexional tangents to the envelope.

It is sometimes stated that when particular solutions occur in the discriminants, they do so to the first power in Δ_c, and cubed in Δ_p. This rule may be combined with those of Art. 64 in the symbolical form: $\Delta_c = EN^2C^3P$, $\Delta_p = ET^2CP^3$, where E, N, C, P, T denote envelope, node-locus, cusp-locus, particular solution, and tac-locus respectively. These rules are useful as suggestions in simple cases, but examples in which they fail are easily constructed (Exs. 3, 4, 6, 13, 14).

We shall now explain the conception of particular solutions and other exceptional loci as boundaries.* We restrict ourselves to the case where $f(x, y, c)$ is a polynomial in x, y, c, and such that corresponding to every pair of real values of x, y we get an equation in c of degree n with, say, m real roots corresponding to real curves, and $(n-m)$ imaginary roots corresponding to imaginary curves. We further stipulate that the roots, which are, of course, functions of x and y, shall vary continuously when x and y do so.

Let a certain curve $B(x, y) = 0$ (not occurring in a multiple form, or made up of a number of simpler curves) be a boundary between two regions, in one of which m has a certain value M and in the other a value $M - 2$. As the point (x, y) travels continuously out of the first region, across the boundary B, into the second, a pair of real unequal roots become less unequal, then equal (on B) and finally (in the second region) conjugate complex. Δ_c, which contains the square of the difference of these roots, must vanish on B and then change sign, as the square of the difference of two conjugate complex

---

* Here and elsewhere I have made considerable use of some valuable suggestions made by Mr. H. B. Mitchell, formerly Professor of Mathematics at Columbia University, New York. However, he must not be held responsible for my treatment, for our points of view are rather different.

roots is negative. $B(x, y)$ must also change sign as (x, y) travels across it. More generally, if m changes from M to $M - 2r$, where r is an odd integer, Δ_c will change sign, and $B(x, y)$ will occur in Δ_c to an odd power (which, however, need not be r; cf. Ex. 14, where $B(x, y)$ occurs cubed, but $r = 1$). If r is an even integer, $B(x, y)$ occurs to an even power. Conversely, if $B(x, y)$ occurs to an odd power, r must be odd. However, if $B(x, y)$ occurs to an even power so that Δ_c does not change sign, r need not be even; it may be zero, as in Ex. 13, where B is an envelope which is crossed by all the curves of the family. In such cases the envelope must occur to an even power, contrary to the rule $\Delta_c = EN^2C^3P$. Similar considerations apply to Δ_p, on replacing the number of real curves through a point by the number of real directions through it. A specially interesting case is that of Clairaut's form (Ex. 9). An inflexional tangent to the envelope corresponds to two equal roots in p, and so leads to $\Delta_p = 0$. As for Clairaut's form $\Delta_c = \Delta_p$, $\Delta_c = 0$ also.

An alternative geometrical method * of investigating singular solutions is to replace p by z, thus converting the differential equation into the algebraic equation of a surface. Similarly, in the complete primitive c may be replaced by z. This method requires a good knowledge of the geometry of surfaces.

The difficulties in the theory of singular solutions are great even for differential equations with coefficients which are polynomials in x and y. When the coefficients are transcendental functions, with singularities of various degrees of complexity, the difficulties are greatly increased.†

Examples for solution

[We shall use C.P., Diff. Eq., Δ_c, Δ_p, and S.S., to denote respectively complete primitive, differential equation, c-discriminant, p-discriminant and singular solution. Δ_c and Δ_p have been obtained from the formulae given above, but numerical factors have been omitted.

The student should draw rough graphs (without calculating exact values of x and y) which will show the form of a few members of each family of curves and their position relative to the loci given by the discriminants.]

(1) Given the C.P. $y(x + c) + c^2 = 0$, obtain the Diff. Eq.
$$x^2p^2 + y(2x - y)p + y^2 = 0,$$
also $\Delta_c = y(4x - y)$, $\Delta_p = y^3(4z - y)$.

---

* *Encyklopädie der Mathematischen Wissenschaften*, III. D 8, or Goursat's *Cours d'Analyse Mathématique*, Vol. II. 4th ed., Art. 435.

† M. J. M. Hill, *Proc. Lond. Math. Soc.*, Series 2, Vol. 17, 1918, p. 149.

[The C.P., for non-zero values of c, represents a family of rectangular hyperbolas. $y = 0$ is an asymptote of all these hyperbolas, and also part of the particular integral $xy = 0$ obtained from the C.P. by putting $c = 0$. $y = 4x$ is an envelope (a S.S.). The rules $\Delta_c = EN^2C^3P$, $\Delta_p = ET^2CP^3$ hold good. The plane can be divided into four regions, in two of which the number of real curves of the family through any point is two, while in the other two regions the number is zero. The boundaries between these regions are the loci given by the discriminants, and both occur to odd powers. This agrees with our theory of boundaries, for in this case $M = 2$, $M - 2r = 0$, so $r = 1$, which is odd.]

(2) Given the C.P. $y = c(x - c)^2$, obtain the Diff. Eq.

$$p^3 - 4xyp + 8y^2 = 0,$$

also $\Delta_c = y(27y - 4x^3)$, $\Delta_p = y^3(27y - 4x^3)$.

[As mentioned in Art. 160, $y = 0$ is an envelope (a S.S.) and also a particular integral. Moreover, it may be regarded as a tac-locus. $27y = 4x^3$ is an envelope. The second and fourth of these geometrical interpretations, but not the first and third, are suggested by the rules $\Delta_c = EN^2C^3P$, $\Delta_p = ET^2CP^3$.]

(3) Given the C.P. $4y^2 = 3c^2x(x - c)^2$, obtain the Diff. Eq.

$$(2px - y)^4 = 3x^5(2px - 3y)^2,$$

also $\Delta_c = x^3y^4(3x^5 - 64y^2)$, $\Delta_p = x^{27}y^4(3x^5 - 64y^2)$.

[The calculation of the discriminants is rather laborious. $y = 0$ is a node-locus as well as a particular solution. $x = 0$ is a common tangent at the origin to all the curves except that for which $c = 0$. (Cf. Ex. 8.) $3x^5 = 64y^2$ is the envelope. To understand why the various factors in the discriminants occur to odd or even powers we notice that $x = 0$ is a boundary between regions where the number of real curves through any point increases from zero to two, while the envelope is the boundary between regions where this number increases from two to four. On $y = 0$ the four coincide in pairs, but on each side of the positive part, between it and a branch of the envelope, the number is the same, namely, four. The rules $\Delta_c = EN^2C^3P$, $\Delta_p = ET^2CP^3$ fail to suggest the geometrical interpretation of the loci $x = 0$, and $y = 0$.]

(4) Given the C.P. $4y^3 = c(3x - c)^2$, obtain the Diff. Eq.

$$yp^3 - 3xp + 2y = 0,$$

also $\Delta_c = y^3(y^3 - x^3)$, $\Delta_p = y(y^3 - x^3)$.

[The C.P., for non-zero values of c, represents a family of semi-cubical parabolas with cusps on $y = 0$, which is a cusp-locus and also a particular solution. $y^3 = x^3$ is an envelope (a S.S.). The rules $\Delta_c = EN^2C^3P$, $\Delta_p = ET^2CP^3$ suggest that $y = 0$ is a cusp-locus, but they fail to indicate that it is also a particular solution.]

(5) Given the C.P. $y^2 = c(3x - c^2)$, obtain the Diff. Eq.

$$8y^2p^3 - 54xp + 27y = 0,$$

also $\Delta_c = y^4 - 4x^3$, $\Delta_p = y^2(y^4 - 4x^3)$.

[The C.P., for non-zero values of c, represents a family of parabola with $y = 0$ as axis, any point of which is the vertex of two such parabola with their concavities turned opposite ways. $y = 0$ is a tac-locus an also a particular solution. $y^4 = 4x^3$ is an envelope (a S.S.). $y^2 = c(3x - c$ touches the envelope at the points $\{c^2, \pm \sqrt{(2c^3)}\}$, which are imaginar if c is negative, and intersects it at $\{\frac{1}{4}c^2, \pm \frac{1}{2}\sqrt{(-c^3)}\}$, which are imaginar if c is positive. The rules suggest the tac-locus, but not the particula solution.]

(6) Show that for all values of m except 0, the complete primitiv of
$$y^{m-2}p^2 = 1 \text{ is } 4y^m = m^2(x+c)^2.$$

Show that for the three cases, m an odd positive integer greater than $m = 1$, and m an odd negative integer, Δ_c and Δ_p are respectively

$$y^m, \quad y, \quad y^{-m},$$
and
$$y^{m-2}, \quad y, \quad y^{2-m},$$

provided that these discriminants are obtained from equations mult plied by the least power of y necessary to get rid of negative powers.
[$y = 0$ is in the first case a cusp-locus, in the second an envelope (S.S.), and in the third the limiting form of a particular solution, which asymptotic to all curves included in the complete primitive. $c = c$ gives $y^{-m} = 0$, if m is negative, so in general this limiting form of particular integral contains the solution $y = 0$ in a multiple form. $m = -1$, but not otherwise, the particular solution occurs to the powe given by the rules $\Delta_c = EN^2C^3P$, $\Delta_p = ET^2CP^3$. The rules give th powers of the cusp-locus correctly only for $m = 3$.]

(7) Given the C.P. $y = x(x+c)^2$, obtain the Diff. Eq.
$$x^2p^2 - 2xyp + y^2 - 4x^3y = 0,$$
also
$$\Delta_c = xy, \quad \Delta_p = x^5y.$$

Show that $y = 0$ is an envelope (a S.S.), and $x = 0$ a limiting form of particular solution, but not itself a solution.
[The vanishing of the discriminants at the origin, a point commo to all the curves of the family, could have been predicted. For sin at the origin the equation of the family is satisfied for any value of the coefficients of every power of c and also the term independent of vanish there, hence $\Delta_c = 0$, for every term in it vanishes. As the curv have different tangents at the common point, the Diff. Eq. is satisfie there for any value of p, so by an argument similar to that f $\Delta_c, \Delta_p = 0$. (Cf. Ex. 7, p. 79).]

(8) Show that for all non-zero values of c, the curves of the famil $y^2 = x(x+c)^2$ touch $x = 0$ at the origin. Obtain the Diff. Eq.
$$4x^2p^2 - 4xyp + y^2 - 4x^3 = 0,$$
also
$$\Delta_c = xy^2, \quad \Delta_p = x^5.$$

Show that $y = 0$ is a node-locus, while $x = 0$ is a limiting form of particular solution (though not itself a solution), and also a line touchin all the curves, except that for which $c = 0$, at one point. (Such a lin does not satisfy our definition of an envelope.)

[As in Ex. 7, Δ_c must vanish at the origin. Δ_p also vanishes though the curves this time have not different tangents). Cf. Ex. 9, 79.]

(9) Show that for the differential equation (of Clairaut's form)

$$(y - px)^2 = p^3,$$

$$\rho = y^3 (27y - 4x^3) = \Delta_c.$$

[$27y = 4x^3$ is the envelope (a S.S.); $y^2 = 0$ is a particular solution, d represents the inflexional tangent to the envelope. Now through y point three tangents to $27y = 4x^3$ can be drawn. All of these are al for the region in the first quadrant between the curve and $y = 0$, also r the similar region in the third quadrant. For the other regions two e imaginary. For a point on $y = 0$ two are coincident, so $y = 0$ must cur in the discriminants. Similarly, whenever the envelope solution any other differential equation of Clairaut's form possesses inflexional ngents, these occur in the discriminants.]

(10) Given a differential equation

$$f(x, y, p) = 0, \quad \dots\dots\dots\dots\dots\dots\dots\dots\dots(1)$$

duce that

$$\frac{\partial f}{\partial x} + p \frac{\partial f}{\partial y} + \frac{dp}{dx} \frac{\partial f}{\partial p} = 0. \quad \dots\dots\dots\dots\dots\dots\dots\dots\dots(2)$$

Hence show that for any point on a solution given by the p-discrimint, for which

$$\frac{\partial f}{\partial p} = 0, \quad \dots\dots\dots\dots\dots\dots\dots\dots\dots(3)$$

$$\frac{\partial f}{\partial x} + p \frac{\partial f}{\partial y} = 0. \quad \dots\dots\dots\dots\dots\dots\dots\dots\dots(4)$$

Equations (1), (3) and (4) are necessary conditions for a singular lution. For Clairaut's form $f(x, y, p) \equiv y - px - F(p)$, so equation (4) satisfied identically. But in general there is no reason why all three ould possess a simultaneous solution, so in general a differential quation has no singular solution.

[Applying this to Ex. (i) on p. 75, we find the three conditions are

$$p^2 (2 - 3y)^2 = 4(1 - y), \quad 2p(2 - 3y)^2 = 0,$$

$$p\{-6p^2 (2 - 3y) + 4\} = 0.$$

$1 - y = 0$, giving $p = 0$, satisfies all three, but $2 - 3y = 0$ does not tisfy the first.]

(11) [In this example the third definition of a singular solution (Art. 60) is to be used. Ex. 10 holds for all three definitions.]

Show that if a curve exists for every point of which the three quations

$$f(x, y, \lambda) = 0, \quad \frac{\partial f}{\partial \lambda} = 0, \quad \frac{\partial f}{\partial x} + \lambda \frac{\partial f}{\partial y} = 0$$

ave a common solution in λ, then along it

$$\frac{\partial f}{\partial x} dx + \frac{\partial f}{\partial y} dy + \frac{\partial f}{\partial \lambda} d\lambda = 0,$$

and hence $\qquad -\lambda \dfrac{\partial f}{\partial y}\, dx + \dfrac{\partial f}{\partial y}\, dy = 0.$

Hence show that if $\dfrac{\partial f}{\partial y} \neq 0$, $\lambda = p$ and the curve is a singular solution

of the differential equation $f(x, y, p) = 0$, while if $\dfrac{\partial f}{\partial y} = 0$, then $\dfrac{\partial f}{\partial x} = 0$ also.

[This shows that the necessary conditions for a singular solution given in Ex. 10, become sufficient by the addition of the condition $\dfrac{\partial f}{\partial y} \neq 0$. But this last condition is not necessary. In Ex. 2, $\dfrac{\partial f}{\partial y} = 16y - 4x$. This is zero for one envelope $y = 0$, but not for the other, $27y = 4x^3$.]

(12) Show that the locus of the points of inflexion of the curve represented by the complete primitive of equation (1) of Ex. 10 satisfies equation (4) of Ex. 10, and hence will be included in the result obtained by eliminating p between these equations.

Apply this process to the equations of Ex. 7, performing the elimination by Sylvester's method, and obtain $x^6y(4y - x^3) = 0$. [Notice that all the loci of Δ_p are included, as well as the locus of inflexions $4y = x^3$.]

(13) Show that the equations $y^2 = (x - c)^3$, $y = (x - c)^3$, $x^{\frac{2}{3}} + y^{\frac{2}{3}} = c$ all represent families of curves in which neighbouring curves do not intersect in real points, and yet an envelope $y = 0$ exists. (In the third case $x = 0$ is also an envelope.)

Obtain the corresponding Diff. Eqs., $8p^3 = 27y$, $p^3 = 27y^2$, $xp^3 + y = 0$,

$\qquad$ c-discriminants, y^4, y^2, $x^4y^4(x - y)^2(x + y)^2$;

and $\qquad$ p-discriminants, y^2, y^4, x^2y^2.

[Notice that in all these cases the envelope occurs to an even power for the reason given in the discussion of discriminant-loci as boundaries. For the first and third families the envelope is also a cusp-locus, so ordinary rules hold, but this is not so for the second family. The loci $x - y = 0$, $x + y = 0$ are where two imaginary curves, given by negative values of c in the equation of the third family, become coincident.]

(14) Show that $y = (x - c)^4$ represents a family of curves having four point contact with its envelope $y = 0$.

Obtain the corresponding Diff. Eq. $p^4 = 256y^3$, and discriminants $\Delta_c = y^3$, $\Delta_p = y^9$.

[The envelope again occurs to a power higher than the first. This time the power is odd, as it should be, since the number of real curves through any point is two on one side of the envelope, and zero on the other side.]

(15) Show that each of the equations $x^{\frac{1}{2}} + y^{\frac{1}{2}} = c^{\frac{1}{2}}$, $x^{\frac{1}{2}} + y^{\frac{1}{2}} =$ $(x + y - c)^2 = 4xy$, $(x + y - c^2)^2 = 4xy$, represents a family of parabolas with a common axis bisecting the angle $x0y$, and having $x = 0$ and $y =$ as envelopes. Show that the attempt to determine Δ_c fails for the first and second forms (or it may be considered to give $0 = 1$, the equation of the line at infinity, which touches all parabolas), while for the third $\Delta_c = xy$, and for the fourth $\Delta_c = x^2y^2(x - y)^2$.

$[x - y = 0$ is a particular curve corresponding to $c = 0$. In discussing discriminants we should avoid forms like the first and second, in which the terms are not single-valued, and also like the fourth, where different curves correspond to different values of c^2 and not of c itself.]

162. Riccati's equation. This name was originally given to the differential equation *

$$y_1 + by^2 = cx^m,$$

where b, c, and m are constants. For a certain set of particular values of m it can be integrated in finite terms (see Exs. 7-14 below), but in general the solution requires infinite series closely connected with Bessel Functions.†

By a Riccati's equation is now usually understood the generalised form

$$y_1 = P + Qy + Ry^2, \quad \dots\dots\dots\dots\dots\dots(1)$$

where P, Q, and R are functions of x. This equation is of some importance in Differential Geometry.‡

163. Reduction to a linear equation of the second order. Put

$$y = -\frac{u_1}{Ru}, \text{ giving } y_1 = -\frac{u_2}{Ru} + \frac{u_1{}^2}{Ru^2} + \frac{R_1 u_1}{R^2 u}.$$

When we substitute in equation (1) the terms in $u_1{}^2$ disappear.§ Hence, on multiplying up by R^2u, we obtain

$$-Ru_2 + R_1 u_1 = PR^2 u - QRu_1,$$

$$i.e. \quad Ru_2 - (QR + R_1)u_1 + PR^2 u = 0, \quad \dots\dots\dots\dots(2)$$

a linear equation of the second order. In special cases (as in the examples below) this may be integrated in finite terms, but in general solution in series will be required. However, in every case the solution will be of the form

$$u = Af(x) + BF(x),$$

giving

$$y = -\frac{u_1}{Ru} = -\frac{Af_1(x) + BF_1(x)}{R\{Af(x) + BF(x)\}}$$

$$= -\frac{cf_1(x) + F_1(x)}{cRf(x) + RF(x)},$$

where A/B has been replaced by c.

* Suffixes denote differentiations with respect to x.

† For the history of Riccati's equation and its connection with Bessel Functions, see Watson's *Theory of Bessel Functions*, pp. 1-3 and 85-94.

‡ There are 20 references to Riccati in the index of Darboux's *Leçons sur la Théorie Générale des Surfaces*. See also Eisenhart's *Differential Geometry*, pp. 25, 158, 249, 429, and Forsyth's *Differential Geometry*, pp. 20, 383.

§ This property is the real reason for choosing the substitution and enables us to recall it if it is forgotten.

This gives the important result that *the general integral of Riccati's equation is a homographic function of the constant of integration.* Conversely, it is easily shown (as outlined in Ex. 6 below) that we obtain a Riccati's equation by eliminating the arbitrary constant from any equation of the form

$$y = \frac{cg(x) + G(x)}{cf(x) + F(x)}.$$

164. The cross-ratio of any four particular integrals of a Riccati's equation is independent of x. We may take the four integrals to be $p(x)$, $q(x)$, $r(x)$, $s(x)$, which are derived from $\dfrac{cg(x) + G(x)}{cf(x) + F(x)}$ by giving the four special values α, β, γ, δ.

Then $$p - q = \frac{\alpha g + G}{\alpha f + F} - \frac{\beta g + G}{\beta f + F} = \frac{(\alpha - \beta)(gF - fG)}{(\alpha f + F)(\beta f + F)},$$

with similar expressions for the other differences of any two of p, q, r, s. When we form the cross-ratio, all the factors involving functions of x cancel out, and we obtain

$$\frac{(p - q)(r - s)}{(p - s)(r - q)} = \frac{(\alpha - \beta)(\gamma - \delta)}{(\alpha - \delta)(\gamma - \beta)} = C \text{ say,}$$

where C is independent of x.

165. Method of solution when three particular integrals are known. Let these be $q(x)$, $r(x)$, $s(x)$. Then it follows from the last result, with $p(x)$ replaced by y, that the general solution is

$$\frac{\{y - q(x)\}\{r(x) - s(x)\}}{\{y - s(x)\}\{r(x) - q(x)\}} = C,$$

so in this case the general solution has been obtained without *quadratures* (i.e., without integrations).

166. Method of solution when two particular integrals are known. Let these be $q(x)$, $r(x)$.

Then, as $$y_1 = P + Qy + Ry^2,$$
and $$q_1 = P + Qq + Rq^2,$$
$$y_1 - q_1 = (y - q)\{Q + (y + q)R\}.$$
Similarly $$y_1 - r_1 = (y - r)\{Q + (y + r)R\}.$$
Hence $$\frac{y_1 - q_1}{y - q} - \frac{y_1 - r_1}{y - r} = (q - r)R,$$
giving $$\log\frac{y - q}{y - r} = c + \int (q - r)R\,dx,$$

so in this case the general solution requires **one quadrature.**

167. Method of solution when one particular integral is known. Let this be $q(x)$.

The substitution * $y = q(x) + \dfrac{1}{z}$ transforms equation (1) into

$$q_1 - \frac{z_1}{z^2} = P + \left(q + \frac{1}{z}\right)Q + \left(q^2 + \frac{2q}{z} + \frac{1}{z^2}\right)R.$$

But, since $q(x)$ is an integral,

$$q_1 = P + qQ + q^2R.$$

Subtracting and multiplying up by z^2, we get

$$-z_1 = zQ + (2zq + 1)R,$$

or

$$z_1 + (Q + 2qR)z = -R,$$

a linear equation which can be solved by the use of an integrating factor $\exp\left\{\int(Q + 2qR)\,dx\right\}$. The determination of this factor requires one quadrature and the completion of the solution (as in Arts. 18-20) requires another, making two in all.

Examples for solution

In Exs. 1-5 the student should work from first principles, imitating the methods used above. He should not merely quote the results and substitute in them.

(1) By reduction to a linear equation show that the solution of

$$y_1 = -2 - 5y - 2y^2$$
$$2y(ce^{3x} + 1) = -(ce^{3x} + 4).$$

(2) Show that the solution of

$$x^2y_1 + 2 - 2xy + x^2y^2 = 0$$
$$y(x^2 + cx) = 2x + c.$$

(3) Show that $\tan x$ is one integral of $y_1 = 1 + y^2$, and hence obtain the general solution in the form

$$y(c - \tan x) = c\tan x + 1.$$

(4) Show that there are two values of the constant k for which k/x is an integral of $x^2(y_1 + y^2) = 2$, and hence obtain the general solution.

$$[k = 2 \text{ or } -1; \ y(cx^4 - x) = 2cx^3 + 1.]$$

(5) Show that 1, x, x^2 are three integrals of

$$x(x^2 - 1)y_1 + x^2 - (x^2 - 1)y - y^2 = 0,$$

and hence obtain the general solution

$$y(x + c) = x + cx^2.$$

---

* This appears artificial. A more natural (but longer) method is first to put $y = q(x) + u$, which will give an equation of Riccati's form with P replaced by zero. But this is a special case of Bernoulli's equation (Art. 21), and the usual method of solution requires the substitution $1/u = z$. By combining these two substitutions we get that given in the text.

(6) By eliminating the arbitrary constant c from the equation

$$y = \frac{cg(x) + G(x)}{cf(x) + F(x)}$$

obtain the Riccati's equation

$$(gF - Gf)y_1 = (gG_1 - g_1G) + (Gf_1 - G_1f - gF_1 + g_1F)y + (fF_1 - f_1F)y^2.$$

(7) Show that when $m = 0$ Riccati's equation

$$y_1 + by^2 = cx^m$$

can be integrated in finite terms.

$\quad$ [$yk(Ae^{2xk} + 1) = c(Ae^{2xk} - 1)$, where $k = \sqrt{(bc)}$, if bc is positive;

$\qquad\qquad yk = c \tan(A + kx)$, where $k = \sqrt{(-bc)}$, if bc is negative;

$\qquad\qquad y = cx + A$, if $b = 0$;

$\qquad y(bx + A) = 1$, if $c = 0$.]

(8) Show that the substitution $y = z/x$ transforms Riccati's equation into

$$xz_1 - z + bz^2 = cx^{m+2},$$

and hence show that the latter equation can be integrated in finite terms if $m = 0$.   [Use the result of Ex. 7.]

(9) By the substitution $z = yx^a$, transform the equation

$$xz_1 - az + bz^2 = cx^n$$

into $\qquad\qquad x^{1-a}y_1 + by^2 = cx^{n-2a}.$

By the further substitution $X = x^a$ obtain an equation of Riccati's form, with b, c, m replaced by b/a, c/a, $(n - 2a)/a$ respectively.   Hence show that the first equation of this example can be integrated in finite terms if $n = 2a$.

(10) Show that the substitution $z = \dfrac{a}{b} + \dfrac{x^n}{u}$ transforms the first equation of Ex. 9 into one of similar form with a, b, c replaced by $n + a$, c, b respectively.   Hence show that either equation is integrable in finite terms if $n = 2a$ or $n = 2(n + a)$.   By a repetition of this reasoning show that the first equation of Ex. 9 is integrable in finite terms if $n = 2(sn + a)$, where (as also in the following examples) s is zero or any positive integer.

(11) Show that the substitution $z = \dfrac{x^n}{u}$ transforms the equation of Ex. 9 into one of similar form with a, b, c replaced by $n - a$, c, b respectively.   Deduce that either is integrable in finite terms if $n = 2(sn - a)$.

(12) From the results of Exs. 9, 10, and 11 deduce that Riccati's equation is integrable in finite terms if $m + 2 = 2s(m + 2) \pm 2$.

Show that this result is equivalent to $m = -4r/(2r \pm 1)$, where r, like s, is zero or a positive integer, or to $2/(m + 2) =$ an odd integer (positive or negative).

(13) Show that the substitutions $y = \dfrac{1}{bx} + \dfrac{1}{x^2Y}$, $X = x^{m+3}$, transform

Riccati's equation into another of similar form with b, c, m replaced by $c/(m+3)$, $b/(m+3)$, $-(m+4)/(m+3)$ respectively. Deduce that if m is of the form $-4s/(2s-1)$, the transformation replaces s by $(s-1)$. By considering s such transformations show that in this case Riccati's equation is integrable in finite terms.

(14) Show that the substitutions $y = 1/Y$, $X = x^{m+1}$, transform Riccati's equation into another of similar form with b, c, m replaced by $c/(m+1)$, $b/(m+1)$, $-m/(m+1)$ respectively. Deduce (using the result of Ex. 13) that Riccati's equation is integrable in finite terms if m is of the form $-4s/(2s+1)$.

168. Two methods of integrating the total differential equation $P\,dx + Q\,dy + R\,dz = 0$.

We have already (in Chap. XI.) given the necessary and sufficient condition of integrability of this equation, and a general method of obtaining the integral when the condition is satisfied. We shall now give two additional methods. One of these (involving an integrating factor) has the defect that it can be used only for certain homogeneous equations, but for these equations it is perhaps the simplest method available. The other (Mayer's method) is quite general. It requires only one integration, and this gives it a theoretical advantage over the other general method (Art. 117), which requires two. However, the beginner is not advised to use this method, for the single integration required is often more difficult to effect (on account of the lack of symmetry of the expressions involved) than the two integrations required in Art. 117. Moreover, Mayer's method, if applied without careful attention to certain conditions, may give results that are absolutely wrong.

169. Integrating factor for homogeneous equations. Let

$$P\,dx + Q\,dy + R\,dz = 0 \quad\ldots\ldots\ldots\ldots\ldots\ldots\ldots(1)$$

be an integrable equation in which P, Q, R are homogeneous functions of the same degree n in x, y, z, that is to say, in which P, Q, R may be expressed in the forms

$$x^n f(u, v),\ \ x^n g(u, v),\ \ x^n h(u, v)$$

respectively, where $u = y/x$, and $v = z/x$.

Then $\qquad dy = u\,dx + x\,du, \quad dz = v\,dx + x\,dv.$

Hence equation (1) becomes

$$x^n\{f(u, v)\,dx + g(u, v)(u\,dx + x\,du) + h(u, v)(v\,dx + x\,dv)\} = 0,$$

i.e. $\qquad x^n\{(f + ug + vh)\,dx + x(g\,du + h\,dv)\} = 0,$

from which, dividing by $x^{n+1}(f + ug + vh)$, if this expression is not zero, we obtain $\qquad \dfrac{dx}{x} + \dfrac{g\,du + h\,dv}{f + ug + vh} = 0. \quad\ldots\ldots\ldots\ldots\ldots\ldots(2)$

Now since equation (1) is integrable, so is equation (2), either immediately or after multiplication by an integrating factor. But the first term in equation (2) involves only x, and the second term only the variables u and v. One variable is separated from the other two, and this separation, which is the most favourable form for integration, would be destroyed by multiplication by any factor (except a mere constant). Hence there is no need to look for an integrating factor, and equation (2) must be exact as it stands. But apart from the change of variables, equation (2) was derived from equation (1) by division by the factor $x^{n+1}(f + ug + vh)$, which is equal to $Px + Qy + Rz$.

Hence $1/(Px + Qy + Rz)$ is an integrating factor of the integrable homogeneous equation

$$P \, dx + Q \, dy + R \, dz = 0,$$

except when $Px + Qy + Rz = 0$. A similar theorem holds good for the equation

$$P_1 \, dx_1 + P_2 \, dx_2 + \ldots + P_n \, dx_n = 0.$$

Ex. $(y^2 + yz) \, dx + (zx + z^2) \, dy + (y^2 - xy) \, dz = 0.$

Here $Px + Qy + Rz = xy^2 + xyz + xyz + yz^2 + y^2z - xyz$
$$= y(xy + xz + z^2 + yz) = y(x + z)(y + z),$$

so the integrating factor is $1/\{y(x + z)(y + z)\}$.

Multiplying the differential equation by it we obtain

$$\frac{dx}{x+z} + \frac{z \, dy}{y(y+z)} + \frac{(y-x) \, dz}{(x+z)(y+z)} = 0,$$

i.e. $$\frac{dx}{x+z} + \frac{\{(y+z) - y\} \, dy}{y(y+z)} + \frac{\{(y+z) - (x+z)\} \, dz}{(x+z)(y+z)} = 0,$$

or $$\frac{dx}{x+z} + \frac{dy}{y} - \frac{dy}{y+z} + \frac{dz}{x+z} - \frac{dz}{y+z} = 0,$$

or $$\frac{dx + dz}{x+z} + \frac{dy}{y} - \frac{dy + dz}{y+z} = 0,$$

whence $\log (x + z) + \log y - \log (y + z) = \log c,$
giving $y(x + z) = c(y + z).$

If $Px + Qy + Rz = 0$, we put $y = ux$, $z = vx$ and divide out x^{n+1} giving $g \, du + h \, dv = 0$, involving only two variables.

170. Mayer's method. Write the total differential equation in the form

$$dz = P(x, y, z) \, dx + Q(x, y, z) \, dy.$$

It may be proved that if the condition of integrability (Arts. 118 and

119) is satisfied, *and if the functions P and Q are holomorphic in the neighbourhood of a point* (x_0, y_0, z_0), then there exists one solution (and only one) of the differential equation representing a surface passing through this point.* Mayer's method determines this surface by finding the curve of intersection of the surface and a variable plane drawn parallel to the axis of z through the point (x_0, y_0, z_0). The simplest values consistent with the holomorphic condition are taken for x_0 and y_0; *e.g.* 0 and 0, or 0 and 1, or 1 and 1. z_0 occurs in the final result as the arbitrary constant. The procedure will be best understood by a study of the following examples. (Of course these examples can be solved at sight, but if harder ones had been chosen the principle of the method might have been obscured by the details of the complicated integrations which Mayer's method often involves).

Ex. (i). $$dz = 2x\,dx + 4y\,dy. \dotfill (1)$$
The condition of integrability is
$$2x(0-0) + 4y(0-0) - 1(0-0) = 0,$$
which is satisfied. We may take $x_0 = 0$ and $y_0 = 0$, as the functions $2x$ and $4y$ are holomorphic in the neighbourhood of $(0, 0, z_0)$. The plane through this point parallel to the axis of z is given by
$$y = mx, \quad dy = m\,dx. \dotfill (2)$$
From equations (1) and (2),
$$dz = (2 + 4m^2) x\,dx,$$
whence we get $$z - z_0 = (1 + 2m^2) x^2, \dotfill (3)$$
determining the constant of integration by the condition that $z = z_0$ when $x = 0$.

Equation (3) represents a cylinder (with generators parallel to the axis of y) through the curve of intersection of the plane (2) and the surface required.

Eliminating m from equations (2) and (3) we get as the equation of the surface
$$z - z_0 = x^2 + 2y^2.$$

This is the general solution of equation (1), if z_0 is taken to be an arbitrary constant.

Ex. (ii). $$dz = \frac{3z\,dx}{x} - \frac{2z\,dy}{y}. \dotfill (4)$$

The condition of integrability is
$$\frac{3z}{x}\left(\frac{-2}{y} - 0\right) - \frac{2z}{y}\left(0 - \frac{3}{x}\right) - 1(0-0) = 0,$$

which is satisfied. We cannot take $x_0 = 0$, $y_0 = 0$, as this makes the functions $3z/x$ and $2z/y$ infinite. However, $x_0 = 1$, $y_0 = 1$ will do.

---

* Goursat, *Cours d'Analyse Mathématique*, Vol. II., 4th ed., Arts. 385 and 441.

Put $$y = 1 + m(x - 1). \quad \dots\dots\dots\dots\dots\dots\dots\dots\dots(5$$

Equation (4) becomes

$$dz = \frac{3z\,dx}{x} - \frac{2zm\,dx}{1 + m(x - 1)},$$

giving $$\log z - \log z_0 = 3 \log x - 2 \log\{1 + m(x - 1)\},$$

whence $$z\{1 + m(x - 1)\}^2 = z_0 x^3. \quad \dots\dots\dots\dots\dots(6$$

Eliminating m from (5) and (6), we get the solution

$$zy^2 = z_0 x^3.$$

It will be observed that *all* the surfaces of this family pass through the point $(0, 0, z_0)$.

Examples for solution

(1) Show that the attempt to solve Ex. (ii) above, with $(0, 0, z_0$ as the fixed point, breaks down when we try to make the cylinder corresponding to equation (6) pass through that point.

(2) Solve $$y^2\,dz = y\,dx + (y^2 - x)\,dy.$$

[The correct result, choosing the fixed point as $(0, 1, z_0)$ is

$$y(z - z_0) = y(y - 1) + x.$$

The choice of $(0, 0, z_0)$ leads to the incorrect result $z - z_0 = y$.]

(3) Solve $\quad (1 + xy)\,dz = (1 + yz)\,dx + x(z - x)\,dy.$

[Result $z = x + z_0(1 + xy.$]

171. Linear differential equations of the second order. The following discussion (Arts. 171-177) is supplementary to Chaps. IX and X. Suffixes will be used to denote differentiations with respect to x. We shall use $h(x)$, $k(x)$, $j(x)$, $H(x)$, $K(x)$, or sometimes h, k, j H, K, to denote functions of x which are *holomorphic* at the origin (*i.e.* expansible in power series convergent within a sufficiently small circle whose centre is the origin) and which have the further property that they do not vanish at the origin. Their reciprocals also will be holomorphic,* and so will their logarithmic derivates such as

$$h_1(x)/h(x).$$

Whenever we speak of singular points, it is to be understood that these points are isolated, *i.e.* that a circle of sufficiently small radius with any one point as centre will exclude all the others.

172. Regular integrals. It was mentioned on p. 110 that solutions of Frobenius' forms are called *regular* integrals. We shall now consider in more detail what is implied by this. Let us examine the forms of the answers to the examples in Chap. IX. Although we

---

* *Bromwich's Infinite Series*, 2nd ed., Arts. 54 and 84.

distinguished four * cases in process of solution, there were only two essentially different forms of the complete primitive $au + bv$. One integral, say u, was always of the form $x^\alpha h(x)$. The second integral, v, had in some examples a similar form, say $x^\beta k(x)$, as in Arts. 95 and 99; in others, as in Arts. 97 and 98, it had the form

$$x^\alpha \{h(x) \log x + x^s k(x)\},$$

where s was an integer, positive or negative (e.g. 1 in Ex. 1, Art. 97 and -4 in Ex. 1, Art. 98).

We take these forms as the definitions of integrals *regular at the origin* † (of a linear differential equation of the second order), with the slight modification that s is allowed to take also the value zero. This makes no real difference, for if s is zero we can replace the integral $v = x^\alpha \{h(x) \log x + k(x)\}$ by the linear combination of integrals

$$v - \frac{k(0)u}{h(0)} = x^\alpha \left\{ h(x) \log x + k(x) - \frac{k(0)}{h(0)} h(x) \right\},$$

which is of similar form except that $k(x)$ has been replaced by a new holomorphic function of which x is a factor. Similarly in the first form of v, namely $x^\beta k(x)$, we can always suppose α and β unequal, for if not v can be replaced by $v - \dfrac{k(0)}{h(0)} u$, which has $x^{\alpha+1}$ as a factor.

For linear differential equations of the mth order an integral regular at the origin is defined as one of the form

$$x^\alpha \{h(x)(\log x)^r + x^s k(x)(\log x)^{r-1} + \ldots + x^n j(x)\},$$

where $s, \ldots, n$ are zero or any integers (positive or negative), and r can have any of the values $0, 1, 2, \ldots, m-1$. Thus for first-order equations regular integrals cannot involve $\log x$. For the second

---

* In the method of Frobenius for equations of the mth order (*Crelle*, Vol. LXXVI. 1873, pp. 214-224, or Forsyth's *Theory of Differential Equations*, Vol. IV. pp. 78-93, or Ince's *Ordinary Differential Equations*, pp. 396-402), it is convenient for the theoretical treatment to distinguish only two cases, the second of which includes our cases II., III. and IV. To deal with this second case the series with its coefficients as functions of c is multiplied by $f(c+1)f(c+2) \ldots f(c+r)$, where $f(c) = 0$ is the indicial equation, and r is the greatest difference between any two of its roots that belong to a set differing by integers (cf. our method for case III.). In this series and its successive partial differential coefficients with respect to c are substituted respectively the roots, arranged so that the difference between any one and the following is a *positive* integer or zero. However, in solving examples this method often leads to a large amount of unnecessary work, and hence in Chap. IX. we have modified it considerably, particularly in our Case IV.

† Points other than the origin are considered in Art. 175. It is unfortunate that the word *regular* has in Differential Equations a meaning different from that usual in Theory of Functions, where it is equivalent to *holomorphic* (as defined in Art. 171). Thus an expression involving $\log x$ or x^α (where α is not zero or a positive integer) may be an integral regular at the origin, and yet cannot be a function regular at that point.

order the logarithm occurs either linearly or not at all. This may also be deduced from Chap. X. as follows: In Art. 107 both integrals were free from logarithms. In Art. 110 we obtained a second integral by differentiating partially with respect to c a series of the form $x^c \sum\limits_0^\infty a_n x^n$, where the a's were functions of c, and then, after differentiation, replacing c by β. The result (not given in Art. 110) is

$$x^\beta \left\{ \left(\sum_0^\infty a_n(\beta) x^n \right) \log x + \left(\sum_0^\infty \frac{\partial a_n(\beta)}{\partial \beta} x^n \right) \right\},$$

which is of the form

$$x^\alpha \{ h(x) \log(x) + x^s k(x) \}$$

If the first λ of the coefficients $a_n(\beta)$ are zero and also the first μ of the coefficients $\dfrac{\partial a_n(\beta)}{\partial \beta}$, then $\alpha = \beta + \lambda$ and $s = \mu - \lambda$.

It will be noticed that the co-factor of $\log x$ is itself an integral. This may be proved independently. Take the differential equation as

$$y_2 + y_1 P(x) + y Q(x) = 0, \quad \dotfill (1)$$

where $P(x)$ and $Q(x)$ are uniform * (*i.e.* single-valued) in the neighbourhood of the origin.

If in the left-hand side of this equation we substitute for y the integral $x^\alpha \{ h(x) \log x + x^s k(x) \} = u \log x + w$ say, the result must, by definition of an integral, be identically zero. In this result $\log x$ occurs with a co-factor $(u_2 + u_1 P + u Q)$. This and all the other terms in the result, except $\log x$, are the product of x^α and a uniform function, since u and w, and hence also u_1, u_2, w_1, w_2, are products of this kind, while P and Q are uniform. If we could divide the identity by the co-factor of $\log x$, we should obtain the absurd result that the non-uniform function $\log x$ is the quotient of two uniform functions, *i.e.* is itself a uniform function. Hence the division is illegitimate, and this can be due only to the co-factor being zero; *i.e.* u is itself an integral.

A similar theorem holds for the co-factor of the highest power of $\log x$ occurring in a regular integral of an equation (with coefficients uniform in the neighbourhood of the origin) of the mth order. Thus in every case in which there are regular integrals at least one of them must be free from logarithms and of the form $x^\alpha h(x)$.

---

* This differential equation includes as particular cases those considered in Chaps. IX. and X.

173. Fuchs' theorem. *The necessary and sufficient condition that a linear differential equation of the second order, whose coefficients are uniform in the neighbourhood of the origin, should have all its integrals regular at the origin is that the equation should be expressible in the form*

$$x^2 y_2 + x y_1 p(x) + y p(x) = 0,$$

where p and q are holomorphic at the origin.

The discussion of the method of Frobenius (Arts. 106-110) proves that this condition is sufficient. We have now to prove that it is necessary. From Art. 172, at least one integral is of the form $x^\alpha h(x)$. Denote this by $u(x)$. Put $y = u \int z\, dx$, and substitute in equation (1) of Art. 172. The terms involving the sign of integration have a factor $(u_2 + u_1 P + uQ)$ and therefore vanish, as u is an integral, and we get

$$2u_1 z + u z_1 + Puz = 0. \quad \dots\dots\dots\dots\dots\dots(2)$$

Now the integral y may have either of the two forms

$$x^\beta k(x), \quad x^\alpha \{h(x) \log x + x^s k(x)\}.$$

Hence
$$\frac{y}{u(x)} = x^{\beta-\alpha} \frac{k(x)}{h(x)}, \quad \text{or} \quad \log x + x^s \frac{k(x)}{h(x)}$$
$$= x^{\beta-\alpha} H(x), \quad \text{or} \quad \log x + x^s H(x), \text{ say,}$$

so that
$$z = \frac{d}{dx}\left(\frac{y}{u}\right) = x^{\beta-\alpha-1}\{(\beta-\alpha)H + xH_1\},$$

or
$$x^{-1} + x^{s-1}(sH + xH_1).$$

In both cases we can write z in the form * $x^\gamma K(x)$, where $K(x)$ is holomorphic with $K(0) \neq 0$. Hence from equation (2)

$$P = -\frac{z_1}{z} - \frac{2u_1}{u} = -\frac{\gamma}{x} - \frac{K_1}{K} - \frac{2\alpha}{x} - \frac{2h_1}{h} = \frac{p(x)}{x} \text{ say,}$$

where p is holomorphic at the origin.

Also, since $x^\alpha h(x)$ is an integral of equation (1),

$$x^\alpha h_2 + 2\alpha x^{\alpha-1} h_1 + \alpha(\alpha-1)x^{\alpha-2}h + (x^\alpha h_1 + \alpha x^{\alpha-1}h)P + x^\alpha hQ = 0,$$

giving
$$Q = \frac{1}{x^2}\left\{ -\frac{x^2 h_2}{h} - \frac{2\alpha x h_1}{h} - \alpha(\alpha-1) - \left(\frac{xh_1}{h} + \right)p\right\} = \frac{q(x)}{x^2} \text{ say,}$$

where q is holomorphic at the origin.

On multiplying each side of equation (1) by x^2, and replacing xP and x^2Q by p and q respectively, we get the form required by the theorem.

---

* In the first case $\gamma = \beta - \alpha - 1$. In the second case $\gamma = -1$ or $s-1$, according as the integer s is positive or negative.

Example for solution

By eliminating the arbitrary constants from $y = Ax^{\frac{1}{2}} + Bx^{\frac{1}{4}} \log x$, obtain the differential equation

$$8x^2(4 - \log x)y_2 + 2x(8 - \log x)y_1 - y \log x = 0,$$

which is therefore a linear differential equation of the second order having all its integrals regular at the origin, but is not expressible in the form given in Fuchs' theorem.

[This example shows the importance of the stipulation that the coefficients of the differential equation should be *uniform* in the neighbourhood of the origin. In fact, this imposes a severe restriction, for it excludes all complete primitives of the form

$$y = Ax^\beta j(x) + Bx^\alpha \{h(x) \log x + x^s k(x)\},$$

except for the special case where $x^\beta j(x)$ is merely a numerical multiple of $x^\alpha h(x)$.]

174. Ordinary and singular points. It may happen that (unlike the other holomorphic functions h, k, j, H, K) p and q may vanish at the origin. In particular if p is divisible by x and q by x^2, the equation in its original form (1) has P and Q holomorphic at the origin. In this case the origin is said to be an *ordinary* point, and on applying the method of Frobenius we shall obtain an indicial equation with 0 and 1 as roots, leading (as in Art. 99) to an indeterminate coefficient and finally to two linearly independent integrals that are both power series. Neither logarithms nor indices other than positive integers (or zero) can occur. But the indicial equation may have 0 and 1 for roots without the origin being an ordinary point, as in Ex. 2 of Art. 98.

Points which are not ordinary are called *singular*. If at a singular point (in whose neighbourhood the coefficients of the equation are uniform) all the integrals are regular, it is called a *regular* singular point.

These definitions refer to singular points of the differential equation itself, that is, of its coefficients when it is written in the form (1). Our discussion of ordinary points shows that the singularities of the integrals are singularities of the equation, but the converse is not always true. For example, by eliminating the arbitrary constants A and B from $y = Ax^m + Bx^n$, we get

$$x^2 y_2 - (m + n - 1)xy_1 + mny = 0.$$

If m and n are unequal positive integers, or if one is zero and the other a positive integer other than 1, the origin is a singularity of the equation but not of the integrals. When, as here, every integral is holomorphic at a point which is singular for the equation, the singularity is said to be *apparent*. In all other cases the singularity is said to be *real*. At an apparent singularity it is necessary that the roots of the indicial equation should be unequal positive integers, or zero and a positive integer greater than 1. It is also necessary that the smaller root should lead to an indeterminate coefficient (very much as in Art. 99).

Examples for solution

(1) Show that a necessary (but not sufficient) condition for the origin to be an apparent singularity of the equation

$$x^2 y_2 + x y_1 p(x) + y q(x) = 0,$$

where $p(x)$ and $q(x)$ are holomorphic at the origin, is $p(0) =$ a negative integer, while the necessary and sufficient conditions for the origin to be an ordinary point are $p(0) = q(0) = q_1(0) = 0$.

(2) Show that the origin is an apparent singularity of

$$x(1 + x^2) y_2 - y_1 - x^3 y = 0,$$

and obtain the complete primitive

$$y = A\left(1 + \tfrac{1}{8}x^4 - \tfrac{1}{16}x^6 + \tfrac{1}{24}x^8 - \ldots\right) + B\left(x^2 - \tfrac{1}{4}x^4 + \tfrac{1}{6}x^6 - \tfrac{7}{64}x^8 \ldots\right).$$

(3) Show that the origin is a real singularity of $x^2 y_2 + (x^2 - 2) y = 0$, but that all the integrals are free from logarithms.

[The roots of the indicial equations are -1 and 2. The smaller root gives a_3 indeterminate (cf. Art. 99). The infinite series obtained can be summed, giving finally $y = A x^{-1} (\cos x + x \sin x) + B x^{-1} (\sin x - x \cos x).$]

175. Equations of Fuchsian type.

To deal with points other than the origin we make a change of variable, putting $X = x - a$, or $X = x^{-1}$, according as the point to be considered is the finite one $x = a$, or that at infinity $x = \infty$. It follows that for equation (1), if the functions P and Q are holomorphic at every finite point except a limited number $a, b, c, \ldots$, then these are the only possible finite singular points. Thus we can find these points by inspection, by seeing where P and Q fail to be holomorphic, without making a change of variable; *e.g.* if

$$P = \frac{x + 2}{x(x - 3)} \quad \text{and} \quad Q = \frac{x^3 + 10}{x^2 (x - 3)(x - 4)^3},$$

the only possible finite singular points are given by $x = 0, 3, 4$. Moreover, to test whether a singular point $x = a$ is regular, we have only to notice whether $(x - a) P$ and $(x - a)^2 Q$ are both holomorphic at $x = a$. In the example given 0 and 3 are regular singular points, but 4 is irregular, since $(x - 4)^2 Q$ is not holomorphic at $x = 4$, owing to the factor $(x - 4)$ in the denominator.

The point at infinity $x = \infty$ is best dealt with by a change of variable.

If all the singular points of an equation (whose coefficients are everywhere uniform) are regular, the equation is said to be of Fuchsian type.

Examples for solution

(1) Show that, for the Hypergeometric equation

$$x(1 - x)y_2 + \{c - (a + b + 1)x\}y_1 - aby = 0,$$

the only singular points are 0, 1 and ∞, which are regular.

(2) Show that for Legendre's equation

$$(1 - x^2)y_2 - 2xy_1 + n(n + 1)y = 0,$$

the only singular points are 1, -1, and ∞, which are regular.

(3) Show that for Bessel's equation

$$x^2 y_2 + xy_1 + (x^2 - n^2)y = 0,$$

the only singular points are 0 and ∞, of which the first is regular, but not the second.

(4) Show that *Riemann's P-equation* $y = P \left\{ \begin{array}{ccc} a & b & c \\ \alpha & \beta & \gamma & x \\ \alpha' & \beta' & \gamma' \end{array} \right\}$,

i.e. $y_2 + \sum \left(\dfrac{1 - \alpha - \alpha'}{x - a} \right) y_1 + \left\{ \sum \dfrac{\alpha\alpha'(a - b)(a - c)}{x - a} \right\} \dfrac{y}{(x - a)(x - b)(x - c)} = 0$

has a, b, c as regular singular points and all other points, including ∞, as ordinary points, provided that $\alpha + \alpha' + \beta + \beta' + \gamma + \gamma' = 1$.

By change of variable show that α and α' are the roots of the indicial equation corresponding to the point a.

(5) Show that the equations of Exs. 1, 2 and 4, but not 3, are of Fuchsian type.

(6) Show that the following equation is of Fuchsian type:

$$y_2 + \frac{P}{\psi}y_1 + \frac{Q}{\psi^2}y = 0,$$

where ψ is the product of any number, say n, linear factors $(x - a)$ $(x - b)$, $(x - c)$, ... of which no two are equal, and P, Q are polynomial in x of degrees not greater than $(n - 1)$ and $(2n - 2)$ respectively.

176. Characteristic index.

Consider the equation

$$y_2 + x^{-\lambda}p(x)y_1 + x^{-\mu}q(x)y = 0,$$

where λ, μ are positive integers or zero, and p, q are holomorphic functions of x which are not zero when $x = 0$.

If we attempt to solve this equation by the method of Frobenius we get the indicial equation by replacing y by a series of powers of x (starting with x^c), and equating to zero the coefficient of the lowest power of x in the result furnished by the left-hand side of the differential equation. The lowest powers of x from its first, second, and third terms will be respectively $c - 2$, $c - \lambda - 1$, and $c - \mu$. Three cases arise:

(i) if the first of these numbers is not greater than either of the others, the indicial equation is of the second degree;

(ii) if the second of these numbers is less than the first and not greater than the third, the indicial equation is of the first degree. (Cf. Exs. 2 and 4, p. 118);

(iii) if the third of these numbers is the least, then the indicial equation is of zero degree. (Cf. the example at the top of p. 118.)

In case (i) $\lambda \leqslant 1$ and $\mu \leqslant 2$, so by Fuchs' theorem there *must* be *two* regular integrals.

In case (ii) there *may* be *one* regular integral. If, however, as is often the case (cf. Ex. 4, p. 118), the single series obtained is divergent for all values of x, there is no regular integral.

In case (iii) there is no series and hence *no* regular integral.

The *characteristic index* may be defined as the number denoting the case which arises, but starting from zero, *i.e.* 0 for case (i), 1 for case (ii), and 2 for case (iii). It is easy to extend this definition and the discussion of the maximum possible degree of the indicial equation to equations of any order, leading to the conclusion that *a linear differential equation of order m and characteristic index r cannot have more than $m - r$ regular integrals.*

177. Normal and subnormal integrals. We saw in Art. 100 that the method of Frobenius failed to discover an integral with a factor $e^{\frac{1}{x}}$. This is a particular case of a normal integral, defined as one of the form $e^z u$, where z is a polynomial in $1/x$ (in the simplest case a numerical multiple of $1/x$), and u is a function of x such as occurs in a regular integral. *Subnormal integrals* differ from normal integrals only by having x replaced by its square root (or by its cube or other higher root in the case of differential equations of order higher than the second).

A method of obtaining normal or subnormal integrals is shown by the following examples:

Ex. (i). $\qquad y_2 - 2x^{-1}y_1 + x^{-4}(-4 + 2x^2)y = 0.$ ....(1)

Here the indicial equation has no roots and there are no regular integrals (*i.e.* the characteristic index is 2). This is due to the term $-4x^{-4}$ in the coefficient of y.

Put
$$y = e^z u,$$
giving
$$y_1 = e^z(u_1 + z_1 u), \quad y_2 = e^z\{u_2 + 2z_1 u_1 + (z_1{}^2 + z_2)u\}.$$

Equation (1) is transformed, after division by e^z, into

$$u_2 + (-2x^{-1} + 2z_1)u_1 + (-4x^{-4} + 2x^{-2} - 2x^{-1}z_1 + z_1{}^2 + z_2)u = 0. \quad \ldots\ldots\ldots(2)$$

To get rid of the term $-4x^{-4}$, take z_1 as ax^{-2}, where $a = \pm 2$. Equation (2) becomes

$$u_2 + (-2x^{-1} + 2ax^{-2})u_1 + (2x^{-2} - 4ax^{-3})u = 0,$$

which has a characteristic index 1, and so may have a regular integral. Applying the method of Frobenius to find this, we get the simple result $u = x^2$ for both values of a. Multiplying by the exponential factor, we obtain finally the two normal integrals $x^2 e^{-2/x}$ and $x^2 e^{2/x}$.

Ex. (ii). $\qquad y_2 + 4x^{-2}y_1 + x^{-6}(-4 + 6x^2 - 4x^3)y = 0.$

Again there are no regular integrals. Proceeding as in Ex. (i), we obtain

$$u_2 + (4x^{-2} + 2z_1)u_1 + (-4x^{-6} + 6x^{-4} - 4x^{-3} + 4x^{-2}z_1 + z_1{}^2 + z_2)u = 0.$$

To get rid of the term $-4x^{-6}$, take z_1 to contain a term bx^{-3}, where $b = \pm 2$. If $z_1 = ax^{-2} + bx^{-3}$, the coefficient of u will contain no term in x^{-5}, provided that a is chosen so that $4b + 2ab = 0$, i.e. $a = -2$.

The choice $z_1 = -2x^{-2} + 2x^{-3}$ leads to

$$u_2 + 4x^{-3}u_1 - 4x^{-4}u = 0,$$

which has one regular integral, $u = x$.

The other choice, $z_1 = -2x^{-2} - 2x^{-3}$, leads to

$$u_2 - 4x^{-3}u_1 + 8x^{-4}u = 0.$$

This has no regular integral, for the only series obtainable, namely,

$$x^2\left(1 + \tfrac{1}{4}x^2 + \frac{1 \cdot 3}{4^2} x^4 + \frac{1 \cdot 3 \cdot 5}{4^3} x^6 + \ldots\right)$$

is divergent. Hence the original equation has one normal integral, $xe^{(2x^{-1} - x^{-2})}$.

Ex. (iii). $\qquad y_2 + x^{-2}(-1 + 3x)y_1 + x^{-2}y = 0.$

This time the characteristic index is 1. The indicial equation is of the first degree, but (as pointed out in Ex. 4, p. 118) the series obtained is divergent.

Proceeding as before, we get

$$u_2 + (-x^{-2} + 3x^{-1} + 2z_1)u_1 + \{x^{-2} + (-x^{-2} + 3x^{-1})z_1 + z_1{}^2 + z_2\}u = 0.$$

As the troublesome term in the original equation was $-x^{-2}$ in the coefficient of y_1, while the coefficient of y was only such as occurs when the integrals are regular, it might be thought desirable to simplify the coefficient of u by taking $z_1 = \tfrac{1}{2}x^{-2}$. But this will introduce a term in x^{-4} into the coefficient of u, giving an equation with no regular integrals.

Let us try to get another equation with characteristic index 1, in the hope that the corresponding series may converge. Put $z_1 = ax^{-2}$.

The coefficient of u will be free from terms in x^{-4} if $a^2 - a = 0$, *i.e.* $a = 0$ or 1. $a = 0$ gives the original equation, but $a = 1$ gives

$$u_2 + (3x^{-1} + x^{-2})u_1 + (x^{-2} + x^{-3})u = 0,$$

which has the regular integral $u = x^{-1}$, giving the one normal integral $y = x^{-1}e^{-1/x}$.

Ex. (iv). $y_2 + \frac{1}{2}x^{-1}y_1 - x^{-3}y = 0.$

This equation has no regular integrals. Proceeding as before, we get

$$u_2 + (\tfrac{1}{2}x^{-1} + 2z_1)u_1 + (-x^{-3} + \tfrac{1}{2}x^{-1}z_1 + z_1{}^2 + z_2)u = 0.$$

To get rid of the term $-x^{-3}$, take $z_1 = kz^{-3/2}$, where $k = \pm 1$. This gives $u_2 + (\tfrac{1}{2}x^{-1} + 2kx^{-3/2})u_1 - kx^{-5/2}u = 0.$

$u = x^c \sum_0^\infty a_n x^{\frac{1}{2}n}$ will be an integral if

$$a_0(2kc - k) = 0, \text{ so that } c = \tfrac{1}{2},$$

$$a_1\{2k(c + \tfrac{1}{2}) - k\} + a_0\{c(c-1) + \tfrac{1}{2}c\} = 0,$$

i.e. $ka_1 + 0 = 0$, so $a_1 = 0$.

Similarly, $a_n = 0$ for all values of $n > 1$, so $u = x^{\frac{1}{2}}$.
The original equation has the two subnormal integrals

$$x^{\frac{1}{2}}e^{-2x^{-\frac{1}{2}}} \quad \text{and} \quad x^{\frac{1}{2}}e^{2x^{-\frac{1}{2}}}.$$

Examples for solution

Find normal or subnormal integrals of the following equations (1)-(5):

(1) $y_2 + 2x^{-1}y_1 - x^{-4}y = 0.$ [*Ans.* $e^{1/x}$, $e^{-1/x}$.]

(2) $y_2 + x^{-1}y_1 + x^{-4}(1 - \tfrac{1}{4}x^2)y = 0.$
          [*Ans.* $x^{\frac{1}{2}}e^{i/x}$, $x^{\frac{1}{2}}e^{-i/x}$; or $x^{\frac{1}{2}}\cos(1/x)$, $x^{\frac{1}{2}}\sin(1/x)$.]

(3) $y_2 + x^{-2}(-2 + x)y_1 + x^{-4}(1 + x - x^2 + x^4)y = 0.$
          [*Ans.* $ue^{-1/x}$, $ve^{-1/x}$, where u and v are as on p. 115.]

(4) $y_2 - \tfrac{1}{2}x^{-1}y_1 - 4x^{-3}y = 0.$ [*Ans.* $x(1 + \tfrac{1}{4}x^{\frac{1}{2}})e^{-4x^{-\frac{1}{2}}}$, $x(1 - \tfrac{1}{4}x^{\frac{1}{2}})e^{4x^{-\frac{1}{2}}}$.]

(5) $y_2 - x^{-6}(1 + 5x^2)y = 0.$
          [*Ans.* $x^{-1}(1 + \tfrac{1}{2}x^2)e^{\frac{1}{2}x^{-2}}$; $z = -\tfrac{1}{2}x^{-2}$ gives a divergent series.]

(6) Transform Bessel's equation of order zero by the substitution $x = 1/X$, and attempt to find normal integrals of the transformed equation. Show that the series obtained are divergent. Reverting to the original variable, obtain the series

$$e^{-ix}x^{-\frac{1}{2}}\left\{1 - \frac{1^2}{8ix} + \frac{1^2 \cdot 3^2}{2!(8ix)^2} - \frac{1^2 \cdot 3^2 \cdot 5^2}{3!(8ix)^3} + \cdots\right\}$$

and a similar series with the sign of i changed.

[The transformed form of Bessel's equation is given in the answer to Ex. 1, p. 118.

These series, although divergent, are very useful. They are called *asymptotic*. For any given value of x, sufficiently large, they give an approximation whose error can be made reasonably small, though not indefinitely so. See Whittaker and Watson's *Modern Analysis*, 4th ed. Arts. 8·1 – 8·32 and 17·5.]

(7) From Whittaker's confluent hypergeometric equation

$$y_2 + \left(-\tfrac{1}{4} + \frac{k}{x} + \frac{\tfrac{1}{4} - m^2}{x^2} \right) y = 0,$$

obtain (by the process of Ex. 6), the series

$$e^{-\frac{1}{2}x} x^k \left[1 + \sum_{r=1}^{\infty} \frac{\{m^2 - (k - \tfrac{1}{2})^2\} \{m^2 - (k - \tfrac{3}{2})^2\} \ldots \{m^2 - (k - r + \tfrac{1}{2})^2\}}{r!\, x^r} \right].$$

[This series is in general the asymptotic expansion of the function denoted by $W_{k,\,m}(x)$, but if $(k - \tfrac{1}{2} \pm m)$ is a positive integer the series terminates, giving an integral in finite terms. Another series $W_{-k,\,m}(-x)$ can be obtained from $W_{k,\,m}(x)$ by changing the signs of k and x.]

178. The equation of vibrating strings. This is

$$\frac{\partial^2 V}{\partial x^2} = \frac{1}{a^2} \frac{\partial^2 V}{\partial t^2}, \quad \ldots\ldots\ldots\ldots\ldots\ldots(1)$$

where a is a constant.

Put $X = x - at$, $T = x + at$.

Then
$$\frac{\partial V}{\partial x} = \frac{\partial V}{\partial X} \frac{\partial X}{\partial x} + \frac{\partial V}{\partial T} \frac{\partial T}{\partial x} = \frac{\partial V}{\partial X} + \frac{\partial V}{\partial T},$$

and
$$\frac{\partial^2 V}{\partial x^2} = \frac{\partial}{\partial x} \left(\frac{\partial V}{\partial x} \right) = \left(\frac{\partial}{\partial X} + \frac{\partial}{\partial T} \right) \left(\frac{\partial V}{\partial X} + \frac{\partial V}{\partial T} \right)$$

$$= \frac{\partial^2 V}{\partial X^2} + 2 \frac{\partial^2 V}{\partial X \partial T} + \frac{\partial^2 V}{\partial T^2}.$$

Similarly
$$\frac{\partial V}{\partial t} = \frac{\partial V}{\partial X} \frac{\partial X}{\partial t} + \frac{\partial V}{\partial T} \frac{\partial T}{\partial t} = a \left(-\frac{\partial V}{\partial X} + \frac{\partial V}{\partial T} \right),$$

and
$$\frac{\partial^2 V}{\partial t^2} = a^2 \left(\frac{\partial^2 V}{\partial X^2} - 2 \frac{\partial^2 V}{\partial X \partial T} + \frac{\partial^2 V}{\partial T^2} \right).$$

Substituting in equation (1), we get

$$4 \frac{\partial^2 V}{\partial X \partial T} = 0,$$

giving
$$\frac{\partial V}{\partial T} = \phi(T),$$

and
$$V = f(X) + \int \phi(T)\, dT,$$

or
$$V = f(X) + F(T),$$

i.e.
$$V = f(x - at) + F(x + at), \quad \ldots\ldots\ldots\ldots\ldots\ldots(2)$$

where f and F are arbitrary functions.

$f(x - at)$ is unaltered if x is increased by a and t by 1; hence it represents a wave moving along the positive direction of the axis of x with speed a. Similarly $F(x + at)$ represents a wave moving along the same line with the same speed in the opposite direction.

An alternative method of solving equation (1) is to use the general result given in Art. 145, with x, y, z replaced by t, x, V respectively. Writing the equation as

$$\left(\frac{\partial^2}{\partial t^2} - a^2 \frac{\partial^2}{\partial x^2}\right) V = 0,$$

or

$$(D^2 - a^2 D'^2) V = 0,$$

we get the auxiliary equation $m^2 - a^2 = 0$, whose roots are $-a$ and a, leading to

$$V = f(x - at) + F(x + at).$$

179. Particular solutions of the Wave equation. This is

$$\frac{\partial^2 V}{\partial x^2} + \frac{\partial^2 V}{\partial y^2} + \frac{\partial^2 V}{\partial z^2} = \frac{1}{a^2} \frac{\partial^2 V}{\partial t^2}, \quad \dots\dots\dots\dots(3)$$

where a is a constant. It is the three-dimensional analogue of the one-dimensional equation (1). Let us attempt to find a solution similar to (2), but with x, y, z, t instead of x, t.

Try $\quad V = f(lx + my + nz - at) + F(lx + my + nz + at), \quad \dots\dots\dots(4)$

where l, m, n are constants. Equation (3) is satisfied if

$$l^2 + m^2 + n^2 = 1.$$

In this case l, m, n are the actual direction-cosines of a certain line. The first function is unaltered if x, y, z, t are increased by la, ma, na, 1 respectively, so it represents a plane wave (whose normal has direction-cosines l, m, n) moving parallel to itself with speed a. The second function represents a parallel wave moving with the same speed in the opposite direction. Hence equation (4) represents the propagation of plane waves. This is one particular solution of the Wave equation.

To obtain a solution for spherical waves transform equation (3) into spherical polar coordinates. The work is essentially a transformation of Laplace's equation,* and we get

$$\frac{1}{r^2} \frac{\partial}{\partial r}\left(r^2 \frac{\partial V}{\partial r}\right) + \frac{1}{r^2 \sin \theta} \frac{\partial}{\partial \theta}\left(\sin \theta \frac{\partial V}{\partial \theta}\right) + \frac{1}{r^2 \sin^2 \theta} \frac{\partial^2 V}{\partial \phi^2} = \frac{1}{a^2} \frac{\partial^2 V}{\partial t^2}. \quad \dots(5)$$

* See Edwards' *Differential Calculus*, Art. 532, or, for a simpler method using Gauss' theorem, any book on Analytical Statics.

For a solution symmetrical in all directions about the origin, *i.e.* independent of θ and ϕ, this reduces to

$$\frac{1}{r^2}\frac{\partial}{\partial r}\left(r^2\frac{\partial V}{\partial r}\right)=\frac{1}{a^2}\frac{\partial^2 V}{\partial t^2}. \quad\ldots\ldots\ldots\ldots\ldots\ldots(6)$$

By the transformation $U=rV$, we get

$$\frac{\partial U}{\partial r}=V+r\frac{\partial V}{\partial r},$$

and

$$\frac{\partial^2 U}{\partial r^2}=2\frac{\partial V}{\partial r}+r\frac{\partial^2 V}{\partial r^2}=\frac{1}{r}\frac{\partial}{\partial r}\left(r^2\frac{\partial V}{\partial r}\right),$$

so equation (6) becomes, after multiplication by r,

$$\frac{\partial^2 U}{\partial r^2}=\frac{1}{a^2}\frac{\partial^2 U}{\partial t^2},$$

giving

$$U=f(r-at)+F(r+at),$$

i.e.

$$V=\frac{1}{r}\{f(r-at)+F(r+at)\}. \quad\ldots\ldots\ldots\ldots\ldots(7)$$

This represents two spherical waves with the same speed a, one diverging from the origin and the other approaching it. The factor $1/r$ shows that the intensity of the disturbance decreases as the distance from the origin increases.

180. Poisson's (or Liouville's) general solution. This obtains V at any time t at a point P in terms of the mean values over a sphere of centre P and variable radius at of the functions, say g and G, which give the values of V and $\dfrac{\partial V}{\partial t}$ respectively when $t=0$ at any point in space.

Take spherical polar coordinates with P as origin.

Now the mean value $\bar{f}$ of a function $f(r,\ \theta,\ \phi,\ t)$ over a sphere of radius r is given by

$$\bar{f}=\frac{1}{4\pi r^2}\int_0^\pi\int_0^{2\pi}fr^2\sin\theta\,d\theta\,d\phi=\frac{1}{4\pi}\int_0^\pi\int_0^{2\pi}f\sin\theta\,d\theta\,d\phi.$$

Take the mean value over a sphere of radius r of each term of the Wave equation (5). The second term becomes

$$\frac{1}{4\pi}\int_0^\pi\int_0^{2\pi}\frac{1}{r^2}\frac{\partial}{\partial\theta}\left(\sin\theta\frac{\partial V}{\partial\theta}\right)d\theta\,d\phi=\frac{1}{4\pi r^2}\int_0^{2\pi}\left[\sin\theta\frac{\partial V}{\partial\theta}\right]_0^\pi d\phi,$$

and the third

$$\frac{1}{4\pi}\int_0^\pi\int_0^{2\pi}\frac{1}{r^2\sin\theta}\frac{\partial^2 V}{\partial\phi^2}d\theta\,d\phi=\frac{1}{4\pi r^2}\int_0^\pi\left[\frac{1}{\sin\theta}\frac{\partial V}{\partial\phi}\right]_0^{2\pi}d\theta.$$

Both are zero, for sin θ vanishes at both limits, while $\phi = 2\pi$ gives the same value of $\dfrac{\partial V}{\partial \phi}$ as $\phi = 0$ (which is really the same position). The first and fourth terms do not vanish. These give

$$\frac{1}{r^2}\frac{\partial}{\partial r}\left(r^2\frac{\partial \bar{V}}{\partial r}\right) = \frac{1}{a^2}\frac{\partial^2 \bar{V}}{\partial t^2}, \quad \dotfill (8)$$

so that

$$r\bar{V} = f(r-at) + F(r+at), \quad \dotfill (9)$$

$$= f(-at) + F(at) + r\{f'(-at) + F'(at)\} + \tfrac{1}{2}r^2\{f''(-at)$$
$$+ F''(at)\} + \dots . \quad \dotfill (10)$$

If $\bar{V}$ is to be finite at the origin $(r=0)$ for all values of t

$$f(-at) + F(at) = 0,$$

giving

$$f'(-at) = \frac{df(-at)}{d(-at)} = -\frac{d\{-F(at)\}}{d(at)} = F'(at).$$

Hence, from equation (10), using a suffix 0 to denote the result of putting $r = 0$,

$$\bar{V}_0 = f'(-at) + F'(at) = 2F'(at). \quad \dotfill (11)$$

From equation (9),

$$\frac{\partial}{\partial r}(r\bar{V}) = f'(r-at) + F'(r+at),$$

and

$$r\frac{\partial \bar{V}}{\partial t} = -af'(r-at) + aF'(r+at),$$

whence

$$2F'(r+at) = \frac{\partial}{\partial r}(r\bar{V}) + \frac{r}{a}\frac{\partial \bar{V}}{\partial t},$$

for all values of r and t. Putting $t = 0$, and using the initial conditions, we get

$$2F'(r) = \frac{\partial}{\partial r}(r\bar{g}) + \frac{r\bar{G}}{a},$$

whence, giving r the special value at, and using equation (11),

$$\bar{V}_0 = \frac{\partial}{\partial(at)}(at\bar{g}) + t\bar{G}.$$

But $\bar{V}_0$, the average value of V over a sphere of zero radius, is simply V_0.

Thus

$$V_0 = \frac{\partial}{\partial t}(t\bar{g}) + t\bar{G}.$$

It follows from the form of this solution that at any time, t, the value of V at any point P depends only upon the initial disturbance at points on the surface of a sphere of centre P and radius at. In an explosion the initial disturbance is generally confined to a region bounded by a closed surface S. If P is external to this surface and d is the shortest distance from P to S, no effect will be produced at until a time d/a has elapsed, for before then the sphere concerned will go only through regions where there is no initial disturbance. At any time t the Wave-front (the locus of points just reached by the disturbance) is a surface obtained from S by producing all the outward normals a distance at.

Other general solutions of the Wave equation have been given by Kirchhoff * (whose form is of importance in Optics), Whittaker, and Bateman.‡

Example for solution

Verify that

$$V = \int_{-\pi}^{\pi}\int_{-\pi}^{\pi} f(x \sin u \cos v + y \sin u \sin v + z \cos u + at, u, v)\, du\, dv$$

where the function f is such that differentiations under the sign of integration are legitimate, is a solution of the Wave equation. [This is Whittaker's solution.]

181. Other differential equations of Mathematical Physics. These include Laplace's equation

$$\frac{\partial^2 V}{\partial x^2} + \frac{\partial^2 V}{\partial y^2} + \frac{\partial^2 V}{\partial z^2} = 0;$$

Poisson's equation

$$\frac{\partial^2 V}{\partial x^2} + \frac{\partial^2 V}{\partial y^2} + \frac{\partial^2 V}{\partial z^2} = -4\pi\gamma\rho;$$

the equation of the conduction of heat

$$\frac{\partial^2 V}{\partial x^2} + \frac{\partial^2 V}{\partial y^2} + \frac{\partial^2 V}{\partial z^2} = \frac{1}{a}\frac{\partial V}{\partial t};$$

the equation of telegraphy

$$LK\frac{\partial^2 V}{\partial t^2} + KR\frac{\partial V}{\partial t} = \frac{\partial^2 V}{\partial x^2};$$

Schrödinger's equation (of Wave Mechanics)

$$\frac{\partial^2 \psi}{\partial x^2} + \frac{\partial^2 \psi}{\partial y^2} + \frac{\partial^2 \psi}{\partial z^2} + \frac{8\pi^2 m(w - V)\psi}{h^2} = 0,$$

of which, in a particular case, a solution is indicated in the example at the end of this article.

* See Jeans, *Electricity and Magnetism* (5th ed.) Art. 580, or Drude, *Theory Optics* (translated by Mann and Millikan), p. 179. For a physical discussion of another equation connected with wave propagation, see Jeans, Art. 645.

† See Whittaker and Watson, *Modern Analysis* (4th ed.), Art. 18·6.

‡ *Ibid.* p. 402.

These equations may be discussed from two points of view. Treatises on pure mathematics * give a local discussion of the general solutions, but the physicist complains of the great length of the discussion, and of the difficulty of applying these general solutions. On the other hand, treatises on physics use a combination of logic and intuition to obtain solutions (usually particular rather than general) which have a physical meaning, and might never have been reached at all by logic alone.

There is usually little doubt that these results are substantially correct, but any uncertainty, however slight, is repugnant to the pure mathematician. Probably his knowledge of the unreliability of intuition in pure mathematics prevents him from appreciating the valuable and generally reliable part that it has played in physics.

Either point of view requires a very extensive treatment, which cannot be given here.†

[The more elementary equations of mathematical physics have been dealt with in several places in this book, e.g. pp. 24, 28, 29, 36, 46-48, 49-61, 189, 190, 234, 235, 241-247, 250, 251.]

Example for solution

From Schrödinger's equation, with $h/2\pi$ replaced by K, and V given the special form $-e^2/r$, obtain, by changing from Cartesian to spherical polar coordinates, replacing ψ by $r^{-1}U(r)S(\theta, \phi)$, (cf. Art. 179),

$$\left\{\frac{d^2U}{dr^2} + \frac{2m}{K^2}\left(w + \frac{e^2}{r}\right)U\right\}S + \frac{U}{r^2}\left\{\frac{1}{\sin\theta}\frac{\partial}{\partial\theta}\left(\sin\theta\frac{\partial S}{\partial\theta}\right) + \frac{1}{\sin^2\theta}\frac{\partial^2 S}{\partial\phi^2}\right\} = 0.$$

By taking r^lS to be a solution of Laplace's equation (and hence $r^{l+1}S$ a solution of what our last equation becomes when m is replaced by zero), obtain

$$\frac{d^2U}{dr^2} + \left\{\frac{2}{K^2}m\left(w + \frac{e^2}{r}\right) - \frac{l(l+1)}{r^2}\right\}U = 0.$$

Finally, by the substitutions

$$R = \frac{2r}{K}\sqrt{(-2mw)}, \quad k = \frac{e^2}{K}\sqrt{\left(\frac{-m}{2w}\right)},$$

reduce it to Whittaker's confluent hypergeometric equation (Ex. 7, following Art. 177), with U, R, and $(l + \frac{1}{2})$ in place of y, x, m respectively.

[For the physical meaning of this work see Biggs, *Wave Mechanics*.]

---

* e.g. Goursat, *Cours d'Analyse Mathématique*, Vol. III.

† See Riemann-Weber, *Partielle Differentialgleichungen und deren Anwendung auf physikalische Fragen* (the latest edition has been quite transformed, and bears the title *Die Differential- und Integralgleichungen der Mechanik und Physik*); Jeffreys, *Operational Methods in Mathematical Physics* (Heaviside's methods); Picard, *Leçons sur Quelques Types Simples d'Équations aux Derivées Partielles avec les Applications à la Physique Mathématique*; Webster, *Partial Differential Equations of Mathematical Physics*; Bateman, *Partial Differential Equations of Mathematical Physics*; Sneddon, *Elements of Partial Differential Equations*.

182. Numerical approximation. Adams' method. Resuming the subject of Chapter VIII. we shall now give a method * which Prof. Whittaker considers to be the best of all those tested in the Edinburgh Mathematical Laboratory. It may be shortly described as the combined use of Taylor's theorem and of a certain formula, given below, belonging to the Calculus of Finite Differences. Taylor's series is used for increments of x small enough to make the series converge rapidly. After thus obtaining a few (generally four) values of y we have sufficient data to obtain further values from the Difference Formula, thus avoiding the use of Taylor's series for large increments of x. The error in the final result may be estimated by a method explained below.

Ex.   Given the differential equation $x\dfrac{dy}{dx} + y - 2x = 0$, with the initial values $x = 2$, $y = 2\cdot5$, find the values of y corresponding to $x = 2\cdot05$, $2\cdot10$, $2\cdot15$, $2\cdot20$, $2\cdot25$, $2\cdot30$, $2\cdot35$, $2\cdot40$, $2\cdot45$, $2\cdot50$, and estimate the order of the errors in the results.

We shall use h to denote the increment of x, x_n for $(x_0 + nh)$, and y_n for the value of y corresponding to x_n.

The successive differential coefficients of y with respect to x will be denoted by y', y'', y''', ... and their initial values by the suffix $_0$.

To determine the coefficients in the Taylor's series

$$y = y_0 + (x - 2)y_0' + \frac{(x-2)^2}{2!}y_0'' + \frac{(x-2)^3}{3!}y_0''' + \dots\,,$$

put $x = 2$, $y = 2\cdot5$ in the original differential equation and in the results of differentiating it successively. We get

$$xy' + y - 2x = 0, \qquad\qquad y_0' = \tfrac{3}{4};$$
$$xy'' + 2y' - 2 = 0, \qquad y_0'' = 1 - y_0' = \tfrac{1}{4};$$

and so on, leading finally to

$$y = 2\tfrac{1}{2} + \tfrac{3}{4}(x-2) + \tfrac{1}{8}(x-2)^2 - \tfrac{1}{16}(x-2)^3 + \tfrac{1}{32}(x-2)^4 - \tfrac{1}{64}(x-2)^5 + \dots \quad (1)$$

If we put in succession $x = 2\cdot05$, $2\cdot10$, $2\cdot15$, $2\cdot20$ in this series, the numerical value of the last term written there will be, at its greatest,

$$\tfrac{1}{64}(0\cdot2)^5 = 0\cdot000005,$$

so the corresponding values of y will be correct to five places of decimals. Thus we get

$$y_1 = 2\cdot53780, \quad y_2 = 2\cdot57619, \quad y_3 = 2\cdot61512, \quad y_4 = 2\cdot65455.$$

---

* Due to J. C. Adams and described in *Theories of Capillary Action*, by F Bashforth and J. C. Adams. See also Chap. XIV. of *The Calculus of Observations* by E. T. Whittaker and G. Robinson.

John Couch Adams, of Cambridge (1819-1892), is best known by his deduction of the existence of the then unknown planet Neptune from the perturbations o Uranus.

We now use the Difference Formula *

$$y_{n+1} - y_n = q_n + \tfrac{1}{2}\Delta q_{n-1} + \tfrac{5}{12}\Delta^2 q_{n-2} + \tfrac{3}{8}\Delta^3 q_{n-3} + \tfrac{251}{720}\Delta^4 q_{n-4} + \ldots \quad (2)$$

where q_n denotes the value of $h\dfrac{dy}{dx}$ when $x = x_n$, $y = y_n$, so in our example

$$q_n = 0\cdot05(2 - y_n/x_n),$$

Δq_n denotes $q_{n+1} - q_n$,

$\Delta^2 q_n$ denotes $\Delta q_{n+1} - \Delta q_n$, and so on.

Putting $n = 4$, equation (2) gives

$$y_5 = y_4 + q_4 + \tfrac{1}{2}\Delta q_3 + \tfrac{5}{12}\Delta^2 q_2 + \tfrac{3}{8}\Delta^3 q_1 + \tfrac{251}{720}\Delta^4 q_0 + \ldots \quad \ldots \ldots \ldots \ldots (3)$$

Now $$q_0 = 0\cdot05(2 - y_0/x_0) = 0\cdot03750.$$

Similarly

$$q_1 = 0\cdot03810, \quad q_2 = 0\cdot03866, \quad q_3 = 0\cdot03918, \quad q_4 = 0\cdot03967.$$

Hence $\Delta q_0 = q_1 - q_0 = 0\cdot00060$, and so on. For the calculation of these differences it is convenient to write the numbers in the form of the following table:

q	Δq	$\Delta^2 q$	$\Delta^3 q$	$\Delta^4 q$
$q_0 = 0\cdot03750$				
	$0\cdot00060$			
$q_1 = 0\cdot03810$		$-0\cdot00004$		
	$0\cdot00056$		$0\cdot00000$	
$q_2 = 0\cdot03866$		$-0\cdot00004$		$0\cdot00001$
	$0\cdot00052$		$0\cdot00001$	
$q_3 = 0\cdot03918$		$-0\cdot00003$		
	$0\cdot00049$			
$q_4 = 0\cdot03967$				

Let us examine the numerical value of the various orders of differences shown in this table. On passing from Δq to $\Delta^2 q$ we find a decided decrease. But there is only a slight further decrease in $\Delta^3 q$, and none at all in $\Delta^4 q$. This suggests that $\Delta^3 q$ and $\Delta^4 q$ are inaccurate. We therefore disregard them and apply equation (3) in the approximate form.

$$y_5 = y_4 + q_4 + \tfrac{1}{2}\Delta q_3 + \tfrac{5}{12}\Delta^2 q_2$$
$$= 2\cdot65455 + 0\cdot03967 + 0\cdot00025 - 0\cdot00001$$
$$= 2\cdot69446.$$

The error due to taking only four terms of the series may be expected to be distinctly less than the last term retained, and therefore negligible to five places of decimals. On the other hand, although the true value of the first and second terms cannot differ from their respective five-figure

* This is obtained by integrating with respect to r, between the limits 0 and 1, the interpolation formula

$$q_n(x_n + rh) = q_n + r\Delta q_{n-1} + \frac{r(r+1)}{2!}\Delta^2 q_{n-2} + \frac{r(r+1)(r+2)}{3!}\Delta^3 q_{n-3} \cdots .$$

See Whittaker and Robinson's *Calculus of Observations*, p. 365.

approximations by more than 0·000005, these errors may, in an unlucky case, be doubled in Δq and doubled again in $\Delta^2 q$. Even if every term used in the calculation of y_5 had its greatest possible error, and if these errors all occurred with the same sign, the resulting error in y_5 would be less than 0·000025.

We now calculate $q_5 = 0·05(2 - y_5/x_5) = 0·04012$. This can be relied upon as accurate to five places of decimals, as an error of 0·000025 in y_5 would be multiplied by the small number 0·05/2·25, and so become negligible to our order of approximation. Adding the value q_5 to our table we can at once get $\Delta q_4 = 0·00045$, and $\Delta^2 q_3 = -0·00004$, and hence

$$y_6 = y_5 + q_5 + \tfrac{1}{2}\Delta q_4 + \tfrac{5}{12}\Delta^2 q_3$$

$$= 2·69446 + 0·04012 + 0·00022 - 0·00002 = 2·73478.$$

(As the last digit is odd for both Δq_3 and Δq_4, in halving we have to choose between two equally good five-figure approximations. We choose the larger and smaller alternately, so as to prevent an accumulation of errors.)

Proceeding in this way, we obtain the results given in the following table:

y	q	Δq	$\Delta^2 q$
$y_0 = 2·50000$	$q_0 = 0·03750$		
		0·00060	
$y_1 = 2·53780$	$q_1 = 0·03810$		$-0·00004$
		0·00056	
$y_2 = 2·57619$	$q_2 = 0·03866$		$-0·00004$
		0·00052	
$y_3 = 2·61512$	$q_3 = 0·03918$		$-0·00003$
		0·00049	
$y_4 = 2·65455$	$q_4 = 0·03967$		$-0·00004$
		0·00045	
$y_5 = 2·69446$	$q_5 = 0·04012$		$-0·00002$
		0·00043	
$y_6 = 2·73478$	$q_6 = 0·04055$		$-0·00003$
		0·00040	
$y_7 = 2·77554$	$q_7 = 0·04095$		$-0·00003$
		0·00037	
$y_8 = 2·81668$	$q_8 = 0·04132$		$-0·00002$
		0·00035	
$y_9 = 2·85817$	$q_9 = 0·04167$		

$$y_{10} = 2·90001$$

The y's may be expected to have small errors in the last digit. A a matter of fact, the differential equation that we have chosen has th exact solution $y = x + 1/x$. Calculating from this we find an error o 0·00002 in y_5, 0·00001 in y_7, y_8, y_9, y_{10}, and zero in the others.

To obtain greater accuracy we may calculate y_1, y_2, y_3, y_4, to mor places of decimals, say eight. The student should do this. It will b

ound that $\Delta q, \Delta^2 q, \Delta^3 q$ and $\Delta^4 q$ all appear to be reliable, and so capable
f use in the difference formula. The final results are

$$y_0 = 2{\cdot}500{,}000{,}00;$$
$$y_1 = 2{\cdot}537{,}804{,}88;$$
$$y_2 = 2{\cdot}576{,}190{,}48;$$
$$y_3 = 2{\cdot}615{,}116{,}28;$$
$$y_4 = 2{\cdot}654{,}545{,}45;$$
$$y_5 = 2{\cdot}694{,}444{,}42 \text{ (error } -2 \text{ in last digit);}$$
$$y_6 = 2{\cdot}734{,}782{,}58 \text{ (error } -3 \text{ in last digit);}$$
$$y_7 = 2{\cdot}775{,}531{,}88 \text{ (error } -3 \text{ in last digit);}$$
$$y_8 = 2{\cdot}816{,}666{,}61 \text{ (error } -6 \text{ in last digit);}$$
$$y_9 = 2{\cdot}858{,}163{,}23 \text{ (error } -4 \text{ in last digit);}$$
$$y_{10} = 2{\cdot}899{,}999{,}93 \text{ (error } -7 \text{ in last digit).}$$

The last term used in the calculation of y_{10}, namely $\frac{251}{720}\Delta^4 q_5$, has
he value $-0{\cdot}000{,}000{,}09$. The magnitude of this indicates that the
rrors this time (unlike those for the five-figure work) probably occur
rom neglect of the higher differences. To remedy this, we can either
alculate y_5 accurately from the Taylor's series, and use $\Delta^5 q$, or (as is
more usual) diminish the interval sufficiently to ensure that $\Delta^5 q$ may
e negligible to our desired order of approximation.

183. Remes' extension of the method of Arts. 90-93. E. Remes
has given * a systematic method of determining suitable values for
he numbers m and M defined in Art. 92, namely,

Case (i) $m = f(a, b)$, $M = f\{a+h, b+hf(a+h, b+h)\}$, if
$$df/dx > 0, \ \partial f/\partial y > 0;$$

Case (ii) $m = f(a, b)$, $M = f\{a+h, b+hf(a, b)\}$, if
$$df/dx > 0, \ \partial f/\partial y < 0;$$

Case (iii) $m = f\{a+h, b+hf(a+h, b-h)\}$, $M = f(a, b)$, if
$$df/dx < 0, \ \partial f/\partial y > 0 \ ;$$

Case (iv) $m = f\{a+h, b+hf(a, b)\}$, $M = f(a, b)$, if
$$df/dx < 0, \ \partial f/\partial y < 0.$$

These values satisfy the inequalities (7), (8), (9), (10) of p. 107.
Remes shows that if we define R and r by the relations

$$r = \tfrac{1}{2}h\{f(a, b) + f(a+h, b+mh)\}, \ R = \tfrac{1}{2}h\{f(a, b) + f(a+h, b+Mh)\},$$

the inequalities hold also when q is replaced by r and Q by R.

* *Phil. Mag.*, Series 7, Vol. 5, Feb. 1928.

Let Σ' denote $\frac{1}{3}(p+2Q)$ if $\dfrac{\partial f}{\partial y}\dfrac{d^2f}{dx^2}>0$,

but $\frac{1}{3}(P+2q)$ if $\dfrac{\partial f}{\partial y}\dfrac{d^2f}{dx^2}<0$.

Let Σ'' denote $\frac{1}{3}(2p+R)$ if $\dfrac{\partial f}{\partial y}\dfrac{d^2f}{dx^2}>0$,

but $\frac{1}{3}(2P+r)$ if $\dfrac{\partial f}{\partial y}\dfrac{d^2f}{dx^2}<0$.

Then Remes proves that the errors in the approximations Σ' and Σ'' are at least of the fourth and third order respectively (taking the increment to be small of the first order) if $\dfrac{\partial f}{\partial y}\dfrac{df}{dx}\dfrac{d^2f}{dx^2}<0$, but at least of the third and fourth orders respectively if $\dfrac{\partial f}{\partial y}\dfrac{df}{dx}\dfrac{d^2f}{dx^2}>0$. This conclusion depends upon m and M being chosen as explained above. The error in the example on p. 107 was much smaller than would be expected from this result, but this seems to be due to luck in the choice of m and M, which were not obtained in the way stipulated by Remes. In general the methods of Adams or Kutta seem much better.

APPENDIX A

he necessary and sufficient condition that the equation $M\,dx + N\,dy = 0$
should be exact

(a) If the equation is exact,

$$M\,dx + N\,dy = \text{a perfect differential} = df, \text{ say.}$$

So
$$M = \frac{\partial f}{\partial x} \quad \text{and} \quad N = \frac{\partial f}{\partial y};$$

erefore
$$\frac{\partial N}{\partial x} = \frac{\partial^2 f}{\partial x\,\partial y} = \frac{\partial^2 f}{\partial y\,\partial x} = \frac{\partial M}{\partial y},$$

the condition is *necessary*.

(b) Conversely, if $\dfrac{\partial N}{\partial x} = \dfrac{\partial M}{\partial y}$, put $F = \displaystyle\int M\,dx$, where the integration

performed on the supposition that y is constant.

Then
$$\frac{\partial F}{\partial x} = M \quad \text{and} \quad \frac{\partial^2 F}{\partial x\,\partial y} = \frac{\partial^2 F}{\partial y\,\partial x} = \frac{\partial M}{\partial y} = \frac{\partial N}{\partial x}.$$

So
$$\frac{\partial}{\partial x}\left(N - \frac{\partial F}{\partial y}\right) = 0,$$

$$N - \frac{\partial F}{\partial y} = \text{a constant as far as } x \text{ is concerned, that is,}$$
$$\text{a function of } y,$$
$$= \phi(y), \text{ say.}$$

Then
$$N = \frac{\partial F}{\partial y} + \phi(y).$$

Now put
$$f = F + \int \phi(y)\,dy.$$

Then
$$N = \frac{\partial f}{\partial y}.$$

Also $\quad M = \dfrac{\partial F}{\partial x}$ by definition of F

$$= \frac{\partial f}{\partial x}, \text{ since } F \text{ and } f \text{ differ only by a function of } y.$$

Thus $M\,dx + N\,dy = \dfrac{\partial f}{\partial x}\,dx + \dfrac{\partial f}{\partial y}\,dy = df$, **a** perfect **differential.**

So the equation is exact, that is, the condition is *sufficient*.

[Our assumption that $\dfrac{\partial^2 f}{\partial x\,\partial y} = \dfrac{\partial^2 f}{\partial y\,\partial x}$ is justified if f and its first and

econd partial differential coefficients are continuous. See **Lamb's**
nfinitesimal Calculus, 2nd ed., Art. 210; or 3rd ed., Art. 193.]

APPENDIX B

The equation $P(x, y, z)\dfrac{\partial f}{\partial x} + Q(x, y, z)\dfrac{\partial f}{\partial y} + R(x, y, z)\dfrac{\partial f}{\partial z} = 0$, *regarded* a *four-dimensional, has no special integrals.* (See Art. 127.)

Let
$$u(x, y, z) = a,$$
$$v(x, y, z) = b,$$

be any two independent integrals of the equations
$$dx/P = dy/Q = dz/R.$$

Then we easily prove that

$$P\frac{\partial u}{\partial x} + Q\frac{\partial u}{\partial y} + R\frac{\partial u}{\partial z} = 0 \quad\text{......................(1}$$

and
$$P\frac{\partial v}{\partial x} + Q\frac{\partial v}{\partial y} + R\frac{\partial v}{\partial z} = 0. \quad\text{......................(2}$$

The left-hand side of (1) does not contain a, and therefore canno vanish merely in consequence of the relation $u = a$. Hence it mus vanish identically. Similarly equation (2) is satisfied identically.

Now let $f = w(x, y, z)$ be *any* integral of the original partia differential equation, so that

$$P\frac{\partial w}{\partial x} + Q\frac{\partial w}{\partial y} + R\frac{\partial w}{\partial z} = 0. \quad\text{......................(3}$$

This is another identical equation, since f does not occur in it.
Eliminating P, Q, R from (1), (2), (3), we get

$$\frac{\partial(u, v, w)}{\partial(x, y, z)} = 0 \text{ identically.}$$

Hence w is a function of u and v, say
$$w = \phi(u, v).$$

That is, $f = w$ is part of the *General Integral*, and therefore, as $f = u$ is *any* integral, there are no *Special Integrals*.

[The student will notice the importance in the above work of a differential equation being satisfied *identically*. Hill's new classificatio of the integrals of Lagrange's linear equation (*Proc. London Math. Soc* 1917) draws a sharp distinction between integrals that satisfy an equation identically and those which have not this property.]

APPENDIX C

The expression obtained for dz by Jacobi's method of solving a single partial differential equation of the first order (Art. 140) is always integrable.

To prove that $dz = p_1\,dx_1 + p_2\,dx_2 + p_3\,dx_3$

is integrable it is necessary and sufficient to prove that

$$L = M = N = 0, \quad\ldots\ldots\ldots\ldots\ldots\ldots\ldots\ldots\text{(A)}$$

where $\quad L \equiv \dfrac{\partial p_2}{\partial x_3} - \dfrac{\partial p_3}{\partial x_2}, \quad M \equiv \dfrac{\partial p_3}{\partial x_1} - \dfrac{\partial p_1}{\partial x_3}, \quad N \equiv \dfrac{\partial p_1}{\partial x_2} - \dfrac{\partial p_2}{\partial x_1}.$

Now, by adding equations (8), (9), (10) of Art. 140 and using the relation $(F, F_1) = 0$, but *not* assuming the truth of (A), we get

$$L\,\frac{\partial(F, F_1)}{\partial(p_2, p_3)} + M\,\frac{\partial(F, F_1)}{\partial(p_3, p_1)} + N\,\frac{\partial(F, F_1)}{\partial(p_1, p_2)} = 0. \quad\ldots\ldots\ldots\ldots\text{(B)}$$

Similarly $\quad L\,\dfrac{\partial(F_1, F_2)}{\partial(p_2, p_3)} + M\,\dfrac{\partial(F_1, F_2)}{\partial(p_3, p_1)} + N\,\dfrac{\partial(F_1, F_2)}{\partial(p_1, p_2)} = 0 \quad\ldots\ldots\ldots\ldots\text{(C)}$

and $\quad L\,\dfrac{\partial(F_2, F)}{\partial(p_2, p_3)} + M\,\dfrac{\partial(F_2, F)}{\partial(p_3, p_1)} + N\,\dfrac{\partial(F_2, F)}{\partial(p_1, p_2)} = 0. \quad\ldots\ldots\ldots\ldots\text{(D)}$

From equations (B), (C), (D) we see that either $L = M = N = 0$ or $\Delta = 0$, where Δ is the determinant whose constituents are the coefficients of L, M, N in (B), (C), (D).

But these coefficients are themselves the co-factors of the constituents of the determinant

$$J = \frac{\partial(F_2, F, F_1)}{\partial(p_1, p_2, p_3)},$$

and by the theory of determinants $\Delta = J^2$.

Now J cannot vanish,* for this would imply the existence of a functional relation which would contradict the hypothesis of Art. 140 that the p's can be found as functions of the x's from

$$F = F_1 - a_1 = F_2 - a_2 = 0.$$

Hence $\quad\quad \Delta \neq 0;$ therefore $L = M = N = 0.$

---

* All the equations of this appendix are satisfied *identically*.

231

APPENDIX D

Suggestions for further reading

No attempt will be made here to give a complete list of works on differential equations. We shall merely give the names of a very small number of the most prominent *works*, classified in three sections.

I. Chiefly of analytical interest (forming a continuation to Chapter X.).

(*a*) Forsyth: *Theory of Differential Equations* (1890 and later years, Cambridge Univ. Press).

This important work is in six volumes, and is the most exhaustive treatise in English upon the subject. It should not be confused with his more elementary work in one volume (4th ed. 1914, Macmillan).

(*b*) Goursat: *Cours d'Analyse mathématique*, Vols. II. and III. (2nd ed. 1911-15, Gauthier-Villars; English translation published by Ginn).

This deals almost entirely with existence theorems.

(*c*) Schlesinger: *Handbuch der Theorie der linearen Differentialgleichungen* (1895-8, 3 vols., Teubner).

II. Partly analytical but also of geometrical interest.

(*a*) Goursat: *Équations aux dérivées partielles du premier ordre* (1891).

(*b*) Goursat: *Équations aux dérivées partielles du second ordre* (1896-98, 2 vols., Hermann et fils).

(*c*) Page: *Ordinary differential equations from the standpoint of Lie's Transformation Groups* (1897, Macmillan).

This deals with the elements in a highly original manner.

III. Of physical interest (forming a continuation to Chapters III. and IV.).

(*a*) Riemann: *Partielle Differentialgleichungen und deren Anwendung auf physikalische Fragen* (1869, Vieweg).

(*b*) Riemann-Weber: A revised eidtion of (*a*), with extensive additions (1900-01, Vieweg).

(*c*) Bateman: *Differential Equations* (1918, Longmans).

This contains many references to recent researches.

Addenda (published since 1920).

I. (*d*) Ince: *Ordinary Differential Equations* (1927, Longmans).

I. (*e*) Bieberbach: *Differentialgleichungen* (3rd ed., 1930, Springer).

I. (*f*) Levy and Baggott: *Numerical Studies in Differential Equations* Vol. I (1934, Watts).

I. (*g*) Poole: *Linear Differential Equations* (1936, Oxford).

I. (*h*) Coddington and Levinson : *Theory of Ordinary Differential Equations* (1955, McGraw-Hill).

I. (*i*) Collatz: *Numerische Behandlung von Differentialgleichungen* (2nd ed., 1955, Springer).

III. (*d*) McLachlan: *Ordinary Non-linear Differential Equations* (1950 Oxford).

Much research on this subject has been published recently.

III. (*e*) Stoker: *Nonlinear Vibrations* (1950, Interscience).

For other references see the second footnote to Art. 181.

MISCELLANEOUS EXAMPLES ON THE WHOLE BOOK

(1) $\dfrac{dy}{dx} = \dfrac{y^3 + 3x^2 y}{x^3 + 3xy^2}$.   [London.]

(2) $\dfrac{dy}{dx} + 2xy = 2x(1 + x^2)$.   [London.]

(3) $\tan y \dfrac{dy}{dx} + \tan x = \cos y \cos^3 x$.   [London.]

(4) $y = 2x\dfrac{dy}{dx} + \left(\dfrac{dy}{dx}\right)^2$.   [London.]

(5) $(1 - x^2)\dfrac{dy}{dx} - xy = x^2 y^2$.   [London.]

(6) $(D^2 + 4)y = \sin 2x$.   [London.]

(7) $(D^3 - D^2 + 3D + 5)y = x^2 + e^x \cos 2x$.   [London.]

(8) $(x^3 D^3 + x^2 D^2)y = 1 + x + x^2$.   [London.]

(9) $\cos x \sin x \dfrac{dy}{dx} = y + \cos x$.   [London.]

(10) $\left.\begin{array}{l} \dfrac{dx}{dt} = x + y + 2\cos t, \\[2mm] \dfrac{dy}{dt} = 3x - y. \end{array}\right\}$ [London.]

(11) $y = x\left(\dfrac{dy}{dx}\right)^3 + 1$.   [London.]

(12) $y\dfrac{d^2 y}{dx^2} - 2\left(\dfrac{dy}{dx}\right)^2 = y^2$.   [London.]

(13) $(D^4 + 8D^2 + 16)y = x\cos 2x$.   [London.]

(14) $\displaystyle\int x^2\, dy + \int xy\, dx = x^3$.   [London.]

(15) $(y^2 + yz - z)\, dx + (x^2 + xz - z)\, dy + (x + y - xy)\, dz = 0$.   [London.]

(16) $(2x^3 - y^3 - z^3)yz\, dx + (2y^3 - z^3 - x^3)zx\, dy + (2z^3 - x^3 - y^3)xy\, dz = 0$.   [London.]

(17) $xp - yq + (x^2 - y^2) = 0$.   [London.]

(18) $(x + 2y - z)p + (3y - z)q = x + y$.   [London.]

(19) $xp + yq + \dfrac{2(xz - yz + xy)}{4y - x + z} = 0.$ [London

(20) $p(x + p) + q(y + q) = z.$ [London

(21) $r + s = p.$ [London

(22) $z - \frac{1}{2}px - qy = p^2/x^2.$ [London

(23) $r - x = t - y.$ [London

(24) $z = px + qy - sxy.$ [London

(25) $z(rt - s^2) + pqs = 0.$ [London

(26) $x^2 r + 2xys + y^2 t = xy.$ [London

(27) $rq(q + 1) - s(2pq + p + q + 1) + tp(p + 1) = 0.$ [London

(28) $y^3 = xy^2 p + x^4 p^2.$ [Math. Trip

(29) $5y \dfrac{d^2 y}{dx^2} = \left(\dfrac{dy}{dx}\right)^2.$

(30) $\dfrac{d^2 y}{dx^2} - \dfrac{n}{x}\dfrac{dy}{dx} + x^{2n} y = 0.$ [Math. Trip

(31) $(zp + x)^2 + (zq + y)^2 = 1.$ [Math. Trip

(32) Find a solution of the equation $\dfrac{d^2 y}{dx^2} - 3\dfrac{dy}{dx} + 2y = e^{3x}$ which sha

vanish when $x = 0$ and also when $x = \log_e 2.$ [Math. Trip

(33) Solve the equation

$$\frac{d^2 x}{dt^2} + 2\kappa \frac{dx}{dt} + (\kappa^2 + \lambda^2) x = A \cos pt.$$

Show that, for different values of p, the amplitude of the particula
integral is greatest when $p^2 = \lambda^2 - \kappa^2$, and prove that the particula
integral is then

$$(A/2\kappa\lambda) \cos (pt - \alpha), \text{ where } \tan \alpha = p/\kappa. \quad \text{[London}$$

(34) Solve the equation

$$\frac{d^2 y}{dx^2} + \frac{dy}{dx} \tan x + y \cos^2 x = 0$$

by putting $z = \sin x.$

(35) (i) Assuming a solution of $\dfrac{\partial^2 V}{\partial x^2} + \dfrac{\partial^2 V}{\partial y^2} + \dfrac{\partial^2 V}{\partial z^2} = 0$ to be of th

form $F(r + z)$, where $r^2 = x^2 + y^2 + z^2$, obtain the function F; and b
integrating with respect to z, deduce the solution $V = z \log (r + z) - r.$

(ii) Assuming a solution of $\dfrac{\partial V}{\partial t} = a^2 \dfrac{\partial^2 V}{\partial x^2}$ to be of the form $\phi(\xi$

where $\xi = x/\sqrt{t}$, obtain the function ϕ; and deduce a second solutio
by differentiating with respect to $x.$ [London

(36) Obtain a rational integral function V of x, y, z which satisfie
the condition

$$\frac{\partial^2 V}{\partial x^2} + \frac{\partial^2 V}{\partial y^2} + \frac{\partial^2 V}{\partial z^2} = 0,$$

and is such as to have the value Az^4 at points on the surface of a spher
of unit radius with its centre at the origin.   [Math. Trip

(37) Show that a solution of Laplace's equation $\nabla^2 u = 0$ is
$$u = (A \cos n\theta + B \sin n\theta) e^{-\lambda z} J_n(\lambda r),$$
here r, θ, z are cylindrical co-ordinates and A, B, n, λ are arbitrary
nstants. [London.]

(38) Show that $J_n(r)(a_n \cos n\theta + b_n \sin n\theta)$, where r and θ are
lar co-ordinates and a_n and b_n are arbitrary constants, is a solution
the equation
$$\frac{\partial^2 V}{\partial x^2} + \frac{\partial^2 V}{\partial y^2} + V = 0.$$
[London.]

(39) Show how to find solutions in series of the equation
$$\frac{\partial u}{\partial t} = a^2 \frac{\partial^2 u}{\partial x^2},$$
d solve completely for the case in which, when $x = 0$,
$$u = a\frac{\partial u}{\partial x} = C \cosh t.$$
[London.]

(40) Obtain two independent solutions in ascending powers of x of
e equation
$$4\frac{d^2 y}{dx^2} + 9xy = 0;$$
d prove by transforming the variables in the equation, or otherwise
at the complete solution may be written in the form
$$y = Ax^{\frac{1}{2}}J_{\frac{1}{3}}(x^{\frac{3}{2}}) + Bx^{\frac{1}{2}}J_{-\frac{1}{3}}(x^{\frac{3}{2}}),$$
here A and B are arbitrary constants. [London.]

(41) Show that the complete solution of the equation
$$\frac{dy}{dx} + P + Qy + Ry^2 = 0,$$
here P, Q, R are functions of x, can be obtained by the substitution
$= y_1 + 1/z$, if a particular solution, y_1, is known.
Show that, if two particular solutions y_1 and y_2 are known, the
omplete solution is
$$\log \left(\frac{y - y_1}{y - y_2} \right) = \int R(y_2 - y_1)\, dx + \text{const.}$$
Obtain the complete solution of the equation
$$(x^2 - 1)\frac{dy}{dx} + x + 1 - (x^2 + 1)y + (x - 1)y^2 = 0,$$
hich has two particular solutions, the product of which is unity.
[London.]

(42) Show that the differential equation
$$(1 - x^2)\frac{d^2 y}{dx^2} + 2\{b + (a - 1)x\}\frac{dy}{dx} + 2ay = 0$$
as a solution of the form $(1 + x)^p (1 - x)^q$, where p and q are determinate
onstants. Solve the equation completely; and deduce, or prove
therwise, that if $2a$ is a positive integer n, one solution of the equation
a polynomial in x of degree n. [London.]

(43) Verify that $1 - x^2$ is a particular solution of the equation

$$x(1-x^2)^2 \frac{d^2y}{dx^2} + (1-x^2)(1+3x^2)\frac{dy}{dx} + 4x(1+x^2)y = 0,$$

and solve it completely.

By the method of *variation of parameters* or otherwise, solve completely the equation obtained by writing $(1 - x^2)^3$ instead of zero on the right-hand side of the given equation.   [London]

(44) Show that the complete solution of the equation

$$\frac{d^2y}{dx^2} + P\frac{dy}{dx} + Qy = 0,$$

where P, Q are given functions of x, can be found if any solution of the equation

$$\frac{du}{dx} + u^2 + Q - \frac{1}{2}\frac{dP}{dx} - \frac{1}{4}P^2 = 0$$

is known.

Hence, or otherwise, solve the equation

$$(1-x^2)\frac{d^2y}{dx^2} - 4x\frac{dy}{dx} + (x^4-3)y = 0. \qquad \text{[London]}$$

(45) Prove by putting $v = we^{ix}$ that the complete solution of the equation $x\frac{d^2v}{dx^2} - 2n\frac{dv}{dx} + xv = 0$, where n is an integer, can be expressed in the form

$$(A\cos x + B\sin x)f(x) + (A\sin x - B\cos x)\phi(x),$$

where $f(x)$ and $\phi(x)$ are suitable polynomials.   [London]

(46) If u, v are two independent solutions of the equation

$$f(x)y''' - f'(x)y'' + \phi(x)y' + \chi(x)y = 0,$$

where dashes denote differentiation with regard to x, prove that the complete solution is $Au + Bv + Cw$, where

$$w \equiv u\int \frac{vf(x)\,dx}{(uv'-u'v)^2} - v\int \frac{uf(x)\,dx}{(uv'-u'v)^2}$$

and A, B, C are arbitrary constants.

Solve the equation

$$x^2(x^2+5)y''' - x(7x^2+25)y'' + (22x^2+40)y' - 30xy = 0,$$

which has solutions of the form x^n.   [London]

(47) Obtain two independent power-series which are solutions of the equation

$$(x^2-a^2)\frac{d^2y}{dx^2} + bx\frac{dy}{dx} + cy = 0,$$

and determine their region of convergence.   [London]

(48) Prove that the equation

$$x(1-x)\frac{d^2y}{dx^2} + (1-2x)\frac{dy}{dx} - \frac{1}{4}y = 0$$

as two integrals

$$\sum_0^\infty a_n x^n, \qquad \sum_0^\infty a_n \left(\frac{1}{4}\log x + 1 - \frac{1}{2} + \frac{1}{3} - \dots - \frac{1}{2n}\right)x^n,$$

where
$$a_n = \left\{\frac{\Gamma\left(n+\frac{1}{2}\right)}{\Gamma(n+1)}\right\}^2.$$ [London.]

(49) Form the differential equation whose primitive is

$$y = A\left(\sin x + \frac{\cos x}{x}\right) + B\left(\cos x - \frac{\sin x}{x}\right),$$

where A, B are arbitrary constants. [London.]

(50) Obtain the condition that the equation
$$P\,dx + Q\,dy = 0$$

may have an integrating factor which is a function of x alone, and apply the result to integrate

$$(3xy - 2ay^2)\,dx + (x^2 - 2axy)\,dy = 0.$$ [London.]

(51) Show that the equations

$$y - x\frac{dy}{dx} + \frac{2ax^2}{x^2 - y^2}\frac{dy}{dx} = 0,$$

$$x^2 - y^2 + 2(xy + bx^2)\frac{dy}{dx} = 0,$$

have a common primitive, and find it. [London.]

(52) Prove that any solution of the equation

$$P\frac{d^2u}{dx^2} + Q\frac{du}{dx} + Ru = 0$$

is an integrating factor of the equation

$$\frac{d^2}{dx^2}(Pu) - \frac{d}{dx}(Qu) + Ru = 0,$$

and conversely that any solution of the latter equation is an integrating factor of the former.

Hence integrate the first of these equations completely, it being given that
$$\frac{d^2}{dx^2}\left(\frac{P}{Q}\right) + \frac{R}{Q} = 0.$$ [London.]

(53) If the equation $\dfrac{d^2y}{dx^2} + P\dfrac{dy}{dx} + Qy = 0,$

where P and Q are functions of x, admits of a solution
$$y = A\sin(nx + \alpha),$$

where A and α are arbitrary constants, find the relation which connects P and Q. [London.]

(54) Solve the equation $\dfrac{d^2y}{dx^2} - 4y = \dfrac{2y}{(1-x)^2},$

having given that it has two integrals of the form

$$y = \frac{a + bx}{1 - x}e^{kx}.$$ [London.]

(55) Show that the linear differential equation whose solutions are the squares of those of $\dfrac{d^2y}{dx^2} + P\dfrac{dy}{dx} + Qy = 0$

may be written $\left(\dfrac{d}{dx} + 2P\right)\left(\dfrac{d^2y}{dx^2} + P\dfrac{dy}{dx} + 2Qy\right) + 2Q\dfrac{dy}{dx} = 0.$

(56) Show that the total differential equation
$$3x^2(y + z)\,dx + (z^2 - x^3)\,dy + (y^2 - x^3)\,dz = 0$$
satisfies the conditions of integrability, and integrate it.   [London.

(57) The operator $\dfrac{d}{dx}$ being represented by D, show that if X is a function of x and $\phi(D)$ a rational integral function of D,
$$\phi(D)\,xX = x\phi(D)\,X + \phi'(D)\,X.$$

Extend the result to the case in which $1/\phi(D)$ is a rational integral function of D.

Solve the differential equation
$$\frac{d^3y}{dx^3} + 8y = 3x^2 + xe^{-2x}\cos x.\qquad\text{[London.}$$

(58) Show that $3\dfrac{d^2y}{dx^2} + 4x\dfrac{dy}{dx} - 8y = 0$

has an integral which is a polynomial in x.   Deduce the general solution.
        [Sheffield.

(59) Show that, if in the equation $P\,dx + Q\,dy + R\,dz = 0$, P, Q, R are homogeneous functions of x, y, z of the same degree, then one variable can be separated from the other two, and the equation, if integrable, is thereby rendered exact.

Integrate
$$z^3(x^2\,dx + y^2\,dy) + z\{xyz^2 + z^4 - (x^2 + y^2)^2\}(dx + dy)$$
$$+ (x + y)(\{z^4 - z^2(x^2 + y^2) - (x^2 + y^2)^2\}\,dz = 0,$$
obtaining the integral in an algebraic form.   [London.

(60) Show that the equation $P\,dx + Q\,dy + R\,dz = 0$, if integrable, can be reduced to the form $\lambda\,du + \mu\,dv = 0$; where λ/μ is a function of u, v only and $u = $ constant, $v = $ constant are two independent solutions of
$$\frac{dx}{\dfrac{\partial Q}{\partial z} - \dfrac{\partial R}{\partial y}} = \frac{dy}{\dfrac{\partial R}{\partial x} - \dfrac{\partial P}{\partial z}} = \frac{dz}{\dfrac{\partial P}{\partial y} - \dfrac{\partial Q}{\partial x}}.$$

Hence, or otherwise, integrate the equation
$$(yz + z^2)\,dx - xz\,dy + xy\,dz = 0.\qquad\text{[London.}$$

(61) Prove that $z^2 = 2xy$ is not included in
$$x + y + \sqrt{(z^2 - 2xy)} = f(x + y + z^2),$$
which is the general solution of
$$\{2\sqrt{(z^2 - 2xy)} - 2x - 1\}zp + \{1 + 2y - 2\sqrt{(z^2 - 2xy)}\}zq = x - y,$$
but that it is nevertheless a solution of the equation.   [Sheffield.

(62) (i) Show how to reduce Riccati's equation

$$\frac{dy}{dx} = a_0(x) + a_1(x)y + a_2(x)y^2$$

to a linear equation of the second order; and hence or otherwise prove that the cross-ratio of any four integrals is a constant.

(ii) Verify that $\frac{1}{2} + x \tan x$, $\frac{1}{2} - x \cot x$ are integrals of

$$x\frac{dy}{dx} = x^2 - \tfrac{1}{4} + y^2,$$

and deduce the primitive. [London.]

(63) By solving

$$\frac{dx}{dt} = -\omega y,$$

$$\frac{dy}{dt} = \omega x,$$

in the ordinary way, and eliminating t from the result, prove that the point (x, y) lies on a circle.

Also prove this by adding x times the first equation to y times the second.

[The equations give the velocities, resolved parallel to the axes, of point which is describing a circle with angular velocity ω.]

(64) Find the orthogonal trajectories of the curves

$$y^2(a - x) = x^3.$$

Prove that they reduce to the system

$$r^2 = b^2(3 + \cos 2\theta). \qquad \text{[Sheffield.]}$$

(65)
$$\frac{dx}{dt} = ny - mz,$$

$$\frac{dy}{dt} = lz - nx,$$

$$\frac{dz}{dt} = mx - ly,$$

where l, m, n are constants, prove that

$$lx + my + nz,$$
$$x^2 + y^2 + z^2,$$

and
$$\left(\frac{dx}{dt}\right)^2 + \left(\frac{dy}{dt}\right)^2 + \left(\frac{dz}{dt}\right)^2$$

are all constant. Interpret these results.

(66) A plane curve is such that the area of the triangle PNT is n times the area of the segment APN, where PN is the ordinate, NT the subtangent at any point P, and A the origin, which is on the curve; show that its equation is $y^{2m-1} = a^{2m-2}x$.

Show that the volume described by the revolution of the segment APN about the axis of x bears a constant ratio to the volume of the one generated by the revolution of the triangle PNT. [London.]

(67) By using the substitutions $x = r \cos \theta$, $y = r \sin \theta$, or otherwise solve the differential equation

$$(x^2 + y^2)(xp - y)^2 = 1 + p^2.$$

Also find the singular solution, and interpret the results geometrically.   [London

(68) Show that the equation

$$(x^2 + y^2 - 2xpy)^2 = 4a^2y^2(1 - p^2)$$

can be reduced to Clairaut's form by making $y^2 - x^2$ a new dependent variable; solve it and show that the singular solution represents two rectangular hyperbolas. Verify also that this solution satisfies the given equation.   [London

(69) Prove that the curves in which the radius of curvature is equal to the length intercepted on the normal by a fixed straight line are either circles or catenaries.   [London

(70) Solve the equation

$$y = x - 2ap + ap^2,$$

and find the singular solution, giving a diagram.   [London.

(71) A plane curve is such that its radius of curvature ρ is connected with the intercept ν on the normal between the curve and the axis of x, by the relation $\rho\nu = c^2$. Show that, if the concavity of the curve is turned away from the axis of x,

$$y^2 = c^2 \sin^2 \phi + b,$$

where ϕ is the inclination of the tangent to Ox. Obtain the value of x as a function of ϕ in the case $b = 0$; and sketch the shape of the curve.   [London

(72) Show that, if the differential equation of a family of curves be given in bipolar co-ordinates r, r', θ, θ', the differential equation of the orthogonal trajectories is found by writing $r \, d\theta$ for dr, $r' \, d\theta'$ for dr' $- dr$ for $r \, d\theta$, $- dr'$ for $r' \, d\theta'$.

Find the orthogonal trajectories of the curves

$$\frac{a}{r} + \frac{b}{r'} = c,$$

c being the variable parameter.   [London

(73) The normal at a point P of a curve meets a fixed straight line at the point G, and the locus of the middle point of PG is a straight line inclined to the fixed straight line at an angle $\cot^{-1} 3$. Show that the locus of P is a parabola.   [London.

(74) Solve the equation $2(p - 1)y = p^2x$; show that the "p-discriminant" is a solution of the equation, and is the envelope of the family of curves given by the general solution.   [London.

(75) Obtain the differential equation of the involutes of the parabola $y^2 = 4ax$, and integrate it. What is the nature of the singular solution?   [London.

(76) Prove that if the normals to a surface all meet a fixed straight line, the surface must be one of revolution. [London.]

(77) Integrate the partial differential equation

$$px + qy = \sqrt{(x^2 + y^2)}.$$

Give the geometrical interpretation of the subsidiary integrals and of the general integral. [London.]

(78) Integrate the differential equation

$$z(x + 2y)\frac{\partial z}{\partial x} - z(y + 2x)\frac{\partial z}{\partial y} = y^2 - x^2.$$

Find the particular solutions such that the section by any plane parallel to $z = 0$ shall be (i) a circle, (ii) a rectangular hyperbola. [London.]

(79) A family of curves is represented by the equations

$$x^2 + y^2 + 6z^2 = \alpha, \quad 2x^2 + 5y^2 + z^2 + 4xy = \beta,$$

where α, β are parameters.

Prove that the family of curves can be cut orthogonally by a family of surfaces, and find the equation of this family. [London.]

(80) Solve $b(bcy + axz)p + a(acx + byz)q = ab(z^2 - c^2)$,

and show that the solution represents any surface generated by lines meeting two given lines.

(81) (i) Solve $$L\frac{dI}{dt} + RI = E,$$

where L, R, and E are constants.

[This is the equation for the electric current I in a wire of resistance R and coefficient of self-induction L, under a constant voltage E.]

(ii) Determine the value of the arbitrary constant if $I = I_0$ when $t = 0$.

(iii) To what value does I approximate when t is large?

[Ohm's law for steady currents.]

(82) Solve $$L\frac{dI}{dt} + RI = E \cos pt.$$

[The symbols have the same meaning as in the last question, except that the voltage $E \cos pt$ is now periodic instead of being constant. The complementary function soon becomes negligible, *i.e.* the free oscillations of the current are damped out.]

(83) Find the Particular Integral of

$$L\frac{d^2Q}{dt^2} + R\frac{dQ}{dt} + \frac{Q}{C} = E \cos pt.$$

[This gives the charge Q on one of the coatings of a Leyden jar when a periodic electromotive force $E \cos pt$ acts in the circuit connecting the coatings. The Particular Integral gives the charge after the free electrical oscillations have been damped out.]

(84) Show that the equations

$$2\frac{dx}{dt} + 3\frac{dy}{dt} - 16x - 3y = 0, \qquad 7\frac{dx}{dt} - 2x - 3y = 0$$

are satisfied by the trial solution $y = mx$, provided that m is a root of the quadratic

$$\frac{2 + 3m}{7} = \frac{16 + 3m}{2 + 3m},$$

and x is given by

$$7\frac{dx}{dt} - (2 + 3m)x = 0.$$

Hence prove that two sets of solutions of the differential equations are

$$y = 4x = 4Ae^{2t}$$

and

$$y = -3x = -3Be^{-t},$$

so that the general solution is $x = Ae^{2t} + Be^{-t}$,

$$y = 4Ae^{2t} - 3Be^{-t}.$$

(85) Use the method of the last example to solve

$$7\frac{d^2x}{dt^2} + 23x - 8y = 0,$$

$$3\frac{d^2x}{dt} + 2\frac{d^2y}{dt^2} - 13x + 10y = 0.$$

[Equations of this type occur in problems on the small oscillations of systems with two degrees of freedom. The motion given by $y = 2x$ (or by $y = -5x$) is said to be a Principal or Normal Mode of Vibration. Clearly it is such that all parts of the system are moving harmonically with the same period and in the same phase. If $y - 2x$ and $y + 5x$ are taken as new variables instead of x and y, they are called Principal or Normal Coordinates.]

(86) Given that L, M, N, R, S are positive numbers, such that LN is greater than M^2, prove that x and y, defined by

$$L\frac{dx}{dt} + M\frac{dy}{dt} + Rx = 0,$$

$$M\frac{dx}{dt} + N\frac{dy}{dt} + Sy = 0,$$

diminish indefinitely as t increases.

[Show that $x = Ae^{at} + Be^{bt}$ and $y = Ee^{at} + Fe^{bt}$, where a and b are *real* and *negative*. These equations give the free oscillations of two mutually influencing electric circuits. L and N are coefficients of self-induction, M of mutual induction, and R and S are resistances.]

(87) Show (without working out the solutions in full) that the Particular Integrals of the simultaneous equations

$$L\frac{dx}{dt} + M\frac{dy}{dt} + Rx + \int\frac{x\,dt}{c} = E\sin pt,$$

$$M\frac{dx}{dt} + N\frac{dy}{dt} + Sy = 0$$

re unaltered if in the first equation the term $\int \dfrac{x}{c}\, dt$ is omitted and L replaced by $L - \dfrac{1}{cp^2}$.

[This follows at once from the fact that the Particular Integrals are of the form $A \sin (pt - \alpha)$.

These equations give the currents in two mutually influencing circuits when the primary, which contains a condenser of capacity c, is acted upon by an alternating electromotive force. This example shows that the effect of the condenser can be compensated for by increasing the self-induction.]

(88) If
$$L \frac{dx}{dt} + M \frac{dy}{dt} + \frac{1}{c} \int x \, dt = f(t)$$

and
$$M \frac{dx}{dt} + N \frac{dy}{dt} = 0,$$

where $LN - M^2$ is a very small positive quantity, show that the Complementary Function for x represents a very rapid oscillation.

[These equations occur in Rayleigh's theory of the oscillatory discharge of a condenser in the primary circuit of an induction coil with a closed secondary. Notice that the second equation shows that the secondary current is at its maximum when the primary current is at its minimum. See Gray's *Magnetism and Electricity*, Arts. 489 and 490.]

(89) Prove that the Particular Integrals of the simultaneous equations
$$m \frac{d^2x}{dt^2} = - a (x - X) + k \cos pt,$$

$$M \frac{d^2X}{dt^2} = - AX + a (x - X)$$

may be written
$$x = \frac{Bk}{a^2 - bB} \cos pt,$$

$$X = \frac{- ak}{a^2 - bB} \cos pt,$$

where $b = mp^2 - a$ and $B = Mp^2 - (a + A)$.

Hence show that x and X are both infinite for two special values of p.

[These equations give the oscillations of the "elastic double pendulum." Masses m and M are arranged so that they can only move in the same horizontal line. A spring connects M to a fixed point of this line and another spring connects m to M. A periodic force acts upon m, and the solution shows that both masses execute forced vibrations whose amplitude becomes very large for two special values of p. Of course this is the phenomenon of Resonance again. It is important to notice that the values of p that give resonance in this case are not the same as they would be if only one mass were present. This may be applied to the discussion of the "whirling" in a turbine shaft. See Stodola's *Steam Turbine*.]

(90) Show that the solution of the simultaneous equations

$$(\tfrac{1}{3}m + M)\,4a\,\frac{d^2\theta}{dt^2} + 2Mb\,\frac{d^2\phi}{dt^2} = -g\,(m + 2M)\,\theta,$$

$$\frac{4b}{3}\,\frac{d^2\phi}{dt^2} + 2a\,\frac{d^2\theta}{dt^2} = -g\phi,$$

where $m = M$ and $a = b$, may be expressed by saying that θ and ϕ are each composed of two simple harmonic oscillations of periods $2\pi/p_1$ and $2\pi/p_2$, $p_1{}^2$ and $p_2{}^2$ being the roots of the quadratic in p^2,

$$28a^2p^4 - 84agp^2 + 27g^2 = 0.$$

[These equations give the inclinations to the vertical of two rods of masses m and M and lengths $2a$ and $2b$ respectively when they are swinging in a vertical plane as a double pendulum, the first being freely suspended from a fixed point and the second from the bottom of the first. The two oscillations referred to are known as the Principal (or Normal) Oscillations. Similar equations occur in many problems on small oscillations. A detailed discussion of these is given in Routh's *Advanced Rigid Dynamics*, with special reference to the case when the equation in p has equal roots.]

(91)
$$\frac{d^2x}{dt^2} + \kappa\,\frac{dy}{dt} + c^2x = 0,$$

$$\frac{d^2y}{dt^2} - \kappa\,\frac{dx}{dt} + c^2y = 0.$$

[These equations give the motion of the bob of a gyrostatic pendulum which does not swing far from the vertical. Notice that if the initial conditions are such that $B = 0$, we get motion in a circle with angular velocity p, while if $A = 0$, we get motion in a circle with angular velocity q in the opposite sense. (For p, q, A, B see the answers.)

Similar equations hold for the path of revolving ions in the explanation of the Zeeman Effect (the trebling of a line in a spectrum by a magnetic field). See Gray's *Magnetism and Electricity*, Arts 565-569.]

(92) Given
$$\begin{cases} \dfrac{dx}{dt} + ax = 0, \\[2mm] \dfrac{dz}{dt} = by, \\[2mm] x + y + z = c, \end{cases}$$

where a, b, c are constants, obtain a differential equation for z.

Hence prove that if $z = \dfrac{dz}{dt} = 0$ when $t = 0$,

$$z = c + \frac{c}{a-b}\,[be^{-at} - ae^{-bt}].$$

[These equations occur in Physical Chemistry when a substance A forms an intermediate substance B, which then changes into a third

ubstance C. x, y, z are the " concentrations " of A, B, C respectively
t any time t. See Harcourt and Esson, *Phil. Trans.* 1866 and 1867.]

(93) The effect on a simple dynamical system with one degree of
reedom of any other dynamical system to which it is linked can be
epresented by the equation

$$\ddot{x} + 2\mu\dot{x} + n^2 x = X.$$

If the exciting system of waves is maintained steady so that
$X = A \cos pt$, find the value of p for which there is resonance, and prove
that if μ exceeds a certain value there is no resonance. Draw curves
llustrating both cases. [Math. Trip.]

(94) Solve the differential equation

$$\ddot{x} + 2k\dot{x} + n^2 x = 0 \quad \text{when} \quad k^2 < n^2.$$

In the case of a pendulum making small oscillations, the time of a
omplete oscillation being 2 secs. and the angular retardation due to
he air being taken as $\cdot 04 \times$ (angular velocity of pendulum), show that
n amplitude of $1°$ will in 10 complete oscillations be reduced to about
0′. [Take $\log_{10} e = \cdot 4343$.] [Math. Trip.]

(95) The motion of a system depends practically on a single co-
rdinate x; its energy at any instant is expressed by the formula
$m\dot{x}^2 + \frac{1}{2}ex^2$; and the time-rate of frictional damping of its energy is
$k\dot{x}^2$. Prove that the period (τ_0) of its free oscillation is

$$2\pi\left(\frac{e}{m} - \frac{1}{16}\frac{k^2}{m^2}\right)^{-\frac{1}{2}}.$$

Prove that the forced oscillation sustained by a disturbing force of
ype $A \cos pt$ is at its greatest when $p^2 = \dfrac{e}{m} - \dfrac{k^2}{8m^2}$, and that the amplitude
f this oscillation is then $\dfrac{Am\tau_0}{\pi k}$, while its phase lags behind that of the
orce by the amount $\tan^{-1}\dfrac{4mp}{k}$. [Math. Trip.]

(96) Show that the substitution $T = \dfrac{1}{2}\left(\dfrac{ds}{dt}\right)^2$ reduces

$$\frac{d^2 s}{dt^2} + P\left(\frac{ds}{dt}\right)^2 = Q$$

 the linear form

$$\frac{dT}{ds} + 2PT = Q.$$

From

$$(s + a)\frac{d^2 s}{dt^2} + \left(\frac{ds}{dt}\right)^2 = (s - a)g,$$

ith the conditions $\dfrac{ds}{dt} = 0$ and $s = 2a$ when $t = 0$, obtain

$$\left(\frac{ds}{dt}\right)^2 = \frac{2g}{3}(s - 2a)$$

nd

$$\frac{d^2 s}{dt^2} = \frac{g}{3}.$$

[This gives the solution of the dynamical problem: " A uniform chain is coiled up on a horizontal plane and one end passes over a smooth light pulley at a height a above the plane; initially a length $2a$ hangs freely on the other side. Prove that the motion is uniformly accelerated." See Loney's *Dynamics of a Particle and of Rigid Bodies*, p. 131.]

(97) Find a solution of the equation

$$\frac{\partial}{\partial r}\left(r^2 \frac{\partial \phi}{\partial r}\right) + \frac{1}{\sin \theta} \frac{\partial}{\partial \theta}\left(\sin \theta \frac{\partial \phi}{\partial \theta}\right) = 0$$

of the form
$$\phi = f(r) \cos \theta,$$

given that
$$-\frac{\partial \phi}{\partial r} = V \cos \theta \text{ when } r = a$$

and
$$-\frac{\partial \phi}{\partial r} = 0 \text{ when } r = \infty.$$

[ϕ is the velocity-potential when a sphere of radius a moves with velocity V in a straight line through a liquid at rest at infinity. See Ramsey's *Hydromechanics*, Part II. p. 152.]

(98) Find a solution of $\quad \dfrac{\partial^2 y}{\partial t^2} = c^2 \dfrac{\partial^2 y}{\partial x^2}$

which shall vanish when $x = 0$, and reduce to $A \cos (pt + \alpha)$ when $x = b$.

[This gives the form of one portion of a stretched string, fixed at both ends, of which a given point is made to move with the periodic displacement $A \cos (pt + \alpha)$. The portion considered is that between the given point and one of the ends. See Ramsey's *Hydromechanics*, Part II. p. 312.]

(99) Obtain the solution of

$$\frac{\partial^2 \phi}{\partial t^2} = c^2\left(\frac{\partial^2 \phi}{\partial r^2} + \frac{2}{r} \frac{\partial \phi}{\partial r}\right)$$

in the form
$$r\phi = f(ct - r) + F(ct + r).$$

[ϕ is the velocity-potential of a spherical source of sound in air. See Ramsey, p. 345.]

(100) Obtain a solution of

$$\frac{\partial^2 \phi}{\partial x^2} + \frac{\partial^2 \phi}{\partial y^2} = 0,$$

such that $\qquad \partial \phi / \partial y = 0 \text{ when } y = -h$

and ϕ varies as $\quad \cos (mx - nt) \text{ when } y = 0.$

[ϕ is the velocity-potential of waves in a canal of depth h, the sides being vertical. See Ramsey, p. 265.]

(101) Obtain the solution of the simultaneous differential equations

$$\frac{d^2 x}{dt^2} - 2n \frac{dy}{dt} + p^2 x = 0,$$

$$\frac{d^2 y}{dt^2} + 2n \frac{dx}{dt} + p^2 y = 0,$$

ith the initial conditions

$$x = a, \quad y = 0, \quad \frac{dx}{dt} = 0, \quad \frac{dy}{dt} = 0,$$

the form

$$z = \frac{a}{2q} \{(q+n)e^{i(q-n)t} + (q-n)e^{-i(q+n)t}\},$$

here

$$z = x + iy \quad \text{and} \quad q = \sqrt{(p^2 + n^2)}.$$

Show that the solution represents a hypocycloid contained between vo concentric circles of radii a and an/q.

[This example gives the theory of Foucault's pendulum experiment :monstrating the rotation of the earth. See Bromwich, *Proc. London [ath. Soc.* 1914.]

(102) Obtain an approximate solution of Einstein's equation of lanetary motion

$$\frac{d^2u}{d\phi^2} + u = \frac{m}{h^2} + 3mu^2$$

ı the following manner:

(*a*) Neglect the small term $3mu^2$, and hence obtain

$$u = \frac{m}{h^2}\{1 + e\cos(\phi - \varpi)\}, \text{ as in Newtonian dynamics.}$$

(*b*) Substitute this value of u in the small term $3mu^2$, and hence btain

$$\frac{d^2u}{d\phi^2} + u = \frac{m}{h^2} + \frac{3m^3}{h^4} + \frac{6m^3}{h^4}e\cos(\phi - \varpi) + \frac{3m^3e^2}{2h^4}\{1 + \cos 2(\phi - \varpi)\}.$$

(*c*) Neglect all the terms on the right-hand side of this differential quation except $\dfrac{m}{h^2}$ and $\dfrac{6m^3}{h^4} e\cos(\phi - \varpi)$. The term in $\cos(\phi - \varpi)$ must

: retained; it is of the same period as the complementary function, and terefore produces a continually increasing particular integral. [See the *sonance* problem Ex. 36 on p. 46.] Hence obtain

$$u = \frac{m}{h^2}\left\{1 + e\cos(\phi - \varpi) + \frac{3m^2}{h^2}e\phi\sin(\phi - \varpi)\right\}$$

$$= \frac{m}{h^2}\{1 + e\cos(\phi - \varpi - \epsilon)\} \text{ approximately,}$$

here $\epsilon = \dfrac{3m^2}{h^2}\phi$ and ϵ^2 is neglected.

[This result proves that when the planet moves through one revolu-on the perihelion (given by $\phi - \varpi - \epsilon = 0$) advances a fraction of a evolution given by $\dfrac{\epsilon}{\phi} = \dfrac{3m^2}{h^2}$. When numerical values are given to the onstants it is found that Einstein's theory removes a well-known iscrepancy between observed and calculated results on the motion f the perihelion of Mercury. See Eddington, *Report on the Relativity 'heory of Gravitation*, pp. 48-52.]

(103) $L(x, y, x', y')$ is a function of the variables x, y, x', y'. X, Y are defined by the equations

$$X = \frac{\partial L}{\partial x'}, \quad Y = \frac{\partial L}{\partial y'}.$$

If these equations can be solved for x' and y' as functions of X, Y, x, and if $H(X, Y, x, y)$ is the function obtained by expressing

$$Xx' + Yy' - L$$

entirely in terms of X, Y, x, y, then prove that

$$\frac{\partial H}{\partial X} = x' \quad \dotfill (1$$

and

$$\frac{\partial H}{\partial x} = -\frac{\partial L}{\partial x}. \quad \dotfill (2$$

Prove also that the equation

$$\frac{d}{dt}\left(\frac{\partial L}{\partial x'}\right) = \frac{\partial L}{\partial x} \quad \dotfill (3$$

is transformed into

$$\frac{dX}{dt} = -\frac{\partial H}{\partial x}. \quad \dotfill (4$$

[This is the *Hamiltonian transformation* in dynamics. Equation (3 is a typical *Lagrangian equation of motion in generalised co-ordinate.* Hamilton replaces it by the pair of equations (1) and (4). See Routh *Elementary Rigid Dynamics*, Chap. VIII. This transformation shoul be compared with that of Ex. 21 of the miscellaneous set at the end c Chap. XII., where we had two partial differential equations derivabl from each other by the *Principle of Duality*.]

(104) Show that Jacobi's method (Art. 140) applied to *Hamilton* *partial differential equation*

$$\frac{\partial z}{\partial t} + H(x_1, x_2, \dots, x_n, p_1, p_2, \dots, p_n, t) = 0$$

leads to

$$\frac{dx_r}{dt} = \frac{\partial H}{\partial p_r}, \quad \frac{dp_r}{dt} = -\frac{\partial H}{\partial x_r} \quad (r = 1, 2, \dots, n),$$

which are the equations of motion of a dynamical system, in Hamilton' form. [See Whittaker's *Analytical Dynamics*, 2nd ed., Art. 142.]

(105) (i) Prove that if $\quad u(x, y, z) = a$

and $\quad v(x, y, z) = b$

are any two integrals of the system of differential equations

$$\frac{dx}{p(x, y, z)} = \frac{dy}{q(x, y, z)} = \frac{dz}{r(x, y, z)},$$

then $\quad \dfrac{1}{p}\dfrac{\partial(u, v)}{\partial(y, z)} = \dfrac{1}{q}\dfrac{\partial(u, v)}{\partial(z, x)} = \dfrac{1}{r}\dfrac{\partial(u, v)}{\partial(x, y)} = m(x, y, z)$, say.

[m is called a *multiplier* of the system.]

(ii) Show that m satisfies the partial differential equation

$$\frac{\partial}{\partial x}(mp) + \frac{\partial}{\partial y}(mq) + \frac{\partial}{\partial z}(mr) = 0.$$

(iii) If $n(x, y, z)$ is any other multiplier of the system, show that

$$p\frac{\partial}{\partial x}\left(\frac{m}{n}\right) + q\frac{\partial}{\partial y}\left(\frac{m}{n}\right) + r\frac{\partial}{\partial z}\left(\frac{m}{n}\right) = 0,$$

nd hence that $\qquad \dfrac{\partial(m/n,\, u,\, v)}{\partial(x,\, y,\, z)} = 0$ identically,

 that m/n is a function of u and v, and $m/n = c$ is an integral of the
·iginal system of differential equations.

(iv) If $u(x, y, z) = a$ can be solved for z, giving $z = f(x, y, a)$, and
capital letters V, P, Q, R, M denote the functions of x, y, a, obtained
y substituting this value of z in v, p, q, r, m, then prove that
$(x, y, a) = b$ is an integral of $\dfrac{dx}{P} = \dfrac{dy}{Q}$.

Prove also that $\qquad\qquad MP = -\dfrac{\partial V}{\partial y}\dfrac{\partial u}{\partial z}$

nd $\qquad\qquad\qquad\qquad MQ = \dfrac{\partial V}{\partial x}\dfrac{\partial u}{\partial z}$

where $\dfrac{\partial u}{\partial z}$ is to be expressed in terms of x, y, $a\Big)$, so that

$$dV = M(Q\,dx - P\,dy)\Big/\frac{\partial u}{\partial z}.$$

[This suggests that if *any* integral $u = a$ and *any* multiplier m are
nown, then $M(Q\,dx - P\,dy)\Big/\dfrac{\partial u}{\partial z}$ will be a perfect differential, leading
 an integral of the system when a is replaced by $u(x, y, z)$.

For a proof of this theorem see Whittaker's *Analytical Dynamics*,
nd ed., Art. 119. A more general theorem is that if $(n-1)$ integrals
 a system of differential equations

$$\frac{dx_1}{p_1} = \frac{dx_2}{p_2} = \ldots = \frac{dx_n}{p_n} = \frac{dx}{p}$$

·e known and also any multiplier, then another integral can be deter-
ined. This is generally referred to as the theorem of *Jacobi's Last
ultiplier. In Dynamics, where this theorem is of some importance
ee Whittaker, Chap. X.), the last multiplier is unity.]

(v) Show that unity is a multiplier of

$$\frac{dx}{xz - 2y} = \frac{dy}{2x - yz} = \frac{dz}{y^2 - x^2}$$

nd $x^2 + y^2 + z^2 = a$ an integral, say $u(x, y, z) = a$.

Show that in this case

$$M(Q\,dx - P\,dy)\Big/\frac{\partial u}{\partial z} = d\{-\tfrac{1}{2}xy - \sqrt{(a - x^2 - y^2)}\},$$

nd hence obtain the second integral $xy + 2z = b$.

(106) Show that if $y = \int_a^b e^{xt} f(t)\, dt$, where a and b are constants, the

$$x\phi\left(\frac{d}{dx}\right)y + \psi\left(\frac{d}{dx}\right)y = e^{bx}\,\phi(b)f(b) - e^{ax}\,\phi(a)f(a)$$

$$- \int_a^b e^{xt}\{\phi(t)f'(t) + \phi'(t)f(t) - \psi(t)f(t)\}\,$$

Hence prove that y will satisfy the differential equation

$$x\phi\left(\frac{d}{dx}\right)y + \psi\left(\frac{d}{dx}\right)y = 0$$

if

$$\phi(t)f(t) = \exp\left\{\int \frac{\psi(t)}{\phi(t)}\,dt\right\}$$

and

$$e^{bx}\phi(b)f(b) = 0 = e^{ax}\phi(a)f(a).$$

Use this method to obtain

$$y = A\int_{-\infty}^{-1} e^{xt}\frac{dt}{\sqrt{(t^2-1)}} + B\int_{-1}^{1} e^{xt}\frac{dt}{\sqrt{(t^2-1)}}$$

as a solution, valid when $x > 0$, of

$$x\frac{d^2y}{dx^2} + \frac{dy}{dx} - xy = 0.$$

The corresponding solution for the case $x < 0$ is obtained by takin the limits of the first integral as 1 to ∞, instead of $-\infty$ to -1.

[Exs. 106-108 give some of the most important methods of obtainin solutions of differential equations in the form of definite integrals.]

(107) Verify that $\quad v = v_0 + \dfrac{2V}{\sqrt{\pi}}\displaystyle\int_0^{x/2\sqrt{(\kappa t)}} e^{-z^2}\,dz$

is a solution of

$$\frac{\partial v}{\partial t} = \kappa \frac{\partial^2 v}{\partial x^2},$$

reducing, when $t = 0$, to $v_0 + V$ for all positive values of x and to $v_0 -$ for all negative values.

[v is the temperature at time t of a point at a distance x from certain plane of a solid extending to infinity in all directions, on th supposition that initially the temperature had the two different consta values $v_0 + V$ and $v_0 - V$ on the two sides of the plane $x = 0$.

Kelvin used this expression for v in his estimate of the age of th earth (see Appendix D of Thomson and Tait's *Natural Philosophy*). Th discovery that heat is continually generated by the radio-active di integration of the rocks introduces a new complexity into the problem

(108) (a) Show that

$$V = \iint e^{lx+my+nz} f(s, t)\, ds\, dt$$

(the limits being any arbitrary quantities independent of x, y, z) is solution of the linear partial differential equation with consta coefficients

$$F\left(\frac{\partial}{\partial x},\ \frac{\partial}{\partial y},\ \frac{\partial}{\partial z}\right)V = 0$$

l, m, n are any constants or functions of s and t such that
$$F(l, m, n) = 0.$$
Extend the theorem to the case when there are n independent variables x, y, z, ... , and $(n-1)$ parameters s, t, ...

Obtain
$$V = \int \int e^{s(x \cos t + y \sin t + sz)} f(s, t) \, ds \, dt$$
s a solution of
$$\frac{\partial^2 V}{\partial x^2} + \frac{\partial^2 V}{\partial y^2} = \frac{\partial V}{\partial z}. \qquad \text{[H. Todd.]}$$

(b) Show that if $F\left(\dfrac{\partial}{\partial x}, \dfrac{\partial}{\partial y}, \dfrac{\partial}{\partial z}\right) V = 0$ is a *homogeneous* linear artial differential equation with constant coefficients a solution is
$$V = \int f(lx + my + nz, t) \, dt,$$
here the limits are any arbitrary quantities independent of x, y, z, and m, n are any constants or functions of t such that
$$F(l, m, n) = 0.$$
Extend the theorem to the case when there are n independent ariables and $(n-2)$ parameters. [See H. Todd, *Messenger of Mathe-atics*, 1914.]

Obtain
$$V = \int_0^{2\pi} f(x \cos t + y \sin t + iz, t) \, dt$$
s a solution of
$$\frac{\partial^2 V}{\partial x^2} + \frac{\partial^2 V}{\partial y^2} + \frac{\partial^2 V}{\partial z^2} = 0.$$
Vhittaker's solution of Laplace's equation.]

(109) By substituting the trial solution
$$y = a_0 + \frac{a_1}{x} + \frac{a_2}{x^2} + \cdots$$
the differential equation $\dfrac{dy}{dx} + y = \dfrac{1}{x}$,

tain the series
$$y = \frac{0!}{x} + \frac{1!}{x^2} + \frac{2!}{x^3} + \frac{3!}{x^4} + \cdots .$$
Prove that this series is divergent for all values of x.
Obtain the particular integral
$$y = e^{-x} \int_{-\infty}^{x} \frac{e^x}{x} \, dx,$$
id by repeated integration by parts show that
$$e^{-x} \int_{-\infty}^{x} \frac{e^x}{x} \, dx = \frac{0!}{x} + \frac{1!}{x^2} + \frac{2!}{x^3} + \cdots + \frac{n!}{x^{n+1}} + e^{-x} \int_{-\infty}^{x} \frac{(n+1)! e^x}{x^{n+2}} \, dx.$$
Hence prove that if x is negative the error obtained by taking $n+1$ rms of the series instead of the particular integral is less than the imerical value of the $(n+1)^{\text{th}}$ term.
[Such a series is called *asymptotic*. See Bromwich's *Infinite Series*, rts. 130-139; or 2nd ed., Arts. 106-118.]

(110) Show that if the sequence of functions $f_n(x)$ be defined by
$$f_0(x) = a + b(x - c), \text{ where } a, b, c \text{ are constants,}$$

and
$$f_n(x) = \int_c^x (t - x) F(t) f_{n-1}(t) \, dt,$$

then
$$\frac{d^2}{dx^2} f_n(x) = - F(x) f_{n-1}(x).$$

Hence show that $y = \sum_0^\infty f_n(x)$ is a solution of
$$\frac{d^2 y}{dx^2} + y F(x) = 0,$$

provided that certain operations with infinite series are legitimate (for a proof of which see Whittaker and Watson's *Modern Analysis*, p. 189.* They give a proof of the existence theorem for linear differential equations of the second order by this method).

(111) Prove that the solution † of the two simultaneous linear differential equations with constant coefficients
$$f(D)x + F(D)y = 0,$$
$$\phi(D)x + \psi(D)y = 0$$
(where D stands for d/dt), may be written
$$x = F(D) \, V,$$
$$y = -f(D) \, V,$$
where V is the complete primitive of
$$\{f(D)\psi(D) - F(D) \, \phi(D)\} V = 0.$$

Hence show that if the degrees of f, F, ϕ, ψ in D be p, q, r, s respectively, the number of arbitrary constants occurring in the solution will in general be the greater of the numbers $(p + s)$ and $(q + r)$, but $(p + s) = (q + r)$ the number of arbitrary constants may be smaller, and may even be zero as in the equations
$$(D + 1)x + Dy = 0,$$
$$(D + 3)x + (D + 2)y = 0.$$

(112) (a) Prove that if
$$y = u(x),$$
$$y = v(x)$$
are any two solutions of the linear differential equation of the first order
$$P(x)y_1 + Q(x)y = 0,$$
then
$$(vu_1 - uv_1)/u^2 = 0,$$
so that $v = au$, where a is a constant.

(b) Prove that if $\quad y = u(x), \quad y = v(x), \quad y = w(x)$

---

* p. 195 in 3rd and 4th editions.
† This, it may be proved, cannot be the most general solution if it gives the number of different arbitrary constants for x and y together less than for V, as will happen if $f(D)$ and $F(D)$ have a common factor other than a mere constant.

are any three solutions of the linear differential equation of the second
order $$P(x)y_2 + Q(x)y_1 + R(x)y = 0,$$

then $$P\frac{d}{dx}(wv_1 - vw_1) + Q(wv_1 - vw_1) = 0$$

and $$P\frac{d}{dx}(uv_1 - vu_1) + Q(uv_1 - vu_1) = 0.$$

Hence show that $w = au + bv.$

[By proceeding step by step in this manner we may show that a
differential equation of similar form but of the nth order cannot have
more than n linearly independent integrals.]

(113) Let u, v, w be any three functions of x.
Prove that if constants a, b, c can be found so that $y \equiv au + bv + cw$
vanishes identically, then

$$\begin{vmatrix} u & v & w \\ u_1 & v_1 & w_1 \\ u_2 & v_2 & w_2 \end{vmatrix} = 0,$$

while conversely, if this determinant (the *Wronskian*) vanishes, the
functions are not linearly independent.
Extend these results to the case of n functions.
[Consider the differential equation of the second order formed by
replacing u, u_1, u_2 in the determinant by y, y_1, y_2 respectively. Such
an equation cannot have more than two linearly independent integrals.
The *Wronskian* is named after Hoëné Wronski, one of the early
writers on determinants.]

(114) Prove that $z = e^{\frac{1}{2}x(t-1/t)}$ satisfies the partial differential equation
$$t\frac{\partial}{\partial t}\left(t\frac{\partial z}{\partial t}\right) = \frac{1}{4}x^2\left(t + \frac{1}{t}\right)^2 z + \frac{1}{2}x\left(t - \frac{1}{t}\right)z.$$

Hence, if $J_n(x)$ is defined as the coefficient of t^n in the expansion
$$e^{\frac{1}{2}x(t-1/t)} = \sum_{-\infty}^{\infty} t^n J_n(x),$$

prove that $y = J_n(x)$ satisfies Bessel's equation of order n,
$$x^2\frac{d^2y}{dx^2} + x\frac{dy}{dx} + (x^2 - n^2)y = 0.$$

[The operations with infinite series require some consideration.]

(115) If u_x denotes a function of x, and E the operator which changes
u_x into u_{x+1}, prove the following results:
  (i) $Ea^x = a \cdot a^x$, i.e. $(E - a)a^x = 0.$
  (ii) $E^2 a^x = a^2 \cdot a^x.$
  (iii) $E(xa^x) = a(xa^x) + a \cdot a^x$, i.e. $(E - a)(xa^x) = a \cdot a^x.$
  (iv) $(E - a)^2(xa^x) = 0.$
  (v) $(p_0 E^2 + p_1 E + p_2)a^x = (p_0 a^2 + p_1 a + p_2)a^x$, if the p's are constant

(vi) $u_x = Aa^x + Bb^x$ is a solution of the *linear difference equation*

$$p_0 u_{x+2} + p_1 u_{x+1} + p_2 u_x = 0,$$
$$\textit{i.e.} \quad (p_0 E^2 + p_1 E + p_2) u_x = 0,$$

if A and B are arbitrary constants and a and b the roots of the auxiliary equation $p_0 m^2 + p_1 m + p_2 = 0$. (Cf. Art. 25.)

Solve by this method $\quad (2E^2 + 5E + 2) u_x = 0$.

(vii) $u_x = (A + Bx) a^x$ is a solution of $(E^2 - 2aE + a^2) u_x = 0$.
Here the auxiliary equation $m^2 - 2am + a^2 = 0$ has equal roots. (Cf. Art. 34.)

(viii) $u_x = r^x (P \cos x\theta + Q \sin x\theta)$ is a solution of

$$(p_0 E^2 + p_1 E + p_2) u_x = 0$$

if P and Q are arbitrary constants, $p \pm iq$ the roots of the auxiliary equation $\quad p_0 m^2 + p_1 m + p_2 = 0$

and $\quad p + iq = r (\cos \theta + i \sin \theta)$. (Cf. Art. 26.)

Solve by this method $\quad (E^2 - 2E + 4) u_x = 0$.

(ix) The general solution of a linear difference equation with constant coefficients

$$F(E) u_x \equiv (p_0 E^n + p_1 E^{n-1} + \ldots + p_{n-1} E + p_n) u_x = f(x)$$

is the sum of a Particular Integral and the Complementary Function, the latter being the solution of the equation obtained by substituting zero for the function of x occurring on the right-hand side. (Cf. Art. 29.)

(x) $a^x / F(a)$ is a particular integral of

$$F(E) u_x = a^x,$$

provided that $F(a) \neq 0$. (Cf. Art. 35.)

Solve by this method $\quad (E^2 + 8E - 9) u_x = 2^x$.

[For further analogies between difference equations and differential equations, see Boole's *Finite Differences*, Chap. XI.]

(116) Show that by applying the method of Art. 53 to Lagrange's equation

$$y = xF(p) + f(p),$$

we get in general (but not for Clairaut's form, where $F(p) = p$) the complete primitive in the parametric form

$$x = c\phi(p) + \psi(p),$$
$$y = cF(p)\phi(p) + F(p)\psi(p) + f(p).$$

Hence show that if C_1, C_2, C_3 are any three curves included in this primitive, corresponding to the values c_1, c_2, c_3 of c, and $P_1(x_1, y_1)$ $P_2(x_2, y_2)$, $P_3(x_3, y_3)$ points on C_1, C_2, C_3 respectively, such that the tangents at these points are all parallel, then

$$(x_3 - x_1)/(x_3 - x_2) = (c_3 - c_1)/(c_3 - c_2) = (y_3 - y_1)/(y_3 - y_2),$$

i.e. P_1, P_2, P_3 are collinear, and the ratio $P_1P_3 : P_2P_3$ is constant as the points move, each along its own curve, in such a way as to keep the corresponding tangents parallel. [Thus given two curves included in the complete primitive, we can construct geometrically any number of others.]

(117) Prove that a plane curve, such that the length of the radius of curvature at any point is twice the length of the normal intercepted between the curve and a fixed straight line, is either a cycloid, whose base is the straight line, or a parabola, whose directrix is the straight line.
[London.]

(118) A curve possesses the property $\rho = k \tan \psi$, where ρ is the radius of curvature, ψ is the angle the tangent makes with the axis of x, and k is positive. Show that the curve has a branch given by the equations

$$x = k(1 - \cos \theta), \quad y = k\{\log (\sec \theta + \tan \theta) - \sin \theta\},$$

where $0 \leqslant \theta < \frac{1}{2}\pi$, and the origin is taken at the point $\theta = 0$. Show that, if s is the length of the arc measured along this branch from the same point,

$$s = k \log \frac{k}{k - x}.$$
[London.]

(119) Obtain a solution of the equation $\dfrac{\partial^2 u}{\partial t^2} = c^2 \dfrac{\partial^2 u}{\partial x^2}$ in the form $f(x) \sin mt$, which is such that

$$\frac{\partial u}{\partial t} = K, \text{ a constant, when } x = 0 \text{ and } t = 0,$$

$$\frac{\partial u}{\partial x} = 0, \text{ when } x = 0, \text{ for all values of } t.$$
[London.]

(120) Obtain for the equation $\dfrac{\partial^2 z}{\partial x^2} + \dfrac{\partial^2 z}{\partial y^2} = 0$ a solution which satisfies the following conditions:

(i) when $y = 0$, $z = \sin x$;

(ii) when $x = 0$ or π, $z = 0$;

(iii) z does not become infinite anywhere in the region of the plane of x, y in which $y > 0$ and $\pi > x > 0$.
[London.]

(121) By two integrations by parts show that, if P, Q, R are functions of x, and suffixes denote differentiations with respect to x,

$$\int z(Py_2 + Qy_1 + Ry) \, dx = z(Py_1 + Qy) - y(Pz)_1 + \int y\{(Pz)_2 - (Qz)_1 + Rz\} \, dx.$$

Deduce that the two equations

$$Py_2 + Qy_1 + Ry = 0, \quad (Pz)_2 - (Qz)_1 + Rz = 0,$$

are such that any integral of one is an integrating factor of the other. Such equations are said to be adjoint to each other.]

Show that, if D represents the operator d/dx, the equation adjoint t
$$\{D + p(x)\}\{D + q(x)\}y = 0$$
is
$$\{D - q(x)\}\{D - p(x)\}z = 0.$$

Verify this for the equation $y_2 + (x + x^2)y_1 + (2x + x^3)y = 0.$ [Her $p(x) = x$, $q(x) = x^2$.]

General solution of $\dfrac{\partial^2 y}{\partial x^2} = \dfrac{1}{a^2}\dfrac{\partial^2 y}{\partial t^2}$.

Factorising the operator, the equation may be written

$$\left(\frac{\partial}{\partial x} - \frac{1}{a}\frac{\partial}{\partial t}\right)\left\{\left(\frac{\partial}{\partial x} + \frac{1}{a}\frac{\partial}{\partial t}\right)y\right\} = 0 = \left(\frac{\partial}{\partial x} + \frac{1}{a}\frac{\partial}{\partial t}\right)\left\{\left(\frac{\partial}{\partial x} - \frac{1}{a}\frac{\partial}{\partial t}\right)y\right\}.$$

Hence (cf. p. 33) the original equation is satisfied by any integra of either of the two Lagrange linear equations

$$\frac{\partial y}{\partial x} + \frac{1}{a}\frac{\partial y}{\partial t} = 0 \quad \text{and} \quad \frac{\partial y}{\partial x} - \frac{1}{a}\frac{\partial y}{\partial t} = 0.$$

For the first of these the subsidiary equations are (from Art. 123)

$$\frac{dx}{1} = \frac{dt}{1/a} = \frac{dy}{0}.$$

Two independent integrals are
$$y = b, \quad x - at = c.$$

The general integral is
$$y = f(x - at).$$

Similarly the second Lagrange equation gives $y = F(x + at)$. Thes are both integrals of the original differential equation. As it is linea a third integral is

$$y = f(x - at) + F(x + at),$$

containing two arbitrary functions, and no more general solution of a equation of order two can be expected. (Cf. pp. 61 and 218.)

A similar method can be used for the equation of Art. 145.

The Method of Parameters. (C. N. Srinivasiengar.)

If a partial differential equation becomes an identity on substitutin $p = f(x, a)/\phi(z, a)$, $q = F(y, a)/\phi(z, a)$, we can use these expressions i conjunction with $dz = p\,dx + q\,dy$ to obtain the complete integral

$$\int \phi(z, a)\,dz = \int f(x, a)\,dx + \int F(y, a)\,dy + b.$$

For example, the equation $z^2(p + q) = x^2 + y^2$ becomes an identity

$$p = (x^2 + a)/z^2, \quad q = (y^2 - a)/z^2,$$
giving
$$z^3 = x^3 + y^3 + 3ax - 3ay + b.$$

This method will deal with all equations of Standard Forms I an III (Arts. 129 and 131) and some of II (Art. 130).

ANSWERS TO THE EXAMPLES

CHAPTER I

Art. 5

(1) $\dfrac{d^2y}{dx^2} = 4y.$

(2) $\dfrac{d^2y}{dx^2} = -9y.$

(3) $y\dfrac{d^2y}{dx^2} = \left(\dfrac{dy}{dx}\right)^2.$

(4) $y = x\dfrac{dy}{dx} + \left(\dfrac{dy}{dx}\right)^3.$

(5) The tangent to a circle is perpendicular to the line joining the point of contact to the centre.

(6) The tangent at any point is the straight line itself.

(7) The curvature is zero.

Art. 8

(1) $y = a + ax + a\dfrac{x^2}{2!} + a\dfrac{x^3}{3!} + a\dfrac{x^4}{4!} + \ldots = ae^x.$

(2) $y = a + bx - a\dfrac{x^2}{2!} - b\dfrac{x^3}{3!} + a\dfrac{x^4}{4!} + \ldots = a\cos x + b\sin x.$

Miscellaneous Examples on Chapter I

(1) $\dfrac{d^3y}{dx^3} = \dfrac{dy}{dx}.$

(2) $\dfrac{d^3y}{dx^3} - 6\dfrac{d^2y}{dx^2} + 11\dfrac{dy}{dx} - 6y = 0.$

(3) $\dfrac{d^2y}{dx^2} - 2\dfrac{dy}{dx} + 2y = 0.$

(4) $y\log_e\left[\dfrac{dy}{dx} + \sqrt{\left\{1 + \left(\dfrac{dy}{dx}\right)^2\right\}}\right] = x\sqrt{\left\{1 + \left(\dfrac{dy}{dx}\right)^2\right\}}.$

(5) $\dfrac{d^3y}{dx^3} = 0.$

(6) $\left\{1 + \left(\dfrac{dy}{dx}\right)^2\right\}^3 = a^2\left(\dfrac{d^2y}{dx^2}\right)^2,$ *i.e.* $\rho^2 = a^2.$

(7) $(x^2 + y^2)\dfrac{d^2y}{dx^2} = 2\left(x\dfrac{dy}{dx} - y\right)\left\{1 + \left(\dfrac{dy}{dx}\right)^2\right\}.$

(8) $\left\{1 + \left(\dfrac{dy}{dx}\right)^2\right\}\dfrac{d^3y}{dx^3} = 3\left(\dfrac{d^2y}{dx^2}\right)^2\dfrac{dy}{dx}.$

(11) $y = ax + bx^2.$

(12) $y = ae^x + be^{-x}.$

(14) $60°$ and $-60°.$

(15) Differentiate and put $x=1$, $y=2$. This gives $\dfrac{d^2y}{dx^2}$ and hence ρ.

(17) (i) $x+1=0$; (ii) $y^2=x^2+6x+1$.

CHAPTER II

Art. 14

(1) $6x^2+5xy+y^2-9x-4y=c$.   (2) $\sin x \tan y + \sin(x+y)=c$.

(3) $\sec x \tan y - e^x = c$.   (4) $x-y+c=\log(x+y)$.

(5) $x+ye^{x^3}=cy$.   (6) $y=cx$.

(7) $e^y(\sin x+\cos x)=c$.   (8) $x^4y+4cy+4=0$.

(9) $ye^x=cx$.   (10) $\sin x \cos y = c$.

Art. 17

(1) $(x+y)^3=c(x-y)$.   (2) $x^2+2y^2(c+\log y)=0$.

(3) $xy^2=c(x-y)^2$.   (4) $cx^2=y+\sqrt{(x^2+y^2)}$.

(5) $(2x-y)^2=c(x+2y-5)$.   (6) $(x+5y-4)^3(3x+2y+1)=c$.

(7) $x-y+c=\log(3x-4y+1)$.   (8) $3x-3y+c=2\log(3x+6y-1)$

Art. 21

(1) $2y=(x+a)^5+2c(x+a)^3$.   (2) $xy=\sin x+c\cos x$.

(3) $y\log x=(\log x)^2+c$.   (4) $x^3=y^3(3\sin x+c)$.

(5) $y^2(x+ce^x)=1$.   (6) $x=y^3+cy$.   (7) $x=e^{-y}(c+\tan y$

Art. 22

(1) The parabola $y^2=4ax+c$.

(2) The rectangular hyperbola $xy=c^2$.

(3) The lemniscate of Bernoulli $r^2=a^2\sin 2\theta$.

(4) The catenary $y=k\cosh\dfrac{x-c}{k}$.   (5) $xy=c^2$.

(6) $y^{\frac{2}{3}}=x^{\frac{2}{3}}+c^{\frac{2}{3}}$.   (7) $y^p=cx^q$.   (8) $r^2=ce^{\theta^2}$.

(9) $\log r+\frac{1}{2}\theta^2+\frac{1}{3}\theta^3=c$. (10) The equiangular spirals $r=ce^{\pm\theta\tan\alpha}$.

Miscellaneous Examples on Chap. II

(1) $xy=y^3+c$.   (2) $cx^3=y+\sqrt{(y^2-x^2)}$.

(3) $\sin x \sin y+e^{\sin x}=c$.   (4) $2x^2-2xy+3y+2cx^2y=0$.

(5) $cxy=y+\sqrt{(y^2-x^2)}$.   (11) $x^3y^{-2}+2x^5y^{-3}=c$.

(12) $\tan^{-1}(xy)+\log(x/y)=c$.   (14) $(x^2-1+y^4)e^{x^2}=c$.

(15) (i) The Reciprocal Spiral $r(\theta-\alpha)=c$.

    (ii) The Spiral of Archimedes $r=c(\theta-\alpha)$.

(16) The parabola $3ky^2=2x$.   (18) $x=y(c-k\log y)$.

(19) (i) $x^2 + (y-c)^2 = 1 + c^2$, a system of coaxal circles cutting the given system orthogonally.

(ii) $r^2 = ce^{-\theta^2}$.　　　(iii) $n^2 = r\{c + \log(\operatorname{cosec} n\theta + \cot n\theta)\}$.

(20) $\left(x + y\dfrac{dy}{dx}\right)\left(x - y\dfrac{dx}{dy}\right) = a^2 - b^2$.

(21) $\log(2x^2 \pm xy + y^2) + \dfrac{6}{\sqrt{7}}\tan^{-1}\dfrac{x \pm 2y}{x\sqrt{7}} = c$.

CHAPTER III

Art. 28

(1) $y = Ae^{-x} + Be^{-3x}$.　　　　　　(2) $y = A\cos 2x + B\sin 2x$.

(3) $y = Ae^{-3x} + Be^{-4x}$.　　　　　(4) $y = e^{2x}(A\cos x + B\sin x)$.

(5) $s = e^{-2t}(A\cos 3t + B\sin 3t)$.　(6) $s = A + Be^{-4t}$.

(7) $y = Ae^x + Be^{-x} + Ce^{-2x}$.　　(8) $y = 2e^{-x} - e^{-2x}$.

(9) $y = A\cos(2x - \alpha) + B\cos(3x - \beta)$.

(10) $y = A\cosh(2x - \alpha) + B\cosh(3x - \beta)$, or
　　 $y = Ee^{2x} + Fe^{-2x} + Ge^{3x} + He^{-3x}$.

(11) $y = Ae^{-2x} + Be^x\cos(x\sqrt{3} - \alpha)$.

(12) $y = Ae^{2x} + Be^{-2x} + Ee^{-x}\cos(x\sqrt{3} - \alpha) + Fe^x\cos(x\sqrt{3} - \beta)$.

(13) $\theta = \alpha\cos t\sqrt{(g/l)}$.　　　　(14) $k^2 < 4mc$.

(16) $Q = Q_0 e^{-Rt/2L}\left(\cos nt + \dfrac{R}{2Ln}\sin nt\right)$, where $n = \sqrt{\left(\dfrac{1}{LC} - \dfrac{R^2}{4L^2}\right)}$.

Art. 29

(1) $y = e^x(1 + A\cos x + B\sin x)$.　(2) $y = 3 + Ae^x + Be^{12x}$.

(3) $y = 2\sin 3x + A\cos 2x + B\sin 2x$.　　　(4) $a = 2$; $b = 1$.

(5) $a = 6$; $b = -1$.　　(6) $a = -4$; $p = 2$.　　(7) $a = 1$; $b = 2$; $p = 1$.

(8) $a = 2$.　　　　(9) $4e^{3x}$.　　　　(10) $3e^{7x}$.

(11) $-\frac{5}{2}\sin 5x$.　　(12) $\frac{25}{29}\cos 5x - \frac{10}{29}\sin 5x$.　　(13) 2.

Art. 34

(1) $y = A + Bx + (E + Fx)e^{-x}$.

(2) $y = (A + Bx + Cx^2)\cos x + (E + Fx + Gx^2)\sin x$.

(3) $y = (A + Bx)e^x + E\cos x + F\sin x$.

(4) $y = A + Bx + Ce^x + (E + Fx)e^{-\frac{1}{2}x}$.

Art. 35

(1) $y = 2e^{3x} + e^{-3x}(A\cos 4x + B\sin 4x)$.

(2) $y = e^{-px}(A\cos qx + B\sin qx) + e^{ax}/\{(a+p)^2 + q^2\}$.

(3) $y = (A + 9x)e^{3x} + Be^{-3x}$.

(4) $y = A + (B + \frac{1}{2}x)e^x + (C + \frac{1}{2}x)e^{-x}$.

(5) $y = (A + ax/2p) \cosh px + B \sinh px$.

(6) $y = A + (B + Cx - 2x^2)e^{-2x}$.

Art. 36

(1) $y = 2 \sin 2x - 4 \cos 2x + Ae^{-x}$.

(2) $y = 4 \cos 4x - 2 \sin 4x + Ae^{2x} + Be^{3x}$.

(3) $y = 2 \cos x + e^{-4x}(A \cos 3x + B \sin 3x)$.

(4) $y = \sin 20x + e^{-x}(A \cos 20x + B \sin 20x)$.

Art. 37

(1) $y = x^3 - 3x^2 + 6x - 6 + Ae^{-x}$.   (2) $y = 6x^2 - 6x + A + Be^{-2x}$.

(3) $y = 6x + 6 + (A + Bx)e^{3x}$.

(4) $y = x^3 + 3x^2 + Ex + F + (A + Bx)e^{3x}$.

(5) $y = 24x^2 + 14x - 5 + Ae^{-x} + Be^{2x}$.

(6) $y = 8x^3 + 7x^2 - 5x + Ae^{-x} + Be^{2x} + C$.

Art. 38

(1) $y = A \cos x + (B + 2x) \sin x$.   (2) $y = Ae^x + (x + 2)e^{2x}$.

(3) $y = Ae^{2x} + (B + Cx - 20x^2 - 20x^3 - 15x^4 - 9x^5)e^{-x}$.

(4) $y = \{A \sin x + (B - x) \cos x\}e^{-x}$.

(5) $y = (A + Bx - x^3) \cos x + (E + Fx + 3x^2) \sin x$.

(6) $y = A + (B + 3x)e^x + Ce^{-x} + x^2 + E \cos x + (F + 2x) \sin x$.

(7) $y = \{A \sin 4x + (B - x + x^2) \cos 4x\}e^{3x}$.

Art. 39

(1) $y = Ax + Bx^2 + 2x^3$.

(2) $y = 2 + Ax^{-4} \cos (3 \log x) + Bx^{-4} \sin (3 \log x)$.

(3) $y = 8 \cos (\log x) - \sin (\log x) + Ax^{-2} + Bx \cos (\sqrt{3} \log x - \alpha)$.

(4) $y = 4 + \log x + Ax + Bx \log x + Cx (\log x)^2 + Dx (\log x)^3$.

(5) $y = (1 + 2x)^2 [\{\log (1 + 2x)\}^2 + A \log (1 + 2x) + B]$.

(6) $y = A \cos \{\log (1 + x) - \alpha\} + 2 \log (1 + x) \sin \log (1 + x)$.

Art. 40

(1) $y = A \cos (x - \alpha)$; $z = -A \sin (x - \alpha)$.

(2) $y = Ae^{5x} + Be^{3x}$; $z = 6Ae^{5x} - 7Be^{3x}$.

(3) $y = Ae^x + B \cos (2x - \alpha)$; $z = 2Ae^x - B \cos (2x - \alpha)$.

(4) $y = e^x + A + Be^{-2x}$; $z = e^x + A - Be^{-2x}$.

(5) $y = A \cos (x - \alpha) + 4B \cos (2x - \beta) + \cos 7x$;
$z = A \cos (x - \alpha) + B \cos (2x - \beta) - 2 \cos 7x$.

(6) $y = -5Ae^{3x} - 4Be^{4x} + 2e^{-x} + \cos 2x - \sin 2x$;
$z = Ae^{3x} + Be^{4x} + 3e^{-x} + 4 \cos 2x + 5 \sin 2x$.

Miscellaneous Examples on Chapter III

(1) $y = (A + Bx + Cx^2)e^x + 2e^{3x}$.   (2) $y = (A + Bx + 6x^3)e^{-3x/2}$.

(3) $y = Ae^{-3x} + Be^{-2x} + Ce^{-x} + E + 2e^{-2x}(\sin x - 2\cos x)$.

(4) $y = Ae^x + B\cos(2x - \alpha) - 2e^x(4\sin 2x + \cos 2x)$.

(5) $y = (A + Bx + Cx^2)e^{-x} + (E + x + 2x^2)e^{3x}$.

(6) $y = A\sin(x - \alpha) + B\sinh(3x - \beta) - 2\sinh 2x$.

(7) $y = (A + Bx + 5x^2)\cosh x + (E + Fx)\sinh x$.

(8) $y = 3 + 4x + 2x^2 + (A + Bx + 4x^2)e^{2x} + \cos 2x$.

(9) $y = (A + Bx + 3\sin 2x - 4x\cos 2x - 2x^2\sin 2x)e^{2x}$.

(10) $y = A\cos(x - \alpha) + \frac{3}{2} - \frac{1}{2}\cos 2x - \frac{3}{4}x\cos x + \frac{1}{16}\sin 3x$.

(11) $y = A\cos(x - \alpha) + B\cos(3x - \beta) - 3x\cos x + x\cos 3x$.

(12) $y = (A_0 + A_1 x + A_2 x^2 + \dots + A_{a-1}x^{a-1})e^{ax} + a^x/(\log a - a)^a$.

(13) $y = A + B\log x + 2(\log x)^3$.   (14) $y = A + Bx^{-1} + \frac{5}{8}x^2$.

(15) $y = Ax^3 + B\cos(\sqrt{2}\log x - \alpha)$.

(16) $y = A + B\log(x + 1) + \{\log(x + 1)\}^2 + x^2 + 8x$.

(17) $x = Ae^{3t} + Be^{-3t} + E\cos t + F\sin t - e^t$;

     $y = Ae^{3t} + 25Be^{-3t} + (3E - 4F)\cos t + (3F + 4E)\sin t - e^t$.

(18) $x = Ae^{2t} + Be^{-t}\cos(\sqrt{3}t - \alpha)$;

     $y = Ae^{2t} + Be^{-t}\cos(\sqrt{3}t - \alpha + 2\pi/3)$;

     $z = Ae^{2t} + Be^{-t}\cos(\sqrt{3}t - \alpha + 4\pi/3)$.

(19) $x = At + Bt^{-1}$; $y = Bt^{-1} - At$.

(20) $x = At\cos(\log t - \alpha) + Bt^{-1}\cos(\log t - \beta)$;

     $y = At\sin(\log t - \alpha) - Bt^{-1}\sin(\log t - \beta)$.

(27) (i) $(x - 1)e^{2x}$; (ii) $\frac{1}{2}(x^2 - 2x + 1)\sin x + \frac{1}{2}(x^2 - 1)\cos x$.

(31) $y = e^{2x} + Ae^x$.

(32) $y = (\sin ax)/(p^2 - a^2) + A\cos px + B\sin px$.

(33) $y = Ae^{ax} + Be^{bx} + e^{bx}\displaystyle\int xe^{(a-b)x}(\log x - 1)\,dx$.

(35) (iii) $y = A\cos(x - \alpha) - x\cos x + \sin x\log\sin x$.

(37) (i) $k/(2phe)$; (ii) zero.

(38) $y = E\cos nx + F\sin nx + G\cosh nx + H\sinh nx$.

CHAPTER IV

Art. 42

(1) $\dfrac{\partial z}{\partial y} = a\dfrac{\partial z}{\partial x}$.

(2) $\dfrac{\partial^2 z}{\partial x^2} + \dfrac{\partial^2 z}{\partial y^2} = 0$.   (Laplace's equation in two dimensions.)

(3) $\dfrac{\partial^2 z}{\partial x^2} + \dfrac{\partial^2 z}{\partial y^2} = \dfrac{1}{a^2}\dfrac{\partial^2 z}{\partial t^2}$.   (4) $y\dfrac{\partial z}{\partial x} + x\dfrac{\partial z}{\partial y} = 0$.

(5) $b\dfrac{\partial z}{\partial x} + a\dfrac{\partial z}{\partial y} = 2abz.$

(6) $x\dfrac{\partial z}{\partial x} + y\dfrac{\partial z}{\partial y} = nz.$ (Euler's theorem on Homogeneous Functions.)

Art. 43

(1) $\dfrac{\partial^2 z}{\partial x^2} = \dfrac{\partial z}{\partial t}.$ (2) $\dfrac{\partial^2 z}{\partial x^2} + \dfrac{\partial^2 z}{\partial y^2} + \dfrac{\partial^2 z}{\partial t^2} = 0.$ (3) $\dfrac{\partial z}{\partial x} + \dfrac{\partial z}{\partial y} = 1.$

(4) $z = x\dfrac{\partial z}{\partial x} + y\dfrac{\partial z}{\partial y} + \left(\dfrac{\partial z}{\partial x}\right)^2 + \left(\dfrac{\partial z}{\partial y}\right)^2.$

(5) $4z = \left(\dfrac{\partial z}{\partial x}\right)^2 + \left(\dfrac{\partial z}{\partial y}\right)^2.$ (6) $\dfrac{\partial z}{\partial x}\dfrac{\partial z}{\partial y} = 1.$

Art. 45

(1) $y = Ae^{-p^2(x+t)}.$ (2) $z = A\sin px \sin pay.$ (3) $z = A\cos p\,(ax -)$

(4) $V = Ae^{-px+qy}\sin z\sqrt{(p^2+q^2)}$, where p and q are positive.

(5) $V = C\cos(pqx + p^2y + q^2z).$

(6) $V = Ae^{-rt}\sin(m\pi x/l)\sin(n\pi y/l)$, where m and n are any intege
and $rl^2 = \pi^2(m^2+n^2).$

Art. 48

(1) $\dfrac{4}{\pi}(\sin x + \tfrac{1}{3}\sin 3x + \tfrac{1}{5}\sin 5x + \ldots).$

(2) $2(\sin x - \tfrac{1}{2}\sin 2x + \tfrac{1}{3}\sin 3x - \ldots).$

(3) $\dfrac{2}{\pi}\left[\left(\dfrac{\pi^3}{1} - \dfrac{6\pi}{1^3}\right)\sin x - \left(\dfrac{\pi^3}{2} - \dfrac{6\pi}{2^3}\right)\sin 2x + \left(\dfrac{\pi^3}{3} - \dfrac{6\pi}{3^3}\right)\sin 3x\ldots\right].$

(4) $\dfrac{4}{\pi}\left[\dfrac{2}{2^2-1}\sin 2x + \dfrac{4}{4^2-1}\sin 4x + \dfrac{6}{6^2-1}\sin 6x + \ldots\right].$

(5) $\dfrac{2}{\pi}[\tfrac{1}{2}(1+e^\pi)\sin x + \tfrac{2}{5}(1-e^\pi)\sin 2x + \tfrac{3}{10}(1+e^\pi)\sin 3x$
$\qquad\qquad\qquad + \tfrac{4}{17}(1-e^\pi)\sin 4x + \ldots].$

(6) $\dfrac{32}{\pi}\displaystyle\sum_1^\infty \dfrac{1}{n^3}\sin\dfrac{n\pi}{2}\left(4\sin\dfrac{n\pi}{4} - n\pi\cos\dfrac{n\pi}{4}\right)\sin nx.$

(7) (a) (2), (3), and (6); (b) (6).

Miscellaneous Examples on Chapter IV

(2) $\dfrac{\partial^2 V}{\partial x^2} = \dfrac{1}{K}\dfrac{\partial V}{\partial t}.$ (5) $\dfrac{\partial^2 V}{\partial t^2} = \dfrac{a^2}{r^2}\dfrac{\partial}{\partial r}\left(r^2\dfrac{\partial V}{\partial r}\right).$

(7) $V = V_0 e^{-gx}\sin(nt-gx)$, where $g = +\sqrt{(n/2K)}.$

(12) $V = \dfrac{8}{\pi}(e^{-Kt}\sin x + \tfrac{1}{27}e^{-9Kt}\sin 3x + \tfrac{1}{125}e^{-25Kt}\sin 5x + \ldots).$

(13) Replace x by $\pi x/l$, t by $\pi^2 t/l^2$, and the factor $8/\pi$ by $8l^2/\pi^3.$

14) $V = \dfrac{\pi^2}{6} - (e^{-4Kt} \cos 2x + \frac{1}{4}e^{-16Kt} \cos 4x + \frac{1}{9}e^{-36Kt} \cos 6x + ...).$

15) $V = \dfrac{400}{\pi} (e^{-Kt} \sin x + \frac{1}{3}e^{-9Kt} \sin 3x + \frac{1}{5}e^{-25Kt} \sin 5x + ...).$

[Notice that although $V = 100$ for all values of x *between* 0 and π, $V = 0$ for $x = 0$ or π, a discontinuity.]

16) Write $100 - V$ instead of V in the solution of (15).

18) $V = \dfrac{4V_0}{\pi} \{e^{-K\pi^2 t/4l^2} \cos (\pi x/2l) - \frac{1}{3}e^{-9K\pi^2 t/4l^2} \cos (3\pi x/2l) + ...\}.$

19) $y = \dfrac{4m}{\pi} (\sin x \cos vt - \frac{1}{9} \sin 3x \cos 3vt + \frac{1}{25} \sin 5x \cos 5vt - ...).$

CHAPTER V
Art. 52

(1) $(y - 2x - c)(y + 3x - c) = 0.$ (2) $(2y - x^2 - c)(2y + 3x^2 - c) = 0.$

(3) $49(y - c)^2 = 4x^7.$ (4) $(2y - x^2 - c)(2x - y^2 - c) = 0.$

(5) $(2y - x^2 - c)(y - ce^x)(y + x - 1 - ce^{-x}) = 0.$

(6) $(y - e^x - c)(y + e^{-x} - c) = 0.$

Art. 54

(The complete primitives only are given here. It will be seen later that in some cases singular solutions exist.)

(1) $x = 4p + 4p^3; \ y = 2p^2 + 3p^4 + c.$

(2) $x = \frac{1}{2}(p + p^{-1}); \ y = \frac{1}{4}p^2 - \frac{1}{2}\log p + c.$

(3) $(p - 1)^2 x = c - p + \log p; \ (p - 1)^2 y = p^2(c - 2 + \log p) + p.$

(4) $x = \frac{3}{2}p^2 + 3p + 3\log(p - 1) + c; \ y = p^3 + \frac{3}{2}p^2 + 3p + 3\log(p - 1) + c.$

(5) $x = 2\tan^{-1} p - p^{-1} + c; \ y = \log(p^3 + p).$

(6) $x = p + ce^{-p}; \ y = \frac{1}{2}p^2 + c(p + 1)e^{-p}.$

(7) $x = 2p + cp(p^2 - 1)^{-\frac{1}{2}}; \ y = p^2 - 1 + c(p^2 - 1)^{-\frac{1}{2}}.$

(8) $x = \sin p + c; \ y = p \sin p + \cos p.$

(9) $x = \tan p + c; \ y = p \tan p + \log \cos p.$

(10) $x = \log(p + 1) - \log(p - 1) + \log p + c; \ y = p - \log(p^2 - 1).$

(11) $x = p/(1 + p^2) + \tan^{-1} p; \ y = c - 1/(1 + p^2).$ (12) $c = 1.$

CHAPTER VI
Art. 58

(1) C.P. $(y + c)^2 = x^3; \ x = 0$ is a cusp-locus.

(2) C.P. $(y + c)^2 = x - 2; \ $S.S. $x = 2.$

(3) C.P. $x^2 + cy + c^2 = 0$; S.S. $y^2 = 4x^2$.

(4) C.P. $y = \sin(x + c)$; S.S. $y^2 = 1$.

(5) C.P. $(2x^3 + 3xy + c)^2 - 4(x^2 + y)^3 = 0$; $x^2 + y = 0$ is a cusp-locus.

(6) C.P. $c^2 - 12cxy + 8cy^3 - 12x^2y^2 + 16x^3 = 0$; $y^2 - x = 0$ is a cusp-lo

(7) C.P. $c^2 + 6cxy - 2cy^3 - x(3y^2 - x)^2 = 0$; $y^2 + x = 0$ is a cusp-locus.

Art. 65

(1) C.P. $(y + c)^2 = x(x - 1)(x - 2)$; S.S. $x(x - 1)(x - 2) = 0$; $x = 1 - 1$
    is a tac-locus and $x = 1 + 1/\sqrt{3}$ a tac-locus of imaginary po
    of contact.

(2) C.P. $(y + c)^2 = x(x - 1)^2$; S.S. $x = 0$; $x = 1/3$ is a tac-locus; x
    is a node-locus.

(3) C.P. $y^2 - 2cx + c^2 = 0$; S.S. $y^2 = x^2$.

(4) C.P. $x^2 + c(x - 3y) + c^2 = 0$; S.S. $(3y + x)(y - x) = 0$.

(5) C.P. $y - cx^2 - c^2 = 0$; S.S. $x^4 + 4y = 0$; $x = 0$ is a tac-locus.

(6) C.P. $y = c(x - c)^2$; $y = 0$ is a S.S. and also a particular integ
    $27y - 4x^3 = 0$ is a S.S.

(7) Diff. Eq. $p^2y^2 \cos^2 \alpha - 2pxy \sin^2 \alpha + y^2 - x^2 \sin^2 \alpha = 0$;
           S.S. $y^2 \cos^2 \alpha = x^2 \sin^2 \alpha$; $y = 0$ is a tac-locus.

(8) Diff. Eq. $(x^2 - 1)p^2 - 2xyp - x^2 = 0$; S.S. $x^2 + y^2 = 1$;
           $x = 0$ is a tac-locus.

(9) Diff. Eq. $(2x^2 + 1)p^2 + (x^2 + 2xy + y^2 + 2)p + 2y^2 + 1 = 0$;
           S.S. $x^2 + 6xy + y^2 = 4$; $x = y$ is a tac-locus.

10) Diff. Eq. $p^2(1 - x^2) - (1 - y^2) = 0$; S.S. $x = \pm 1$ and $y = \pm 1$.

Art. 67

(1) C.P. $y = cx + c^2$; S.S. $x^2 + 4y = 0$.

(2) C.P. $y = cx + c^3$; S.S. $27y^2 + 4x^3 = 0$.

(3) C.P. $y = cx + \cos c$; S.S. $(y - x \sin^{-1} x)^2 = 1 - x^2$.

(4) C.P. $y = cx + \sqrt{(a^2c^2 + b^2)}$; S.S. $x^2/a^2 + y^2/b^2 = 1$.

(5) C.P. $y = cx - e^c$; S.S. $y = x(\log x - 1)$.

(6) C.P. $y = cx - \sin^{-1} c$; S.S. $\pm y = \sqrt{(x^2 - 1)} - \sin^{-1}\sqrt{(1 - 1/x^2)}$.

(7) $\frac{1}{2}(y - px)^2 = -pk^2$; $2xy = k^2$, a rectangular hyperbola with
    axes as asymptotes.

(8) $(x - y)^2 - 2k(x + y) + k^2 = 0$, a parabola touching the axes.

(9) The four-cusped hypocycloid $x^{\frac{2}{3}} + y^{\frac{2}{3}} = k^{\frac{2}{3}}$.

Miscellaneous Examples on Chapter VI

(1) No S.S.; $x = 0$ is a tac-locus.   (2) $Y = PX + P/(P - 1)$.

(5) $2y = \pm 3x$ represent envelopes, $y = 0$ is both an envelope and
    cusp-locus.

(6) C.P. $xy = yc + c^2$.

(7) C.P. $x = yc + xyc^2$; S.S. $y + 4x^2 = 0$.   (Put $y = 1/Y$; $x = 1/X$.)

(8) (i) Putting $p + x = 3t^3$ we get
$$2x = 3(t^3 - t^2);\quad 40y = 9(5t^6 + 2t^5 - 5t^4) + c.$$
   (ii) C.P. $y^2 + 4c^2 = 1 + 2cx$; S.S. $x^2 - 4y^2 + 4 = 0$; $y = 0$ is a tac-locus.

(11) C.P. $r = a\{1 + \cos(\theta - \alpha)\}$, a family of equal cardioids inscribed in the circle $r = 2a$, which is a S.S. The point $r = 0$ is a cusp-locus and also a S.S.

CHAPTER VII
Art. 70

(1) $y = \log \sec x + ax + b$.   (2) $x = a + y + b \log(y - b)$.

(3) $ay = \cos(ax + b)$.   (4) $x = \log\{\sec(ay + b) + \tan(ay + b)\} + c$.

(5) $y = x^3 + ax \log x + bx + c$.

(6) $y = -e^x + ae^{2x} + bx^{n-2} + cx^{n-3} + \dots + hx + k$.

(7) The circle $(x - a)^2 + (y - b)^2 = k^2$. The differential equation expresses that the radius of curvature is always equal to k.

(9) $\sqrt{(1 + y_1^2)} = ky_2$; the catenary $y - b = k \cosh\{(x - a)/k\}$.

Art. 73

(1) $y = x(a \log x + b)$.   (2) $y = ax \cos(2 \log x) + bx \sin(2 \log x)$.

(3) $y = x(a \log x + b)^2$.   (4) $y = x^2(a \log x + b)^2$.

Art. 74

(1) $y = \pm \coth \dfrac{x - c}{\sqrt{2}}$.   (2) $y = -\log(1 - x)$.   (3) $y = \sin^{-1} x$.

(4) $t = \dfrac{1}{a}\sqrt{\left(\dfrac{h}{2g}\right)}\left\{h \cos^{-1}\sqrt{\dfrac{x}{h}} + \sqrt{(xh - x^2)}\right\}$.

(5) (i) The conic $u = \mu/h^2 + (1/c - \mu/h^2)\cos\theta$;
   (ii) $cu = \cos\theta\sqrt{(1 - \mu/h^2)}$ or $\cosh\theta\sqrt{(\mu/h^2 - 1)}$, according as $\mu \lessgtr h^2$.

Art. 75

(1) $y = a(x^2 + 1) + be^{-x}$.   (2) $y = a(x - 1) + be^{-x}$.

(3) $y = a(x - 1) + be^{-x} + x^2$.   (4) $y = 1 + e^{-x^2/2}$.   (5) $y = e^{2x}$.

Art. 77

(2) $y = x^3 + ax - b/x$.   (3) $y = (x^2 + ax)e^x + bx$.

(4) $y = e^{2x} + (ax^3 + b)e^x$.   (5) $y = ax^3 + bx^{-3}$.

(6) $y = ax^2 + b \sin x$.

Art. 80

(1) $y = (a - x)\cos x + (b + \log \sin x)\sin x$.

(2) $y = \left\{a - \log \tan\left(\dfrac{\pi}{4} + x\right)\right\}\cos 2x + b \sin 2x$.

(3) $y = \{a - e^{-x} + \log(1 + e^{-x})\}e^x + \{b - \log(1 + e^x)\}e^{-x}.$

(4) $y = ax + bx^{-1} + (1 - x^{-1})e^x.$ (5) $y = ae^x + (b - x)e^{2x} + ce^{3x}.$

Miscellaneous Examples on Chapter VII

(1) $y = ae^{x/b} - b.$ (2) $y = a + \log(x^2 + b).$

(3) $y = \dfrac{2x^{n+1}}{(n+1)!} + 2a\dfrac{x^n}{n!} + a^2\dfrac{x^{n-1}}{(n-1)!} + bx^{n-2} + cx^{n-3} + \ldots + hx + k.$

(4) $y = -3^{2-n}\cos\{3x - \tfrac{1}{2}\pi(n-2)\} + a\cos x + b\sin x + cx^{n-3} + \ldots + hx + k.$

(5) $y = ax + b\log x.$ (6) $y = ae^x + b(x^2 - 1)e^{2x}.$

(7) $y = a\cos nx + b\sin nx + \dfrac{x}{n}\sin nx - \dfrac{1}{n^2}\cos nx \log \sec nx.$

(8) $y(2x + 3) = a\log x + b + e^x.$

(9) (i) $y = \pm\sqrt{(ax + b)}$; (ii) $y = \pm\sqrt{(a\log x + b)}.$

(10) $y = (a\cos x + b\sin x + \sin 2x)e^{x^2}.$

(12) $y = x^2 z.$ (14) $I = -\tfrac{1}{4}.$

(17) (i) $y = ae^{x^2} + be^{-x^2} - \sin x^2.$ (Put $z = x^2$).

(ii) $y(1 + x^2) = a(1 - x^2) + bx.$ (Put $x = \tan z$).

(18) $\dfrac{d^2y}{dz^2} - 2y = 2(1 - z^2)$; $y = \sin^2 x + A\cosh(\sqrt{2}\sin x + \alpha).$

(19) $y = a\cos\{2(1 + x)e^{-x}\} + b\sin\{2(1 + x)e^{-x}\} + (1 + x)e^{-x}.$

CHAPTER VIII

Art. 83

(1) $y = 2 + x + x^2 - \tfrac{1}{3}x^4 - \tfrac{2}{15}x^5$; exact solution $y = 2 + x + x^2.$

(2) $y = 2x - 2\log x - \tfrac{1}{3}(\log x)^3$; exact value $y = x + \dfrac{1}{x}.$

(3) $y = 2 + x^2 + x^3 + \tfrac{3}{20}x^5 + \tfrac{1}{10}x^6$;

$z = 3x^2 + \tfrac{3}{4}x^4 + \tfrac{6}{5}x^5 + \tfrac{3}{28}x^7 + \tfrac{3}{40}x^8.$

(4) $y = 5 + x + \tfrac{1}{12}x^4 + \tfrac{1}{6}x^6 + \tfrac{2}{63}x^7 + \tfrac{1}{72}x^9$;

$z = 1 + \tfrac{1}{3}x^3 + x^5 + \tfrac{2}{9}x^6 + \tfrac{1}{8}x^8 + \tfrac{11}{324}x^9 + \tfrac{7}{264}x^{11}.$

(5) y has the same value as in Ex. 4.

Art. 87

(1) 2·19.   (2) 2·192.   (3) (a) 4·12, (b) 4·118.

(4) Errors 0·0018; 0·00017; 0·000013;

Upper limits 0·0172; 0·00286; 0·000420.

Art. 89

1·1678487; 1·16780250; 1·1678449.

CHAPTER IX
Art. 95

(1) $u = \left\{1 - \dfrac{x}{2!} + \dfrac{x^2}{4!} - \ldots\right\} = \cos\sqrt{x}; \quad v = x^{\frac{1}{2}}\left\{1 - \dfrac{x}{3!} + \dfrac{x^2}{5!} - \ldots\right\} = \sin\sqrt{x}.$

(2) $u = \left\{1 - 3x + \dfrac{3x^2}{1\,.\,3} + \dfrac{3x^3}{3\,.\,5} + \dfrac{3x^4}{5\,.\,7} + \dfrac{3x^5}{7\,.\,9} + \ldots\right\}; \quad v = x^{\frac{1}{2}}(1-x).$

(3) $u = \left\{1 + \dfrac{1}{3}x + \dfrac{1\,.\,4}{3\,.\,6}x^2 + \dfrac{1\,.\,4\,.\,7}{3\,.\,6\,.\,9}x^3 + \ldots\right\} = (1-x)^{-\frac{1}{3}};$

$v = x^{7/3}\left\{1 + \dfrac{8}{10}x + \dfrac{8\,.\,11}{10\,.\,13}x^2 + \dfrac{8\,.\,11\,.\,14}{10\,.\,13\,.\,16}x^3 + \ldots\right\}.$

(4) $u = x^n\left\{1 - \dfrac{1}{4(1+n)}x^2 + \dfrac{1}{4\,.\,8(1+n)(2+n)}x^4 \right.$

$$\left. - \dfrac{1}{4\,.\,8\,.\,12(1+n)(2+n)(3+n)}x^6 + \ldots\right\}.$$

To get v from u change n into $-n$. If u is multiplied by the constant $\dfrac{1}{2^n\Gamma(n+1)}$, the product is called Bessel's function of order n and is denoted by $J_n(x)$.

Art. 96

(1) and (4), all values of x.   (2) and (3), $|x| < 1$.

Art. 97

(1) $u = \left\{1 + x + \dfrac{2}{4}x^2 + \dfrac{2\,.\,5}{4\,.\,9}x^3 + \dfrac{2\,.\,5\,.\,10}{4\,.\,9\,.\,16}x^4 + \ldots\right\};$

$v = u\log x + \{-2x - x^2 - \tfrac{14}{27}x^3\ldots\}.$

(2) $u = \left\{1 - \dfrac{1}{2^2}x^2 + \dfrac{1}{2^2\,.\,4^2}x^4 - \dfrac{1}{2^2\,.\,4^2\,.\,6^2}x^6 + \ldots\right\};$

$v = u\log x + \left\{\dfrac{1}{2^2}x^2 - \dfrac{1}{2^2\,.\,4^2}(1+\tfrac{1}{2})x^4 + \dfrac{1}{2^2\,.\,4^2\,.\,6^2}(1+\tfrac{1}{2}+\tfrac{1}{3})x^6 - \ldots\right\}.$

u is called Bessel's function of order zero and is denoted by $J_0(x)$.

(3) $u = \left\{1 - 2x + \dfrac{3}{2!}x^2 - \dfrac{4}{3!}x^3 + \ldots\right\};$

$v = u\log x + \left\{2(2-\tfrac{1}{2})x - \dfrac{3}{2!}(2+\tfrac{1}{2}-\tfrac{1}{3})x^2 + \dfrac{4}{3!}(2+\tfrac{1}{2}+\tfrac{1}{3}-\tfrac{1}{4})x^3 - \ldots\right\}.$

(4) $u = x^{\frac{1}{2}}\left\{1 + \dfrac{1\,.\,3}{4^2}x^2 + \dfrac{1\,.\,3\,.\,5\,.\,7}{4^2\,.\,8^2}x^4 + \dfrac{1\,.\,3\,.\,5\,.\,7\,.\,9\,.\,11}{4^2\,.\,8^2\,.\,12^2}x^6 + \ldots\right\};$

$v = u\log x + 2x^{\frac{1}{2}}\left\{\dfrac{1\,.\,3}{4^2}(1+\tfrac{1}{3}-\tfrac{1}{2})x^2 \right.$

$$\left. + \dfrac{1\,.\,3\,.\,5\,.\,7}{4^2\,.\,8^2}(1+\tfrac{1}{3}-\tfrac{1}{2}+\tfrac{1}{5}+\tfrac{1}{7}-\tfrac{1}{4})x^4 + \ldots\right\}.$$

Art. 98

(1) $u = x^{-2}\left\{ -\dfrac{1}{2^2 \cdot 4}x^4 + \dfrac{1}{2^3 \cdot 4 \cdot 6}x^6 - \dfrac{1}{2^3 \cdot 4^2 \cdot 6 \cdot 8}x^8 \right.$

$$\left. + \dfrac{1}{2^3 \cdot 4^2 \cdot 6^2 \cdot 8 \cdot 10}x^{10} - \ldots \right\};$$

$$v = u\log x + x^{-2}\left\{ 1 + \dfrac{1}{2^2}x^2 + \dfrac{1}{2^2 \cdot 4^2}x^4 - \dfrac{11}{2^2 \cdot 4^2 \cdot 6^2}x^6 \right.$$

$$\left. + \dfrac{31}{2^2 \cdot 4^2 \cdot 6^2 \cdot 8^2}x^8 \ldots \right\}.$$

(2) $u = x + 2x^2 + 3x^3 + \ldots = x(1-x)^{-2}$;

      $v = u\log x + 1 + x + x^2 + \ldots = u\log x + (1-x)^{-1}$.

(3) $u = \{1 \cdot 2x^2 + 2 \cdot 3x^3 + 3 \cdot 4x^4 + \ldots\}$;

      $v - u = u\log x + \{-1 + x + 3x^2 + 5x^3 + 7x^4 + \ldots\}$.

(4) $u = \{2x + 2x^2 - x^3 - x^4 + \tfrac{5}{4}x^5 \ldots\}$;

      $v = u\log x + \{1 - x - 5x^2 - x^3 + \tfrac{11}{3}x^4 \ldots\}$.

Art. 99

(1) $y = a_0\{1 - x^2 - \tfrac{1}{3}x^4 - \tfrac{1}{5}x^6 \ldots\} + a_1 x = a_0\left\{1 - \tfrac{1}{2}x\log\dfrac{1+x}{1-x}\right\} + a_1 x$.

(2) $y = a_0\left\{1 - \dfrac{n(n+1)}{2!}x^2 + \dfrac{n(n-2)(n+1)(n+3)}{4!}x^4 - \ldots\right\}$

$$+ a_1\left\{x - \dfrac{(n-1)(n+2)}{3!}x^3 + \dfrac{(n-1)(n-3)(n+2)(n+4)}{5!}x^5 - \ldots\right\}.$$

[For solutions in powers of $1/x$ see No. 7 of the Miscellaneous Examples at the end of Chapter IX.]

(3) $y = a_0\left\{1 - \dfrac{1}{3 \cdot 4}x^4 + \dfrac{1}{3 \cdot 4 \cdot 7 \cdot 8}x^8 - \dfrac{1}{3 \cdot 4 \cdot 7 \cdot 8 \cdot 11 \cdot 12}x^{12} + \ldots\right\}$

$$+ a_1\left\{x - \dfrac{1}{4 \cdot 5}x^5 + \dfrac{1}{4 \cdot 5 \cdot 8 \cdot 9}x^9 - \dfrac{1}{4 \cdot 5 \cdot 8 \cdot 9 \cdot 12 \cdot 13}x^{13} + \ldots\right\}.$$

(4) $y = a_0\{1 - \tfrac{1}{4}x^2 - \tfrac{1}{12}x^3 + \tfrac{5}{96}x^4 \ldots\} + a_1\{x - \tfrac{1}{6}x^3 - \tfrac{1}{24}x^4 + \tfrac{1}{24}x^5 \ldots\}$.

Art. 100

(1) $z^4\dfrac{d^2y}{dz^2} + z^3\dfrac{dy}{dz} + (1 - n^2z^2)y = 0$.   (2) $y = ax^2(1+2x)$.

(3) $y = x^2(1+2x)\left\{a + b\displaystyle\int x^{-2}(1+2x)^{-2}e^{\frac{1}{x}}\,dx\right\}$.

(5) ze^{-z} and $\left[ze^{-z}\log z + z^2\left\{1 - \dfrac{1}{2!}\left(1 + \dfrac{1}{2}\right)z + \dfrac{1}{3!}\left(1 + \dfrac{1}{2} + \dfrac{1}{3}\right)z^2 - \ldots\right\}\right]$,

      where $z = 1/x$.

Miscellaneous Examples on Chapter IX

(1) $u = x^{-\frac{1}{3}} \left\{ 1 + \dfrac{3}{3!} x + \dfrac{9}{6!} x^2 + \dfrac{27}{9!} x^3 + \dots \right\}$;

$\quad v = \quad \left\{ \dfrac{1}{1!} + \dfrac{3}{4!} x + \dfrac{9}{7!} x^2 + \dfrac{27}{10!} x^3 + \dots \right\}$;

$\quad w = x^{\frac{1}{3}} \left\{ \dfrac{1}{2!} + \dfrac{3}{5!} x + \dfrac{9}{8!} x^2 + \dfrac{27}{11!} x^3 + \dots \right\}$.

(2) $u = \left\{ 1 + \dfrac{1}{1^2} x + \dfrac{1}{1^2 \cdot 2^2} x^2 + \dfrac{1}{1^2 \cdot 2^2 \cdot 3^2} x^3 + \dots \right\}$;

$\quad v = u \log x + 2 \left\{ -\dfrac{1}{1^2} x - \dfrac{1}{1^2 \cdot 2^2} \left(1 + \dfrac{1}{2} \right) x^2 \right.$

$\left. \qquad\qquad - \dfrac{1}{1^2 \cdot 2^2 \cdot 3^2} \left(1 + \dfrac{1}{2} + \dfrac{1}{3} \right) x^3 - \dots \right\}$;

$\quad w = u \, (\log x)^2 + 2 \, (v - u \log x) \log x$

$\qquad\qquad + \left\{ 6x + \left(\dfrac{6}{1^4 \cdot 2^2} + \dfrac{8}{1^3 \cdot 2^3} + \dfrac{6}{1^2 \cdot 2^4} \right) x^2 + \dots \right\}$.

CHAPTER XI

Art. 113

(1) $x/a = y/b = z$; straight lines through the origin.

(2) $lx + my + nz = a$; $x^2 + y^2 + z^2 = b$; circles.

(3) $y = az$; $x^2 + y^2 + z^2 = bz$; circles.

(4) $x^2 - y^2 = a$; $x^2 - z^2 = b$; the intersections of two families of rectangular hyperbolic cylinders.

(5) $x - y = a(z - x)$; $(x - y)^2 (x + y + z) = b$.

(6) $x^2 + y^2 + z^2 = a$; $y^2 - 2yz - z^2 = b$; the intersections of a family of spheres with a family of rectangular hyperbolic cylinders.

(7) $\sqrt{(m^2 + n^2)}$. (8) The hyperboloid $y^2 + z^2 - 2x^2 = 1$.

(9) $(x^2 + y^2)(k \tan^{-1} y/x)^2 = z^2 r^2$. (10) $1/x = 1/y + 1/2 = 1/z + 2$.

Art. 114

(1) $y - 3x = a$; $5z + \tan(y - 3x) = be^{5x}$.

(2) $y + x = a$; $\log \{z^2 + (y + x)^2\} - 2x = b$.

(3) $xy = a$; $(z^2 + xy)^2 - x^4 = b$. (4) $y = ax$; $\log(z - 2x/y) - x = b$.

Art. 116

(1) $x^2 + y^2 + z^2 = c^2$; spheres with the origin as centre.

(2) $x^2 + y^2 + z^2 = cx$; spheres with centres on the axis of x, passing through the origin. (3) $xyz = c^3$.

(4) $yz + zx + xy = c^2$; similar conicoids with the origin as centre.

(5) $x - cy = y \log z$.

(6) $x^2 + 2yz + 2z^2 = c^2$; similar conicoids with the origin as centre.

Art. 117

(1) $y = cx \log z$.　　　(2) $x^2 y = cze^z$.　　　　　(3) $(x + y + z^2) e^{x^2} = c$.

(4) $y(x + z) = c(y + z)$.　　　　　(5) $(y + z)/x + (x + z)/y = c$.

(6) $ny - mz = c(nx - lz)$. The common line is $x/l = y/m = z/n$.

Art. 120

(3) $z = ce^{2x}$.　　　　　　　　　(4) $x^2 z + 4 = 0$.

Miscellaneous Examples on Chapter XI

(1) $y = ax$; $z^2 - xy = b$.　　　(2) $x^3 y^3 z = a$; $x^3 + y^3 = bx^2 y^2$.

(3) $y + z = ae^x$; $y^2 - z^2 = b$.　　(4) $y = \sin x + cz/(1 + z^2)$.

(5) $x^2 + xy^2 + x^2 z = t + c$.　　(6) $f(y) = ky$; $x^k = cy^z$.

(8) $dx/x = dy/2y = dz/3z$.　　(9) $y + z = 3e^{x-3}$; $y^2 - z^2 = 3$.

(10) (i) $x^2 + y^2 + z^2 = c(x + y + z)$;　(ii) $x^2 - xy + y^2 = cz$;

　(iii) $y^2 - yz - xz = cz^2$.

(14) $xy = ce^z \sin w$.

CHAPTER XII

Art. 123

(1) $\phi(x/z, y/z) = 0$.　　　　　(2) $\phi(lx + my + nz, x^2 + y^2 + z^2) = 0$.

(3) $\phi\{y/z, (x^2 + y^2 + z^2)/z\} = 0$.　　(4) $\phi(x^2 - y^2, x^2 - z^2) = 0$.

(5) $\phi\{(x - y)^2 (x + y + z), (x - y)/(z - x)\} = 0$.

(6) $\phi\{x^2 + y^2 + z^2, y^2 - 2yz - z^2\} = 0$.

(7) $\phi[y - 3x, e^{-5x}\{5z + \tan(y - 3x)\}] = 0$.

(8) $\phi\{y + x, \log(z^2 + y^2 + 2yx + x^2) - 2x\} = 0$.

(9) $y^2 = 4xz$.　　　　　(10) $a(x^2 - y^2) + b(x^2 - z^2) + c = 0$.

(12) $\phi(x^2 + y^2, z) = 0$; surfaces of revolution about the axis of z.

Art. 126

(1) $\phi(z + x_1, x_1 + x_2, x_1 + x_3) = 0$.

(2) $\phi(z, x_1^2 x_2^{-1}, x_1^3 x_3^{-1}, x_1^4 x_4^{-1}) = 0$.

(3) $\phi(z - x_1 x_2, x_1 + x_2 + x_3, x_2 x_3) = 0$.

(4) $\phi(2z + x_1^2, x_1^2 - x_2^2, x_1^2 - x_3^2) = 0$.

(5) $\phi(4\sqrt{z} - x_3^2, 2x_3 - x_2^2, 2x_2 - x_1^2) = 0$; special integral $z = 0$.

(6) $\phi\{z - 3x_1, z - 3x_2, z + 6\sqrt{(z - x_1 - x_2 - x_3)}\} = 0$; special integral $z = x_1 + x_2 + x_3$.

Art. 129

(1) $z = (2b^2 + 1)x + by + c$.

(2) $z = x \cos \alpha + y \sin \alpha + c$.

(3) $z = ax + y \log a + c$.

(4) $z = a^3 x + a^{-2} y + c$.

(5) $z = 2x \sec \alpha + 2y \tan \alpha + c$.

(6) $z = x(1 + a) + y(1 + 1/a) + c$.

Art. 130

(1) $az = (x + ay + b)^2$.

(2) $z = \pm \cosh \{(x + ay + b)/\sqrt{(1 + a^2)}\}$.

(3) $z^2 - a^2 = (x + ay + b)^2$ or $z = b$.

(4) $z^2(1 + a^3) = 8(x + ay + b)^3$.

(5) $(z + a)e^{x + ay} = b$.

(6) $z = be^{ax + a^2 y}$.

Art. 131

(1) $3z = \pm 2(x + a)^{\frac{3}{2}} + 3ay + 3b$.

(2) $2az = a^2 x^2 + y^2 + 2ab$.

(3) $az = ax^2 + a^2 x + e^{ay} + ab$.

(4) $(2z - ay^2 - 2b)^2 = 16ax$.

(5) $z = a(e^x + e^y) + b$.

(6) $az = a^2 x + a \sin x + \sin y + ab$.

Art. 133

(1) $z = -2 - \log xy$.

(2) $3z = xy - x^2 - y^2$.

(3) $8z^3 = -27x^2 y^2$.

(4) $zx = -y$.

(5) $z = 0$.

(6) $z^2 = 1$.

(7) $z = 0$.

Art. 136

(1) $4z = -y^2$.

(4) A particular case of the general integral, representing the surface generated by characteristics passing through the point $(0, -1, 0)$.

Miscellaneous Examples on Chapter XII

(1) $z = ax + by - a^2 b$; singular integral $z^2 = x^2 y$.

(2) $zx = ax + by - a^2 b$; singular integral $z^2 = y$.

(3) $\phi\{xy, (z^2 + xy)^2 - x^4\} = 0$.

(4) $z = 3x^3 - 3ax^2 + a^2 x + 2y^4 - 4ay^3 + 3a^2 y^2 - a^3 y + b$.

(5) $z = ax_1 + b \log x_2 + (a^2 + 2b)x_3^{-1} + c$.

(6) $z = \phi\{(x_1 + x_3)/x_2, x_1^2 - x_3^2\}$.

(7) $3a(x + ay + b) = (1 + a^3) \log z$, or $z = b$. $z = 0$ is included in $z = b$, but it is also a singular integral.

(8) $z(1 + a^2 + b^2) = (x_1 + ax_2 + bx_3 + c)^2$.

(9) $\phi(z \div e^{4x_1}, z \div e^{4x_2}, z \div e^{4x_3}) = 0$.

(10) $z = ax - (2 + 3a + \frac{1}{2}a^2)y + b$.

11) $z^2 = ax^2 - (2 + 3a + \frac{1}{2}a^2)y^2 + b$.

(12) $z^2 = (1 + a^2)x^2 + ay^2 + b$.

13) $z = a \tan(x + ay + b)$, or $z = b$. $z = 0$ is a singular integral, but it is also included in $z = b$.

14) $z^2 = ax^2 + by^2 - 3a^3 + b^2$. Singular integral $z^2 = \pm 2x^3/9 - y^4/4$.

15) $z = x + y - 1 \pm 2\sqrt{\{(x - 1)(y - 1)\}}$.

(16) $z^2 - xy = c$.

(17) $\phi(z/x, z/y) = 0$; cones with the origin as vertex.

(18) $x^2 + y^2 + z^2 = 2x \cos \alpha + 2y \sin \alpha + c$; spheres with centres on the given circle. The general integral gives other solutions.

(19) $xyz = c$. (This is the singular integral. The complete integral gives the tangent planes.)

(20) The differential equation $(z - px - qy)(1 - 1/p - 1/q) = 0$ has no singular integral, and the complete integral represents planes. Every integral included in the general integral represents the envelope of a plane whose equation contains only one parameter, that is, a developable surface.

CHAPTER XIII
Art. 139

(1) $y^2\{(x - a)^2 + y^2 + 2z\} = b.$ (2) $z^2 = 2ax + a^2 y^2 + b.$

(3) $z = ax + be^y (y + a)^{-a}.$ (4) $z^2 = 2(a^2 + 1)x^2 + 2ay + b.$

(5) $z = ax + 3a^2 y + b.$ (6) $(z^2 + a^2)^3 = 9(x + ay + b)^2.$

(7) $z = x^3 + ax \pm \frac{2}{3}(y + a)^{3/2} + b.$ (8) $z = ax + by + a^2 + b^2.$

Art. 141

(1) $z = a_1 x_1 + a_2 x_2 + (1 - a_1{}^3 - a_2{}^2) x_3 + a_3.$

(2) $z = a_1 x_1 + a_2 x_2 \pm \sin^{-1}(a_1 a_2 x_3) + a_3.$

(3) $z = a_1 \log x_1 + a_2 \log x_2 \pm x_3 \sqrt{(a_1 + a_2)} + a_3.$

(4) $2z = a_1 x_1{}^2 + a_2 x_2{}^2 + a_3 x_3{}^2 - 2(a_1 a_2 a_3)^{1/3} \log x_4 + a_4.$

(5) $2(a_1 a_2 a_3)^{1/3} \log z = a_1 x_1{}^2 + a_2 x_2{}^2 + a_3 x_3{}^2 + 1.$

(6) $4a_1 z = 4a_1{}^2 \log x_3 + 2a_1 a_2 (x_1 - x_2) - (x_1 + x_2)^2 + 4a_1 a_3.$

(7) $(1 + a_1 a_2) \log z = (a_1 + a_2)(x_1 + a_1 x_2 + a_2 x_3 + a_3).$

(8) $z = -(a_1 + a_2) x_1 + (2 \quad - a_2) x_2 + (-a_1 + 2a_2) x_3$
$\qquad - \frac{1}{2}(x_1{}^2 + x_2{}^2 + x_3{}^2) \pm \frac{2}{3}\{x_1 + x_2 + x_3 - 2a_1{}^2 + 2a_1 a_2 - 2a_2{}^2\}^{3/2} + a_3.$

Art. 142

(1) $z = \pm (x_1 + x_2)^2 + \log x_3 + a.$ (2) No common integral.

(3) $z = x_1{}^2 + x_2{}^2 + x_3{}^2 + a$, or $z = x_1{}^2 + 2x_2 x_3 + a.$

(4) $z = a(x_1 + 2x_2) + b \log x_3 + 2ab \log x_4 + c.$

(5) $z = a(3x_1 + x_2{}^3 - x_3{}^3) + b.$ (6) No common integral.

(7) $z = a(x_1 - x_4) + b(x_2 - x_3) + c$, or $z = a(x_1 - 2x_2) + b(2x_3 - x_4) + c.$

(8) $z = \phi(3x_1 + x_2{}^3 - x_3{}^3).$

(9) $z = \phi(x_1 - x_4, x_2 - x_3)$, or $z = \phi(x_1 - 2x_2, 2x_3 - x_4).$

Miscellaneous Examples on Chapter XIII

(1) $z^2 = a_1 \log x_1 - a_1 a_2 \log x_2 + a_2 \log x_3 + a_3.$

(2) No common integral.

(3) $z = a_1 \log x_1 + a_2 x_2 + (a_1 + a_2) x_3 \pm \sqrt{\{a_1 (a_1 + 2a_2) x_4{}^3\}} + a_3.$

(4) $0 = a_1 \log x_1 + a_2 x_2 + (a_1 + a_2) x_3 \pm \sqrt{\{a_1 (a_1 + 2a_2) z^3\}} + 1.$

(5) $2 \log z = c \pm (x_1^2 + x_2^2 + x_3^2).$ (6) $z^3 = x_1^3 + x_2^3 + x_3^3 + c.$

(7) $4z + x_1^2 + x_2^2 + x_3^2 = 0.$ (10) $z = \phi(x_1 x_2, \, x_2 + x_3 + x_4, \, x_4 x_5).$

(11) (iii) $3z = x_1^3 - 3x_1 x_2 + c.$

CHAPTER XIV
Art. 144

(1) $z = x^3 + x f(y) + F(y).$ (2) $z = \log x \log y + f(x) + F(y).$

(3) $z = -\dfrac{1}{x^2} \sin xy + y f(x) + F(x).$ (4) $z = x^3 y^3 + f(y) \log x + F(y).$

(5) $z = \sin(x + y) + \dfrac{1}{y} f(x) + F(y).$ (6) $z = -xy + f(x) + e^{xy} F(x).$

(7) $z = (x^2 + y^2)^2 - 1.$ (8) $z = y^2 + 2xy + 2y + ax^2 + bx + c.$

(9) $z = (x^2 + y^2)^2.$ (10) $z = x^3 y^3 + y(1 - x^3).$

Art. 145

(1) $z = F_1(y + x) + F_2(y + 2x) + F_3(y + 3x).$

(2) $z = f(y - 2x) + F(2y - x).$ (3) $z = f(y + x) + F(y - x).$

(4) The conicoid $4x^2 - 8xy + y^2 + 8x - 4y + z + 2 = 0.$

Art. 146

(1) $z = f(2y - 3x) + x F(2y - 3x).$ (2) $z = f(5y + 4x) + x F(5y + 4x).$

(3) $z = f(y + 2x) + x F(y + 2x) + \phi(y).$ (4) $z(2x + y) = 3x.$

Art. 147

(1) $z = x^4 + 2x^3 y + f(y + x) + x F(y + x).$

(2) $z = 6x^2 y + 3x^3 + f(y + 2x) + F(2y + x).$ (3) $V = -2\pi x^2 y^2.$

Art. 148

(1) $z = e^{x+2y} + f(y + x) + x F(y + x).$

(2) $z = x^2(3x + y) + f(y + 3x) + x F(y + 3x).$

(3) $z = -x^2 \cos(2x + y) + f(y + 2x) + x F(y + 2x) + \phi(y).$

(4) $z = x e^{x-y} + f(y - x) + F(2y + 3x).$

(5) $V = (x + y)^3 + f(y + ix) + F(y - ix).$

(6) $z = 2x^2 \log(x + 2y) + f(2y + x) + x F(2y + x).$

Art. 149

(1) $z = x \sin y + f(y - x) + x F(y - x).$

(2) $z = x^4 + 2x^3 y + f(y + 5x) + F(y - 3x).$

(3) $z = \sin x - y \cos x + f(y - 3x) + F(y + 2x).$

(4) $z = \sin xy + f(y + 2x) + F(y - x).$

(5) $z = \frac{1}{2} \tan x \tan y + f(y+x) + F(y-x)$.

(6) $y = x \log t + t \log x + f(t+2x) + F(t-2x)$.

Art. 150

(1) $z = f(x) + F(y) + e^{3x} \phi(y+2x)$.

(2) $z = e^{-x}\{f(y-x) + xF(y-x)\}$.

(3) $V = \Sigma A e^{h(x+ht)}$.

(4) $z = f(y+x) + e^{-x} F(y-x)$.

(5) $z = \Sigma A e^{h(x+hy)} + \Sigma B e^{k(x+2ky)}$.

(6) $V = \Sigma A e^{n(x \cos \alpha + y \sin \alpha)}$.

(7) $z = e^x\{f(y+2x) + \Sigma A e^{k(y+2kx)}\}$.

(8) $z = 1 + e^{-x}\{(y-x)^2 - 1\}$.

Art. 151

(1) $z = \frac{1}{2} e^{2x-y} + e^x f(y+x) + e^{2x} F(y+x)$.

(2) $z = 1 + x - y - xy + e^x f(y) + e^{-y} F(x)$.

(3) $z = \frac{1}{82}\{\sin(x-3y) + 9 \cos(x-3y)\} + \Sigma A e^{k(y+kx)}$.

(4) $z = x + f(y) + e^{-x} F(y+x)$.

(5) $y = -e^{x+z} + \Sigma A e^{x \sec \alpha + z \tan \alpha}$.

(6) $z = e^{2x}\{x^2 \tan(y+3x) + xf(y+3x) + F(y+3x)\}$.

Art. 152

(1) $y^2 r - 2ys + t = p + 6y$.

(2) $pt - qs = q^3$.

(3) $r + 3s + t + (rt - s^2) = 1$.

(4) $pq(r-t) - (p^2 - q^2)s + (py - qx)(rt - s^2) = 0$.

(5) $2pr + qt - 2pq(rt - s^2) = 1$.

(6) $qr + (zq - p)s - zpt = 0$.

Art. 154

(1) $z = f(y + \sin x) + F(y - \sin x)$.

(2) $z = f(x+y) + F(xy)$.

(3) $y - \psi(x+y+z) = \phi(x)$, or $z = f(x) + F(x+y+z)$.

(4) $z = f(x + \tan y) + F(x - \tan y)$.

(5) $z = f(x^2 + y^2) + F(y/x) + xy$.

(6) $y = f(x+y+z) + xF(x+y+z)$.

(7) $3z = 4x^2 y - x^2 y^4 - 6 \log y - 3$.

Art. 157

(1) $p + x - 2y = f(q - 2x + 3y)$; $\lambda = -\frac{1}{2}$.

(2) $p - x = f(q - y)$; $\lambda = \infty$.

(3) $p - e^x = f(q - 2y)$; $\lambda = \infty$.

(4) $p - y = f(q + x)$; $p + y = F(q - x)$; $\lambda = \pm 1$.

(5) $p - y = f(q - 2x)$; $p - 2y = F(q - x)$; $\lambda = -1$ or $-\frac{1}{2}$.

(6) $px - y = f(qy - x)$; $\lambda = -x$ or $-y$.

(7) $zp - x = f(zq - y)$; $\lambda = z/pq$.

Art. 158

(1) $z = ax + by - \frac{1}{2}x^2 + 2xy - \frac{3}{2}y^2 + c$;

$z = \frac{1}{2}x^2(1 + 3m^2) + (2 + 3m)xy + nx + \phi(y + mx)$

$= 2xy - \frac{1}{2}(x^2 + 3y^2) + nx + \psi(y + mx)$.

(2) $z = \frac{1}{2}(x^2 + y^2) + ax + by + c$; $z = \frac{1}{2}(x^2 + y^2) + nx + \psi(y + mx)$.

(3) $z = e^x + y^2 + ax + by + c$; $z = e^x + y^2 + nx + \psi(y + mx)$.

(4) $x = \frac{1}{2}(\alpha - \beta)$; $y = \frac{1}{2}\{\psi'(\beta) - \phi'(\alpha)\}$; $z = xy + \frac{1}{2}\{\phi(\alpha) - \psi(\beta)\} + \beta y$.

(5) $x = \beta - \alpha$; $y = \phi'(\alpha) - \psi'(\beta)$; $z = xy - \phi(\alpha) + \psi(\beta) + \beta y$.

(6) $z + y/m + mx - n \log x = \phi(x^m y)$;   the other method fails.

(7) $z^2 = x^2 + y^2 + 2ax + 2by + c$; $z^2 = x^2 + y^2 + 2nx + \psi(y + mx)$.

(8) $2z = y^2 - x^2$.

Miscellaneous Examples on Chapter XIV

(1) $z = x^2 y^2 + xf(y) + F(y)$.   (2) $z = e^{x+y} + f(x) + F(y)$.

(3) $yz = y \log y - f(x) + yF(x)$.

(4) $z = f(x + y) + xF(x + y) - \sin(2x + 3y)$.

(5) $z = f(y + \log x) + xF(y + \log x)$.   (6) $z = x + y + f(xy) + F(x^2 y)$.

(7) $z = \log(x + y) . f(x^2 - y^2) + F(x^2 - y^2)$.

(8) $4z = 6xy - 3x^2 - 5y^2 + 4ax + 4by + c$;

     $4z = 6xy - 3x^2 - 5y^2 + 2nx + 2\psi(y + mx)$.

(9) $3z = 3c \pm 2(x + a)^{3/2} \pm 2(y + b)^{3/2}$.

10) $mz + \sin y + m^2 \sin x - mnx = m\phi(y + mx)$.

11) $2x = \alpha - \beta$; $2y = \psi'(\beta) - \phi'(\alpha)$;

     $2z = 3x^2 - 6xy - 7y^2 + \phi(\alpha) - \psi(\beta) + 2\beta y$.

12) $z = x^3 + y^3 + (x + y + 1)^2$.   (13) $z = x^2 - xy + y^2$.

20) $px + qy = f(p^2 + q^2)$; $py - qx = F(q/p)$.

Miscellaneous Examples on the Whole Book

(1) $(x^2 - y^2)^2 = cxy$.   (2) $y = x^2 + ce^{-x^2}$.

(3) $2 \sec x \sec y = x + \sin x \cos x + c$.   (4) $(xy + c)^2 = 4(x^2 + y)(y^2 - cx)$.

(5) $1 + xy = y(c + \sin^{-1} x)\surd(1 - x^2)$.   (6) $y = (A - \frac{1}{4}x) \cos 2x + B \sin 2x$.

(7) $y = \dfrac{x^2}{5} - \dfrac{6x}{25} + \dfrac{28}{125} + \dfrac{1}{16} \, xe^x(\sin 2x - \cos 2x) + Ae^{-x} + Be^x \cos(2x + \alpha)$.

(8) $y = A + Bx + Cx \log x + \log x + \frac{1}{2}x(\log x)^2 + \frac{1}{2}x^2$.

(9) $y + \sec x = c \tan x$.

10) $x = Ae^{2t} + Be^{-2t} - \frac{2}{5}(\cos t - \sin t)$; $y = Ae^{2t} - 3Be^{-2t} - \frac{6}{5} \cos t$.

11) $x^{2/3} = (y - 1)^{2/3} + c$;   S.S. $y = 1$.   (12) $y = a \operatorname{cosec}(b - x)$.

13) $y = \left(A + Bx + \dfrac{x^2}{64}\right) \sin 2x + \left(E + Fx - \dfrac{x^3}{96}\right) \cos 2x$.

14) $2xy = 3x^2 + c$.   (15) $z + xy = c(x + y - xy)$.

16) $x^3 + y^3 + z^3 = cxyz$.   (17) $z = f(xy) - \frac{1}{2}x^2 - \frac{1}{2}y^2$.

18) $(x - y)e^{(x-z)/(x-y)} = f\{(x - 3y + z)/(x - y)^2\}$.

19) $(z + x)^2 = (z + 2y)f(y/x)$.

20) $z = ax + by + a^2 + b^2$;   singular integral $4z + x^2 + y^2 = 0$.

21) $z = e^x f(x - y) + F(y)$.

(22) $z = ax^2 + by + 4a^2$; singular integral $16z + x^4 = 0$.

(23) $z = f(x+y) + F(x-y) + \frac{1}{6}(x^3 + y^3)$.

(24) $z = xf(y) + yF(x)$.   **(25)** $cz = (x+a)(y+b)$.

(26) $z = \frac{1}{2}xy + f(y/x) + xF(y/x)$.   **(27)** $z = f(z+x) + F(z+y)$.

(28) $y(x+c) = c^2 x$; singular solutions $y = 0$ and $y + 4x^2 = 0$.

(29) $ay^4 = (x+b)^5$.   **(30)** $y = A \cos\left(\dfrac{x^{n+1}}{n+1}\right) + B \sin\left(\dfrac{x^{n+1}}{n+1}\right)$.

(31) $x^2 + y^2 + z^2 = 2(x \cos \alpha + y \sin \alpha + c)$.   **(32)** $y = e^x - \frac{3}{2}e^{2x} + \frac{1}{2}e^{3x}$.

(33) $x = e^{-\kappa t}(a \cos \lambda t + b \sin \lambda t) + C \cos(pt - \alpha)$,

where $C = A/\sqrt{\{(\kappa^2 + \lambda^2 - p^2)^2 + 4\kappa^2 p^2\}}$, $\tan \alpha = 2\kappa p/(\kappa^2 + \lambda^2 -$

and a and b are arbitrary constants.

(34) $y = A \cos(\sin x) + B \sin(\sin x)$.

(35) (i) $F = A \log(r+z) + B$;

(ii) $\phi = A \int e^{-\xi^2/4a^2} d\xi + B$; $\dfrac{\partial \phi}{\partial x} = \dfrac{A}{\sqrt{t}} e^{-x^2/4a^2 t}$.

(36) $V = A\{\frac{1}{5} + \frac{2}{7}(3z^2 - r^2) + \frac{1}{35}(35z^4 - 30z^2 r^2 + 3r^4)\}$,

where $r^2 = x^2 + y^2 + z^2$.

(39) $u = C\left(1 + \dfrac{x}{a} + \dfrac{x^4}{4!\,a^4} + \dfrac{x^5}{5!\,a^5} + \ldots\right) \cosh t$

$\qquad + C\left(\dfrac{x^2}{2!\,a^2} + \dfrac{x^3}{3!\,a^3} + \dfrac{x^6}{6!\,a^6} + \dfrac{x^7}{7!\,a^7} + \ldots\right) \sinh t$.

(41) $y - x = c(xy - 1)e^{-x}$.

(42) $y = (1+x)^{a-b}(1-x)^{a+b}\{A + B \int (1+x)^{-a+b-1}(1-x)^{-a-b-1} dx\}$.

If $2a$ is an integer, the integral can be evaluated putting $z = (1+x)/(1-x)$.

(43) (i) $y = (1-x^2)(A + B \log x)$; (ii) $y = (1-x^2)(x + A + B \log x)$.

(44) $(1 - x^2)y = \left(a + b \int e^{-x^2} dx\right)e^{\frac{1}{2}x^2}$. [Put $\log y = \int (u - \frac{1}{2}P)\,dx$. $u =$

a solution of the differential equation in u.]

(45) $f(x) = 1 - \dfrac{(2n-2)}{(2n-1)}\dfrac{x^2}{2!} + \dfrac{(2n-2)(2n-4)(2n-6)}{(2n-1)(2n-2)(2n-3)}\dfrac{x^4}{4!} - \ldots$;

$\phi(x) = x - \dfrac{(2n-2)(2n-4)}{(2n-1)(2n-2)}\dfrac{x^3}{3!} + \ldots$.

(46) $y = Ax^5 + Bx^3 + E(x^2 + 1)$, replacing $C/6$ by E.

(47) $u = 1 + \dfrac{c}{2!}\left(\dfrac{x}{a}\right)^2 + \dfrac{c\{c + 2(b+1)\}}{4!}\left(\dfrac{x}{a}\right)^4$

$\qquad + \dfrac{c\{c + 2(b+1)\}\{c + 4(b+3)\}}{6!}\left(\dfrac{x}{a}\right)^6 + \ldots$;

$v = \left(\dfrac{x}{a}\right) + \dfrac{\{c+b\}}{3!}\left(\dfrac{x}{a}\right)^3 + \dfrac{\{c+b\}\{c + 3(b+2)\}}{5!}\left(\dfrac{x}{a}\right)^5 + \ldots$;

both converge within the circle $|x| = |a|$.

(49) $x^2 \dfrac{d^2y}{dx^2} = (2 - x^2)\, y.$

(50) $\dfrac{1}{Q}\left\{\dfrac{\partial P}{\partial y} - \dfrac{\partial Q}{\partial x}\right\}$ must be a function of x alone; $x^3 y - a x^2 y^2 = c.$

(51) $x^2 + y^2 + 2bxy = 2ax.$

(52) $uve^w = a\displaystyle\int v^2 e^w\, dx + b,$ where $v = Q/P$ and $w = \displaystyle\int v\, dx.$

(53) $Pn \cot(nx + \alpha) + Q = n^2.$

(54) $y(1 - x) = A(3 - 2x)e^{2x} + B(1 - 2x)e^{-2x}.$

(56) $x^3 + yz = c(y + z).$

(57) $y = Ae^{-2x} + e^x(B \cos x\sqrt{3} + C \sin x\sqrt{3})$
$\qquad + \tfrac{3}{8}x^2 + \tfrac{1}{24649}e^{-2x}\{157x(6 \cos x + 11 \sin x)$
$\qquad\qquad\qquad + 3(783 \cos x - 56 \sin x)\}.$

(58) $y = (3 + 4x^2)\{A + B\displaystyle\int (3 + 4x^2)^{-2}e^{-\frac{2}{3}x^2}\, dx\}.$

(59) $z^4(x + y)^4(x^2 + y^2 + z^2) = c(x^2 + y^2 - z^2).$ (60) $xz = c(y + z).$

(62) (i) Put $y = -\dfrac{1}{u a_2(x)}\dfrac{du}{dx};$ (ii) $y - \dfrac{1}{2} = \dfrac{x(c + \tan x)}{1 - c \tan x}.$

     [See Ex. 41 for method.]

(65) If a particle P moves so that its velocity is proportional to the radius vector OP and is perpendicular to OP and also to a fixed line OK, then it will describe with constant speed a circle of which OK is the axis.

(67) $r^2 \sin 2(\theta + \alpha) = 1;$ singular solution $r^4 = 1.$

(68) $y^2 - x^2 = cx + 2a^2 \pm a\sqrt{(4a^2 - c^2)};$ singular solution $y^2 - x^2 = \pm 2ay.$

(70) $4a(y - c) = (x - c)^2;$ singular solution $y = x - a.$

(71) $x + a = c \cos\phi + c \log\tan\tfrac{1}{2}\phi.$ (72) $a \cos\theta + b \cos\theta' = k.$

(74) $2cy = (x + c)^2;$ singular solution $y(y - 2x) = 0.$

(75) $x + py + ap^2 = 0;$ $(y + ap)\sqrt{(p^2 + 1)} = c + a \sinh^{-1} p,$
$x\sqrt{(p^2 + 1)} + p(c + a \sinh^{-1} p) = 0.$

     There is no singular solution. The p-discriminant $y^2 = 4ax$ represents the cusp-locus of the involutes.

(77) $y = ax,\ z = b + \sqrt{(x^2 + y^2)};$ $z = \sqrt{(x^2 + y^2)} + f(y/x).$

     The subsidiary integrals represent a family of planes through the axis of z and a family of right circular cones with the axis of z as axis; the general integral represents a family of surfaces each of which contains an infinite number of the pairs of straight lines in which the planes and cones intersect.

(78) $x^2 + y^2 + z^2 = f\{x^2 + y^2 + (x + y)^2\};$ $x^2 + y^2 + z^2 = c^2;$ $z^2 = xy + c.$

(79) $(2x - y)^7 = c^5 z(x + 2y).$

(80) $(ax - by)/(z + c) = f\{(ax + by)/(z - c)\}.$

(81) (i) $I = E/R + Ae^{-Rt/L}$; (ii) $A = I_0 - E/R$; (iii) $I = E/R$.

(82) $I = a \cos (pt - \epsilon) + Ae^{-Rt/L}$, where $a = E/\sqrt{(R^2 + L^2 p^2)}$, $\tan \epsilon = Lp/R$, and A is arbitrary.

(83) $Q = a \sin (pt - \epsilon)$, where $\tan \epsilon = (CLp^2 - 1)/pCR$ and $a = EC/\sqrt{\{(CLp^2 - 1)^2 + p^2 C^2 R^2\}}$.

(85) $x = A \cos (t - \alpha) + B \cos (3t - \beta)$; $y = 2A \cos (t - \alpha) - 5B \cos (3t - \beta)$.

(86) a and b are the roots of $\lambda^2 (LN - M^2) + \lambda (RN + LS) + RS = 0$.

(91) $x = A \cos (pt - \alpha) + B \cos (qt - \beta)$, $y = A \sin (pt - \alpha) - B \sin (qt - \beta)$ where $2p = \sqrt{(4c^2 + \kappa^2)} + \kappa$, $2q = \sqrt{(4c^2 + \kappa^2)} - \kappa$.

(92) $\dfrac{d^2 z}{dt^2} + (a + b) \dfrac{dz}{dt} + abz = abc$.

(93) $p = \sqrt{(n^2 - 2\mu^2)}$ makes the amplitude of the particular integral a maximum, provided $2\mu^2$ does not exceed n^2.

(94) $x = Ae^{-kt} \cos (pt - \epsilon)$ where $p = \sqrt{(n^2 - k^2)}$.

(97) $\phi = \frac{1}{2} Va^3 r^{-2} \cos \theta$.   (98) $y \sin (pb/c) = A \sin (px/c) \cos (pt + \alpha)$.

(100) $\phi = C \cosh m(y + h) \cos (mx - nt)$.

(115)   (vi) $u_x = A(-2)^x + B(-\frac{1}{2})^x$;

      (viii) $u_x = 2^x \left(P \cos \dfrac{\pi x}{3} + Q \sin \dfrac{\pi x}{3} \right)$;

      (x) $u_x = A(-9)^x + B + \dfrac{2^x}{11}$.

(119) $u = \dfrac{K}{m} \cos \dfrac{mx}{c} \sin mt$.

(120) $z = e^{-y} \sin x$.

Note on alternative forms of answers.

In several examples a slight variation in the method of solution may lead to a different form of the complete primitive. Thus in Ex. 3, Art. 70, the answer given is $ay = \cos (ax + b)$, but the student may equally well obtain $ay = \sin (ax + b)$, or $ay = \sinh (ax + b)$. If in the first form b is replaced by $(b - \frac{1}{2}\pi)$ we obtain the second, while if in the second a and b are replaced by ai and bi respectively, we obtain the third after division by i. Other forms may be obtained by replacing a by $1/a$.

In the answer to Ex. 4, Art. 116, c^2 may be replaced by $-c^2$, or c or $-c$. In general an arbitrary constant must be supposed to have all values, real, imaginary, or complex, and may be replaced by any function of a new arbitrary constant.

Where pairs of integrals are needed, alternative pairs often arise very naturally. Thus the answers to Exs. 5 and 6, Art. 113, may be replaced by

$$y - z = a(y - x), \quad (y - z)^2 (x + y + z) = b,$$

and by

$$x^2 + y^2 + z^2 = a, \quad x^2 + 2y^2 - 2yz = b,$$

spectively.   In this set of examples the pairs $u = a$, $v = b$ may be
placed by $f(u, v) = a$, $F(u, v) = b$, where f and F are any two inde-
endent functions of u and v.

Alternative answers are to be found for several of the examples on
artial differential equations, e.g. $\dfrac{\partial z}{\partial x} \sin \alpha = \dfrac{\partial z}{\partial y} \cos \alpha$ for Ex. 3, Art. 42,
nd $z^2(a - y^2) = (x + b)^2$ for Ex. 2, Art. 139 (see note on p. 171).

Note on limiting solutions

In addition to the " complete primitives ", some differential equa-
ons have *limiting solutions* (which are *not* singular solutions) obtained
y letting an arbitrary constant become infinite.   For example, the
omplete primitive of Ex. 4, Art. 14, is $x - y + c = \log (x + y)$.   When
$\rightarrow - \infty$ we get the limiting solution $x + y = 0$.   Similarly the complete
rimitive of Ex. 2, Art. 70, is $x = a + y + b \log (y - b)$. By taking $a/b \rightarrow + \infty$
e get the limiting solution $y = b$.

A discussion of such solutions and of their geometrical representation
given in my article " The incompleteness of ' complete ' primitives
f differential equations " (*Mathematical Gazette*, XXIII, 1939, p. 49)
hich also contains some new results concerning singular solutions of
e total differential equation $P\,dx + Q\,dy + R\,dz = 0$.

INDEX

(The numbers refer to the pages)

(The numbers refer to the pages)

(The numbers refer to the pages)

(*The numbers refer to the pages*)